OPTIMISATION OF DISASTER FORECASTING AND PREVENTION MEASURES IN THE CONTEXT OF HUMAN AND SOCIAL DYNAMICS

NATO Science for Peace and Security Series

This Series presents the results of scientific meetings supported under the NATO Programme: Science for Peace and Security (SPS).

The NATO SPS Programme supports meetings in the following Key Priority areas: (1) Defence Against Terrorism; (2) Countering other Threats to Security and (3) NATO, Partner and Mediterranean Dialogue Country Priorities. The types of meeting supported are generally "Advanced Study Institutes" and "Advanced Research Workshops". The NATO SPS Series collects together the results of these meetings. The meetings are co-organized by scientists from NATO countries and scientists from NATO's "Partner" or "Mediterranean Dialogue" countries. The observations and recommendations made at the meetings, as well as the contents of the volumes in the Series, reflect those of participants and contributors only; they should not necessarily be regarded as reflecting NATO views or policy.

Advanced Study Institutes (ASI) are high-level tutorial courses to convey the latest developments in a subject to an advanced-level audience.

Advanced Research Workshops (ARW) are expert meetings where an intense but informal exchange of views at the frontiers of a subject aims at identifying directions for future action.

Following a transformation of the programme in 2006 the Series has been re-named and re-organised. Recent volumes on topics not related to security, which result from meetings supported under the programme earlier, may be found in the NATO Science Series.

The Series is published by IOS Press, Amsterdam, and Springer Science and Business Media, Dordrecht, in conjunction with the NATO Public Diplomacy Division.

Sub-Series

A.	Chemistry and Biology	Springer Science and Business Media
B.	Physics and Biophysics	Springer Science and Business Media
C.	Environmental Security	Springer Science and Business Media
D.	Information and Communication Security	IOS Press
E.	Human and Societal Dynamics	IOS Press

http://www.nato.int/science
http://www.springer.com
http://www.iospress.nl

Sub-Series E: Human and Societal Dynamics – Vol. 52 ISSN 1874-6276

Optimisation of Disaster Forecasting and Prevention Measures in the Context of Human and Social Dynamics

Edited by

Ion Apostol

Deputy Minister of Ecology and Natural Resources of the Republic of Moldova

David L. Barry

Director, DLB Environmental, Cranleigh, Surrey, United Kingdom

Wilhelm G. Coldewey

University of Münster, Institute of Geology and Palaeontology,
Department of Applied Geology, Germany

and

Dieter W.G. Reimer

UWIK-CONSULTING, Bonn, Germany

IOS Press

Amsterdam • Berlin • Tokyo • Washington, DC

Published in cooperation with NATO Public Diplomacy Division

Proceedings of the NATO Advanced Research Workshop on Optimisation of Disaster Forecasting and Prevention Measures in the Context of Human and Social Dynamics
Chisinau, Moldova
7–10 April 2008

ISBN 978-1-58603-948-6
Library of Congress Control Number: 2008944037

Publisher
IOS Press BV
Nieuwe Hemweg 6B
1013 BG Amsterdam
Netherlands
fax: +31 20 687 0019
e-mail: order@iospress.nl

Distributor in the UK and Ireland
Gazelle Books Services Ltd.
White Cross Mills
Hightown
Lancaster LA1 4XS
United Kingdom
fax: +44 1524 63232
e-mail: sales@gazellebooks.co.uk

Distributor in the USA and Canada
IOS Press, Inc.
4502 Rachael Manor Drive
Fairfax, VA 22032
USA
fax: +1 703 323 3668
e-mail: iosbooks@iospress.com

PRINTED IN THE NETHERLANDS

Preface

Following the very successful Workshop (ARW) in April 2007, which was structured on the basis of the ever-increasing frequency and severity of natural disasters in the region, the second ARW was convened on 7–10 April 2008 in Chisinau, Moldova. This was aimed at further supplementing the efforts to transfer technology and knowledge and so help decrease the vulnerability of the population to both natural and man-made disasters.

As the Moldova–NATO Individual Partnership Action Plan (IPAP) has foreseen, the follow-up ARW tried to unify the efforts of the scientific community in creating a greater understanding of the various threats to society and the environment. Thus, this ARW had the task of further evaluating accumulated European theoretical knowledge and practical experience in the relevant fields of concern so that practical recommendations can be developed for the prevention and mitigation of disasters.

The agenda consisted of about 30 presentations (from ten countries), and discussions, that addressed a wide range of disaster-management regimes. The principal themes focused (for a series of typical disaster scenarios) on how these disasters can affect both the human and natural environments. Accordingly, the presentations and syndicate discussions covered the following areas of concern: natural disasters such as earthquakes, landslides, and floods; man-made disasters such as accidents at mining and tailings dams; nuclear/radiological facilities; transport accidents involving hazardous materials; fires; and environmental contamination. Monitoring and the assessment of health and environmental pollution risks, as well as the communication of these risks to the public, were also discussed.

The essence of the various themes centred on the integrated techniques for predicting, measuring and assessing the various physical, environmental, health and social risks, and how these risks might be prevented or at least mitigated. The ARW again recognised the complex inter-relationships between several of the key factors that must be involved, to varying degrees of sophistication, in the overall management of the range of hazards and their associated risks. These factors include: monitoring; risk and other modelling exercises; control measures (such as licensing); public liaison and information management (including education); and cost-benefit assessments.

The scientific content of the presentations, and the subsequent written papers, were thus focused on: risk assessment as part of national policies regarding protection of man and environment; the need for strong co-operation at international and national levels; using a cost–benefit approach; information sharing and networking; and vulnerability as a moderating factor in risk assessment. The presentations appeared to be very useful especially to those partner countries that are developing their legal framework in civil emergency planning as well as in environmental protection. (Some participating countries, such as Moldova, the Ukraine and Georgia, are aligning their legal frameworks to EU directives and other international standards.)

The ARW contributions reflected the extensive experience in the participating countries (namely, Armenia, Austria, Bulgaria, Georgia, Germany, Kazakhstan, Moldova, Romania, Turkey, United Kingdom, and the Ukraine, together with a further written paper from the Netherlands) in the field of combating natural and man-made disasters, as well as how their secondary impacts should be assessed and adapted to the specific conditions in the Republic of Moldova.

In the opinion of the ARW participants there is a continuing need to convene similar more dedicated follow-up ARWs, with the aim of gaining a greater understanding of further specific topics, such as environmental and health monitoring, drought conditions, and the role of land-use planning, in mitigating the effects of natural disasters and preventing man-made disasters.

Contents

Theme 3. Mining/Industrial Hazards/Risks

Theme 4. Health/Radiological Hazards/Risks

Theme 5. Hazard/Risk Communication/Public Participation

Theme 1

Land-Based Hazards/Risks

Optimisation of Disaster Forecasting and Prevention Measures in the Context of Human and Social Dynamics
I. Apostol et al. (Eds.)
IOS Press, 2009.
doi:10.3233/978-1-58603-948-6-3

Forecasting and Preventing Disasters from Natural or Man-made Fires in Forest Areas

Lăcrămioara BOZ [1], Lavinia TOFAN [2], Ovidiu TOMA [3*]

[1] *Faculty of Law, Centre of Studies – Focşani, George Bariţiu University of Braşov, Lunii Street no. 6, BRASOV / ROMANIA, Tel. (+40 268) 319 806; Fax (+40 268) 319 948; E-mail: ramramira@yahoo.com; http:// www.universitateagbaritiu.ro*

[2] *Faculty of Chemical Engineering, Gh. Asachi Technical University of Iaşi, Bd. D. Mangeron, nr. 71 A, IASI/ ROMANIA, Tel (+40 232) 278680; E-mail: lavinia_tofan@yahoo.com*

[3*] *Faculty of Biology, Department of Molecular and Experimental Biology, Alexandru Ioan Cuza University of Iasi, Bd. Carol I , no. 20 A , 700505 IASI / ROMANIA Tel. (+40 232) 201630 ; Fax (+40 232) 201472; http://www.bio.uaic.ro; E-mail: otoma@uaic.ro*

Abstract. This paper is intended as an "alarm signal" for the preservation of the two priceless valuables, human life and the environment. It is essential for this approach to empower the idea of rendering everybody responsible for all factors in the sense that they should extend to individuals' efforts as well as to team efforts in order to preserve life and the environment. This is achieved by complying with prevention regulations for fire hazards in forest areas and also by proper behaviour concerning the warning, evacuation and saving people in a forest area on fire. In addition, the article aims to render public opinion sensitive enough to understand that forest fires are disasters that can be avoided, or at least limited, and carrying forward positive results is entirely in man's power.

Keywords. disasters, forest fires, education, forecasting, prevention

Introduction

Disasters always occur and they threaten the world population more and more often and with increasing power. Besides, those due to natural causes, such as earthquakes, unstable ground, avalanches, drought, floods, hurricanes, and tornados, there are also on-going man-made disasters caused by nuclear accidents, the testing of certain types of ammunition used in military operations, pollution under various forms, massive forest clearing, and the setting on fire of thousands of hectares of forests. All these aggravate the phenomenon of global warming and its major effects on the irreversible damage on the ozone layer [1-7].

Natural disasters are due to inevitable natural phenomena and they are impossible to prevent, while disasters due to irresponsible or indifferent human behaviour can be limited or even eradicated. Losses of any kind from disasters are huge: human lives, material damage and environmental deterioration. Some disasters could be prevented so that the negative effects can be minimal [8-12].

1. General Considerations

The deterioration of the environment caused by man can not only cause the destruction of ecological equilibrium, but also implies a response on the part of the environment – which is qualitatively altered – against humans. The new environmental conditions thus created are less favourable to a healthy life.

In the following examples the emphasis will be laid on disasters caused by both nature and man whose consequences, regardless their origin, are of huge impact, that is, for instance, forest fires (Photo 1), which can be considered a real crime against the environment.

Photo 1 – Forest fire [13]

Modern society, continuously challenged by change, acknowledges the priceless value of forests as a refuge and way of relaxation, and for its property of ensuring life-needed oxygen. Unfortunately, not all beneficiaries have regard for these sanctuaries of nature. The price of their reckless and indifferent acts most often leads to real catastrophes. Creating a fire in the forest and leaving the place without having extinguished it or doing it improperly, throwing a cigarette at random, uncovering a fire without warning the authorities about its existence or, even worse, committing arson in a forest area, are actions that can turn forests into torches.

The fire is an event that escapes one's control as far as its spreading, intensity and duration are concerned and that is why it requires the professional intervention of fire-fighters who have proper equipment and techniques, as well as the tactics, necessary to approach such emergencies (Photo 2).

Photo 2 – The intervention of firemen [13]

A fire will break out whenever three conditions are met: air, a source of heat (flame), and combustible matter (solid, liquid or gas). If one of these conditions is no longer fulfilled, the fire is extinguished, as the flame is no longer fed. The source of heat may be natural (such as lightening or volcanic lava) or man-made. Most cases of forest fire hazard are due to drought and very hot summer temperatures (known as canicula) – the forest, because of its wooden matter (i.e. trees), may be considered as combustible in its entirety. The fire can burn both live vegetation, such as trees, branches and leaves, and similar dead vegetation. A forest fire is classified as such when it causes a minimum area of 1ha to be destroyed.

The causes of forest fires are natural or anthropogenic. The influence of natural factors is explained by weather conditions and the characteristics of a particular type of vegetation. Summer periods, characterized by drought and strong winds, are favourable to the breaking out of fires, as the wind makes the soil dry faster and increases the risk that the fire could propagate a long distance. The heat dries the vegetation and, by evaporation, the volatile essences that ensure the propagation of fire are released into the atmosphere. Lightning and the incandescent projectiles from erupting volcanoes are also natural factors originating fires (Photo 3).

Photo 3 – The fight with fire [13]

The human factor is also a cause of fires because of improper behaviour as far as prevention regulations are concerned, that is to say owing to some people's behaviour (such as smoking or using fires for cooking) during entertainment activities. Besides carelessness and indifference to forest fires, what is also especially dangerous is the intentional act of setting on fire – arson – a forest for reasons that could not possibly extenuate the gravity of such a crime.

In recent years, the news on disasters due to forest fires has been continuously brought to public attention by mass media. In Portugal, a country with activities in the wood industry at a European level, had losses from fires that amounted to €1 billion and, of course, with catastrophic effects on human life and environmental damage. The Greek authorities called the devastating fire of 2007 "Europe's unprecedented catastrophe". Despite all the efforts and forces engaged in this terrible fight, the fire laid waste a large part of Greece and the scientists stated that the loss of the forest area equalled the effect on the atmosphere of the overnight doubling of the number of cars, and that it is also an essential loss for the natural cooling system.

In Peru, a fire quickly propagated to the forested sides of the Urubamba valley and endangered Machu Picchu, the ancient Inca city. The firemen intervened with great difficulty because of the uneven ground and the wind, and thick smoke prevented the helicopters and planes from spreading extinguishing substances. In Oregon, USA, over 200,000ha of forest burnt in 2002; in Belgrade and Kosovo, Serbia, over 300,000ha of forest burnt and the fires broke out in several places simultaneously. In 2003 in Australia, fire destroyed an area three times larger than Great Britain and quickly spread because of the canicula and the gusts of wind; it burnt to the ground the houses of the inhabitants, even killing people, despite the desperate efforts of thousands of firemen sent out on a mission to stop the fire.

The examples may continue and, as shown, all areas in the world are potentially affected. A forest fire, regardless of its cause, is devastating and very hard to beat, as the circumstances are always difficult: uneven ground, places out of reach, intervention forces which are hardly enough when compared with the intensity and extent of the fire, favouring conditions such as heat, drought, high wind, lack of water, and deficiency of equipment (such as special-need vehicles, helicopters and planes).

In Romania in 2007, every region of the country faced forest fires (Photo 4).

Photo 4 – "Slaughtered nature" [13]

Yearly almost 350ha of forest are lost in fires in Romania, the damage being most significant if the long-term effects are also taken into account: the extinction of flora and fauna which, needless to say, sometimes include rare species, the alteration of both environment and landscape, and the loss of human lives.

2. Control Materials and Methods

The intervention of firemen, besides the use of special-need vehicles (Photo 5), also involves the use of tools such as shovels, big brooms, sandbags, and rubber rugs to extinguish the fire on the ground.

Photo 5 – The use of special-need vehicles [13]

Because of the catastrophe that happened in Greece, with more than 65 victims and over 200,000ha of forest burnt to ashes, and all the other forest fires in Europe, in the autumn of 2007 the European Union discussed the creation of a European Civil Protection Force designed to intervene in the case of a disaster on the territory of any Member State.

3. Results and Discussions

Based on the information so far presented, several question arise naturally:

- What should be done?
- What are the steps to be taken in order to avoid such catastrophes? and
- What are the rules with which one must comply in order to reduce fire hazard?

In order to prevent forest fires from happening it is utterly important to convey recommendation-type information concerning the behaviour of citizens (adults as well as minors) and their education should focus on the way in which disasters can be prevented.

The first step in this direction is to comply with the specific safety procedures:

- the training of personnel involved in any activity in forest areas on wood exploitation;
- maintenance of forests with stress on safety regulations in the event of fire;
- the organization of patrol intense-rhythm activities during drought periods;
- good maintenance of forest roads so as to allow quick and easy movement of the special-need vehicles and intervention forces (Photo 5);
- ensuring water supplies by building accesses and arranging car platforms;
- appropriating forest ranges with means of first intervention (such as sand, shovels and big brooms); and
- planning pleasure spots and placing panels showing the main regulations for preventing and extinguishing forest fires.

Of course, the most important measure is that of educating people on what the prevention regulations are and how one is expected to properly behave in case of forest fire, namely:

- remain calm;
- if a member of a group (tourists etc.) then do not panic the others;
- leave the area immediately;
- inform immediately the inhabitants about the fire;
- call 112 – Dispatcher's Office for Emergencies and give the most accurate information possible; inform the authorities if there are any group of tourists or forest workers in the area in order to locate and evacuate them as well;
- if someone's clothes take fire, roll the person on the ground and stifle his or her clothes;
- if the air becomes unbreathable due to thick smoke, crawl (on one's knees and elbows); and help children, the old and the disabled to be evacuated first.

A forest fire always exceeds the capacity of one or two persons to extinguish it and that is why they should not waste time in trying to do so, but should inform the intervention forces as soon as possible. Also, they must warn all the people met on their way out of that area in order to give them the chance to save themselves by leaving the place immediately.

4. Conclusions

To prevent a forest fire or to make all the efforts to inform the intervention forces in due time is a duty of honour of every citizen, regardless of his or her citizenship, ethnicity or reason for being in that place – as inhabitant or tourist.

It is everyone's right and duty to watch over a healthy environment for both present and future generations.

References

[1] Boz Lăcrămioara, George Bariţiu University of Braşov, Romania, *Educatie şi prevenire*, Proceedings of the scientific meeting with international participation "SIGPROT 2005", 8th Edition, Alexandru Ioan Cuza Police Academy, Faculty for Firemen, Printech Publishing, 2005, 144 -149 Bucharest, Romania

[2] Boz Lăcrămioara, Dan Marciuc, George Bariţiu University of Braşov, Romania, *Concursuri de desene*, "Pompierii români" review, 7/2005/, Bucharest, Romania

[3] Boz Lăcrămioara, Dan Marciuc, George Bariţiu University of Braşov, Romania, *Rolul conlucrării cu O.N.G.-urile în educaţiaşsi prevenirea incendiilor*, "Pompierii ieşeni" review, 2005, 4th year, no. 4, Iasi, Romania

[4] Boz Lăcrămioara, George Bariţiu University of Braşov, Romania, *Educaţia şi prevenirea incendiilor în unităţile de învăţământ*", Proceedings of the scientific meeting with international participation "SIGPROT 2006", 9th Edition, Alexandru Ioan Cuza Police Academy, Faculty for Firemen, Printech Publishing, 2006, 47-53, Bucharest, Romania

[5] Boz Lăcrămioara, George Bariţiu University of Braşov, Romania, *Împreună pentru siguranţa copiilor*, "Salvatorii ieşeni" review, 2006, ", 5th year 5, no. 2, Iasi, Romania

[6] Boz Lăcrămioara, George Bariţiu University of Braşov, Romania, *Situaţii de urgenţă – educaţie şi prevenire în mediul rural*, Proceedings of the scientific meeting with international participation, "SIGPROT 2007", 10th Edition, Alexandru Ioan Cuza Police Academy, Faculty for Firemen, Printech Publishing, 2007, Bucharest, Romania

[7] Boz Lăcrămioara, George Bariţiu University of Braşov, Romania, *Vacanţă în siguranţă*, "Salvatorii ieşeni" review, 2007, 6th year, no. 2, Iasi, Romania

[8] Toma, Ovidiu, Alexandru Ioan Cuza University of Iasi, Romania, *Consortium Regional de Recherche Moldova - pour la Monitorisation et Protection d'Environnement – pour une meilleure gestion de la biodiversité*, Conférence internationale, sous le haut patronage de Monsieur Jacques Chirac, Président de la République française, et de Monsieur Koïchiro Matsuura, Directeur général de l'UNESCO, "Biodiversite: science et gouvernance", 2005, UNESCO, Paris, France

[9] Toma Ovidiu, Alexandru Ioan Cuza University of Iasi, Romania, *The University Regional Consortium (Moldavia) for Environment Monitoring and Protection – as a premise for the optimisation of living conditions and life*, the World Conference on Ecological Restoration "Ecological Restoration – A Global Challenge", 2005, Zaragoza, Spain

[10] Toma Ovidiu, Alexandru Ioan Cuza University of Iasi, Romania, *The Moldavian University Regional Consortium for Environment Monitoring and Protection – as a premise for the optimisation of living conditions and life and for student service improvement*, 1st Biennial Conference of the International EcoHealth and Ecology, "EcoHealth ONE: Forging Collaboration between Ecology and Health", University of Wisconsin, 2006, Wisconsin-Madison, USA

[11] Toma Ovidiu, Alexandru Ioan Cuza University of Iasi, Romania, *Recherche régionale pour la monitorisation et protection de l'environnement, gestion biodiversité*, journees scientifiques "Recherche et développement durable: approches, méthodologies, stratégies d'action et de formation", Centre de Recherches et de Transferts Technologies de l'Université Abdelhamid IBN BADIS-Chemin des Cretes-Mostaganem, 2006, Mostaganem, Algeria

[12] Toma Ovidiu, Alexandru Ioan Cuza University of Iasi, Romania, *The University Regional Research Consortium (Moldavia) for Environment Monitoring and Protection – as a premise for the optimisation of living conditions because of the prevention of natural and human ecological catastrophes*, NATO Security through Science Book, IOS Press, 2007, 1, Amsterdam, Holland

[13] *** www.igsu.ro (photos)

Optimisation of Disaster Forecasting and Prevention Measures in the Context of Human and Social Dynamics
I. Apostol et al. (Eds.)
IOS Press, 2009.
doi:10.3233/978-1-58603-948-6-11

Multiple Risk Assessment for Various Natural Hazards for Georgia

T. CHELIDZE, N. TSERETELI, E. TSERETELI, L. KALDANI, J. DOLIDZE, O. VARAZANASHVILI and D. SVANADZE
Department of Seismology and Experimental Physics, Institute of Geophysics, Alecidze Nr 1, 0171, Tbilisi, Georgia

Abstract. The critical importance of accurate mapping and assessment of natural hazards and risks is discussed, which hazards and risks relate not only to seismic events but also to floods and avalanches, and their consequences. Accurate mapping is crucial in the early warning process and there is evidence of considerable discrepancies in previous mapping efforts. These discrepancies can have major economic effects on, for example, investment potential and insurance factors. Thus, hazards must be more accurately defined and risk assessments must be based on multi-risk calculation methods.

Keywords. Georgia, earthquake, avalanche, landslide, debris flow, flash flood, population density, multiple risks, risk discrepancies

Introduction

The sustainable development of the Southern Caucasus (SC) region depends critically on the correct assessment of natural hazards that are characteristic for different areas of this mountainous region: earthquakes, landslides, debris flows, flash-floods and floods, and avalanches. The destruction caused in recent decades by strong seismic events (such as Spitak, Racha, Tbilisi and, Baku) and other natural hazards, has seriously affected the national economies of the SC countries. The rate of risks associated with these hazards increases every year due to the appearance of new complicated technological features such as oil and gas pipelines, communication lines, large dams, power stations, and chemical factories. The GIS-technology, together with space images, allows exact mapping of such risks and the assessment of integrated effects. For example, earthquakes induce many secondary effects that may cause even larger damage than the event itself. Combining maps of seismic hazard with maps of landslide-prone areas, lakes, and large engineering features, gives the chance to evaluate integrated hazards and risks. Exact cartography of hazards is very important for planning investments and insurance activities, as well as for providing for the safety of the population of the region.

The topic is in full agreement with main priorities of Hyogo Framework of Action, namely to identify, assess and monitor disaster risks and enhance early warning. The framework also calls for the promotion of: regional programmes, including technical cooperation; capacity enhancement; the development of methodologies and standards for hazard and vulnerability monitoring and assessment; the sharing of information;

and the effective mobilization of resources. This work was initiated by the specifically devised programme of *EUR-OPA: GIS Mapping of Integrated Major Hazards in the Southern Caucasus* as the early warning tool and *From Hazards to Risks – Comparative analysis of Assessment Techniques in the South Caucasus region*. According to the programme scientists from Georgia compiled GIS-based maps at a scale of 1:000 000 for five hazards (namely, earthquakes, landslides, debris flows, avalanches and flash-floods) because they cause the largest mortality and economic losses.

1. Disaster Cartography as an Early Warning Tool

At present the concept of early warning is considered as one of the main directions in disaster management and the importance of such systems was confirmed by the recent Sumatra earthquake and tsunami. At the same time it is clear that an early warning is possible only for some specific hazards, such as tsunamis, hurricanes and storms, when the source and propagation details are known exactly. For most disasters (such as earthquakes, landslides, volcano eruptions, debris flows, etc.) that information is partially or totally absent. The early warning of strong seismic events, based on the difference of propagation velocities of elastic waves and triggered electromagnetic signal, gives too short (from several to dozen of seconds) a warning time for realization of preventive activities and is quite expensive. That is why we think that the concept of early warning systems should include the probabilistic assessment/mapping of hazards and the recurrence period. This approach allows the implementation of activities which reduce considerably the losses and casualties. To develop that approach it is necessary to have the statistical information on disasters as well as effective monitoring system.

2. Mapping of Mass-Movement Potential on the Territory of Georgia: Criteria for Destabilization

The landslide and debris flow static zoning of Georgia is based on the integrated analysis of main factors, which upset the balance of forces and destroy the existing state of equilibrium. There are three main factors: (i) the state and properties of rocks, which defines the sensitivity of geological formations to impacting forces; (ii) the geometry and slope of the terrain (i.e. topography); and (iii) the climatic characteristics of the territory. For landslide-hazard mapping the criterion of "Landslide potential" was used and the intensity of landslide processes on the given area of a definite climatic zone, is characterized by the following two coefficients:

(i) coefficient of areal damage (K_p) which is the ratio of (a) area damaged by landslides F_p to (b) the entire area (F) of a given homogeneous geological space:

$$K_p = F_p/F$$

(ii) the "density" of landslide events (D) namely, the number of landslides (N) normalized to the same area F:

$$D = N/F$$

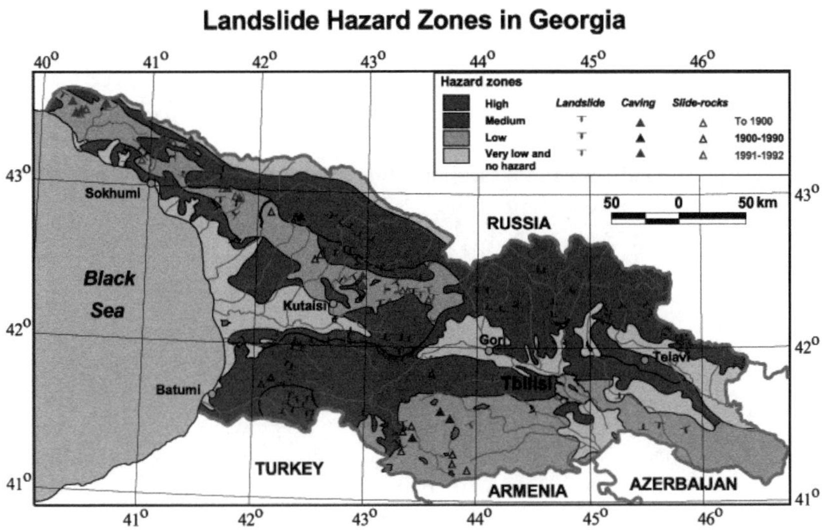

Figure 1. Landslide hazard zones in Georgia.

Both these coefficients vary in the range from 0 to 1. Their average values gives the integrated characteristic of landslide hazard (maximum is 1). Landslide hazard zones are delineated according to following rules (Fig. 1):

high >0.5
medium 0.1–0.5
low 0.001–0.1
no or very low ≤ 0.001.

Besides these stationary factors, mass-movements can be stimulated by short-term perturbations, such as earthquakes, intensive meteorological impact and man-made effects.

The time-dependence in the static maps can be introduced by taking into account the deviation of meteorological parameters from the long term average (LTA), mainly considering the intensity and duration of deviation. Accordingly, the following three situations can be distinguished: 'stable', 'normal' and 'extreme'. The 'stable' situation is expected when the precipitation and air humidity is less than the long-term average. The situation is 'normal' (i.e. background state) if the deviation is small; say if the precipitation exceeds the long term average by no more than 100mm per year. The 'extreme' situation is expected when precipitation exceeds the LTA by 200–400 mm.

For assessment of debris flow hazards (Fig. 2) a combination of several parameters is used, namely:

(i) ratio of total length of debris-flow generating rivers ($\sum l$) to the total length of a given river basin (L):

$$K_{s1} = \sum l / L;$$

(ii) ratio of number of generating rivers of debris flow basin ($\sum n$) to the number of rivers of a given basin, where debris flow events were not registered (n):

$$K_{s2} = \sum n / n;$$

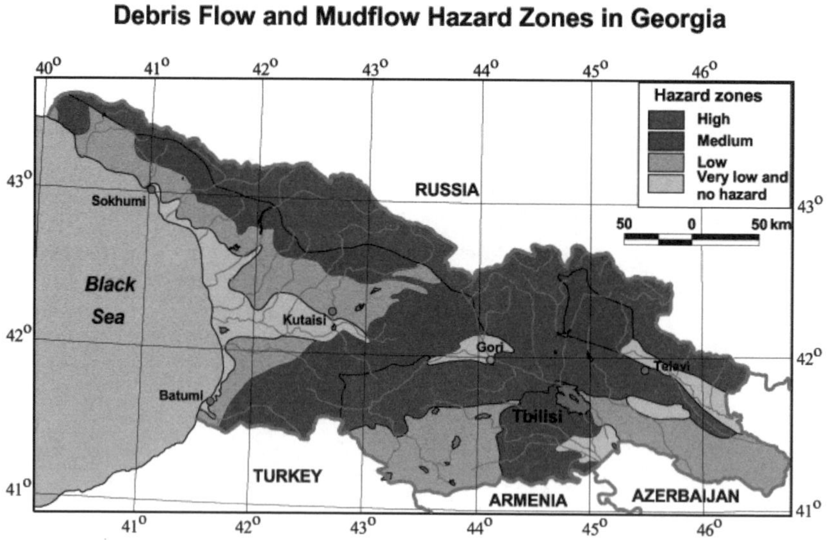

Figure 2. Debris flow and mudflow hazard zones in Georgia.

(iii) ratio of total area of debris-flow generating rivers ($\sum f$) to the total area a given river basin (F):

$$K_{s3} = \sum f/F; \text{ and}$$

(iv) ratio of total area of active debris-flow sources that feed generating rivers ($\sum s$) to the total area of these river basins (F):

$$K_{s4} = \sum s/F.$$

3. Mapping of Flash-Flood and Flood Hazards in Georgia

The orography (i.e. the nature of the mountainous terrain) of the territory is of decisive value in assessing flash-flood and flood hazards. The flooding of a river basin can be caused by:

(i) melting of snow cover, especially when the air temperature is rising fast and there is intensive rain;
(ii) heavy showers in the summer/autumn period;
(iii) incessant autumn rains, covering large part of a river basin; and
(iv) intensive winter rains of short duration in the seaside areas of the Black Sea.

For the South Caucasus the most typical hazards are rivers with springtides, rivers with high waters in the warm period of a year, and rivers with flood flows. Maximal water discharge during such anomalous events can be almost 30 times larger than the average annual water discharge.

The critical values of precipitation per 12 hours that cause disastrous water flows, and flooding in rivers and dry ravines are:

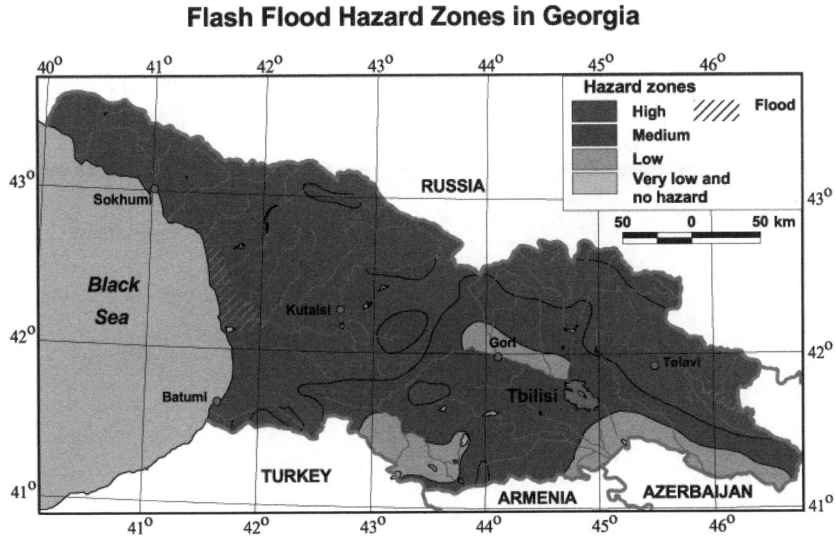

Figure 3. Flash flood hazard zones in Georgia.

(i) in the seaside regions of Western Georgia >130 mm;
(ii) in the central and western part of Colchis lowland and adjoining mountains slopes >100 mm;
(iii) in the remaining part of Western Georgia, on the Southern slopes of Larger Caucasus >80 mm; and
(iv) in the remaining part of Eastern Georgia – 60 mm.

Using these critical values, the recurrence rates of disastrous heavy rains are calculated and corresponding flash-flood hazard maps are compiled (Fig. 3).

High >16%: recurrence once in less than 6 years;
Medium 8–16%: recurrence once in 6–12 years;
Low 4–8%: recurrence once in 12–25 years; and
No or very low <4%: recurrence once in 25 years.

4. Avalanche Zoning of Georgia

For avalanche hazard zoning of Georgia the following two quantitative parameters are used:

- “density” of avalanche sources, namely, the number of avalanche sources per kilometre of valley (n/L); and
- recurrence rate of avalanche events, namely, the number of avalanches generated by a given source in ten years ($n_i/100$).

For mapping of sources the results of numerous expeditions carried out for many years are used and for the assessment of recurrence the data is obtained from meteorological stations. These data allow development of theoretical prognostic method.

The following gradation of avalanche hazard is developed: *high, medium, low and no hazard* (Fig. 4). A ‘high’ hazard grade corresponds to the following criteria: the

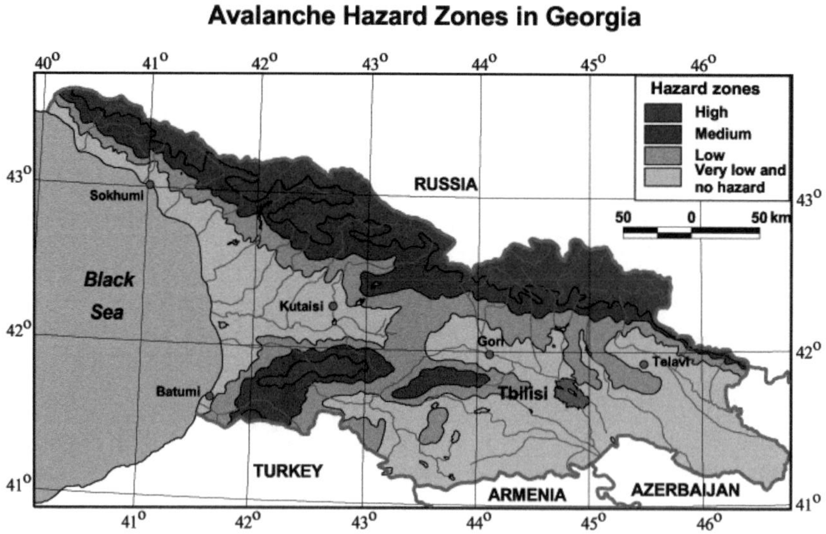

Figure 4. Avalanche hazard zones in Georgia.

Table 1. Distribution of high, medium and low avalanche hazards

Avalanche hazard	Georgia	
	F (km^2)	%
High (H)	10,700	15
Medium (M)	12,000	17
Low (L)	12,900	19
Σ (H+M+L)	35,600	51
No hazard	34,100	49
Σ	69,700	100

number of avalanche sources per kilometre of valley exceeds 5 ($n/L > 5$), and the recurrence rate of avalanche events exceeds 10 in 10 years ($n_i/100 > 10$). A 'medium' avalanche hazard is recognized in areas where the number of avalanche sources per kilometre of valley is less than 5 ($n/L < 5$) and/or the recurrence rate of avalanche events is less than 10 in ten years ($n_i/100 < 10$). The avalanche hazard is 'low' if the number of avalanche sources per one kilometre of valley is less than 1 ($n/L < 1$) and the recurrence rate of avalanche events is less than 1 in 10 years ($n_i/100 < 1$). "No or very low" hazard zone is free of avalanches (no hazard). In Georgia the area prone to avalanche hazard covers 33% of the territory. Table 1 shows the distribution of high, medium and low avalanche hazard in the region.

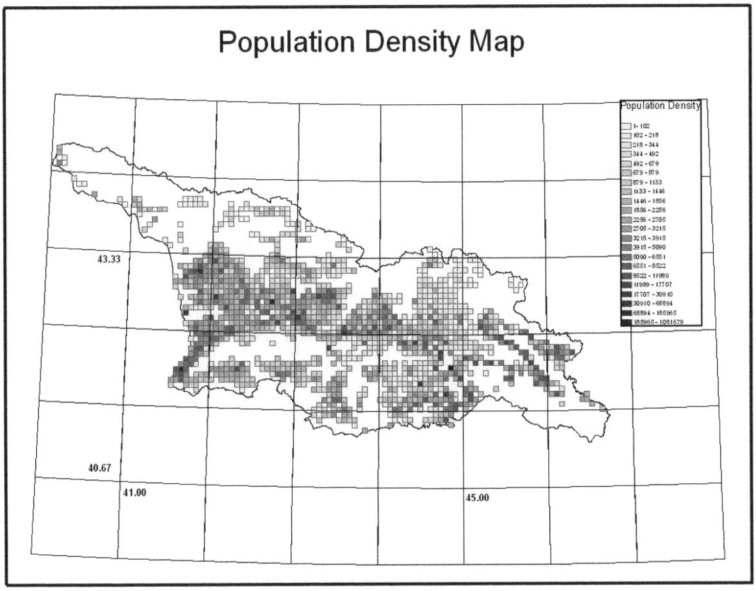

Figure 5. Population density map.

5. Principles of Multi-Risk Calculations for Natural Hazards: The Case of Georgia

A risk index related to total economic losses was estimated for Georgia and the technique used can be applied in the future to the whole territory of the South Caucasus. Five types of hazard (earthquake, landslide, debris flow, avalanche and flash flood) were considered for the period 1980–2005. The distribution of risk index by hazard type is based on the method developed by the Columbia University group during compilation of the Map of Global Natural Disaster Risk Hotspots (Synthesis Report – Natural Disaster Hotspots: A Global Risk Analysis, 2005).

The investigated territory of Georgia was divided on a $2.5' \times 2.5'$ latitude-longitude grid, with each grid considered as a unit area. According to the mentioned method, the risk assessment is based on two data sets: population (Fig. 5), and total Gross Domestic Products (GDP) per unit area (Fig. 6). For Georgia the GDP is not available for unit areas but only for sub-national units and territorial entities. There are some data that show the share in total GDP of territorial entities from the corresponding district centre. The national macro-economic parameters were calculated by the Ministry of Economical Development of Georgia in 2005. This allowed the estimation of GDP for district centre units. These estimates were applied to population densities using the description of population in 2002 as supplied by the Statistics Department of the Ministry of Economical Development of Georgia.

The following quantities were calculated:

1. The total economic losses, estimated from 1980–2005 for each hazard – E_i.
2. The total Gross Domestic Product (GDP) for the areas affected by the i-th hazard was estimated for the period 2002 – GDP_i.

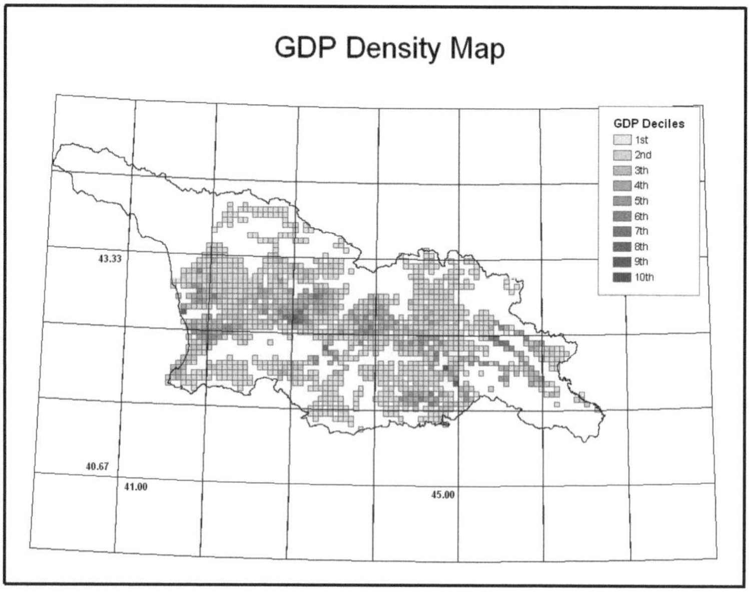

Figure 6. GDP density map.

3. The regional economic loss rate R_i was computed for the i-th hazard:

   $$R_i = E_i / GDP_i$$

4. For each grid cell that falls into a hazard zone, the local expected economic loss was calculated:

   $$E_{ij} = R_i * GDP_j$$

   where GDP_j is taken per unit area. As mentioned above, we applied GDP of district centre units to the map of population density. Rural areas also were not taken into account.
5. The poverty headcounts (PH) in districts was estimated by UNDP in Georgia (Poverty mapping in Georgia. 2003). PH measures the share of the population for which consumption or income is less than the poverty line. As GDP for Georgia is applied to the population density, we decided to take into account the PH for districts multiplied the GDP of districts by the parameter $(1 - PH_n)$, where 'n' denote the districts.
6. The degree of hazard was expressed, in our case, in terms of frequency for earthquakes (i.e. regional economic loss rate multiplied by hazardous event frequency Ω) and the exposed area. The accumulated economic loss in the grid cell is: $E_{ij} = R_i * \Omega_j * GDP_j$. For other natural hazards the degree of hazards was expressed instead of the frequency (as we do not have frequency for those hazards) in terms of the gradation of corresponding hazard maps, the gradations of 'no or very low', 'low', 'medium' and 'high' were assigned values of 1, 2, 3, and 4 respectively.
7. The resulting weighted economic loss (EL) from hazard I is:

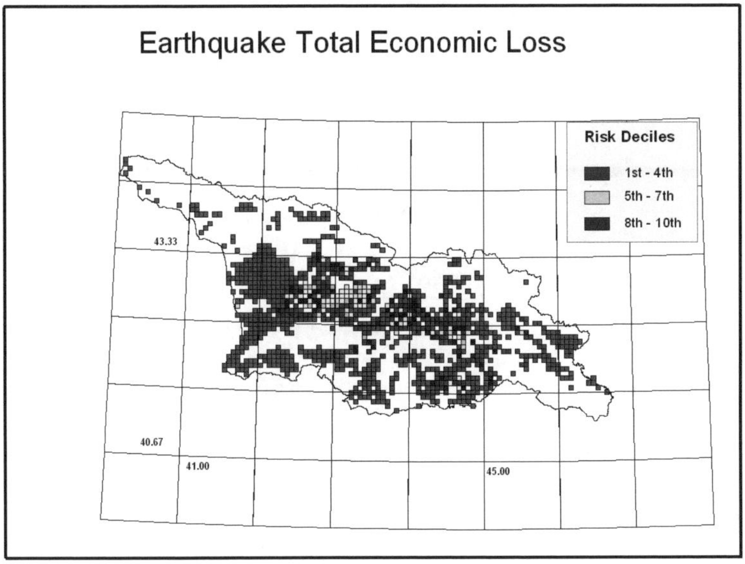

Figure 7. Earthquake total economic loss.

$$E_{ij}^{*} = E_{ij} * E_{i} / \sum_{i=1}^{v} E_{ij},$$

where ν is the number of grid cells in the area exposed to hazard i.

8. The economic loss – multi-hazard disaster risk index (RI) can be sum of the single-hazard economic losses (in our case for five hazard) in the grid cell:

$$\mathrm{RI} = \sum_{i}^{5} E^{*}_{ij}.$$

The risk indices for these five hazards are expressed in deciles. The top three deciles of cells were chosen for calculation of the summary multi-hazard risk index (Figs 7–12). As mentioned above the risk index was calculated for the period 1980–2005. (Due to the difficult political situation the risk index was not calculated for the Abkhazeti region.) It is evident that the risk index map differs considerably from the hazard maps, as it takes into consideration such additional parameters as GDP and population density.

6. Conclusions

There are big discrepancies in hazard and risk assessments, in particular, for the Caucasus region in different World Disaster Maps (for example, between Global Natural Disaster Hotspots Map and the Map of Global Seismic Hazard Assessment Program GSHAP and World Map of Natural Disasters of Munich Group). According to the Hotspot Map, the Southern Caucasus is prone only to hydro-meteorological hazards while

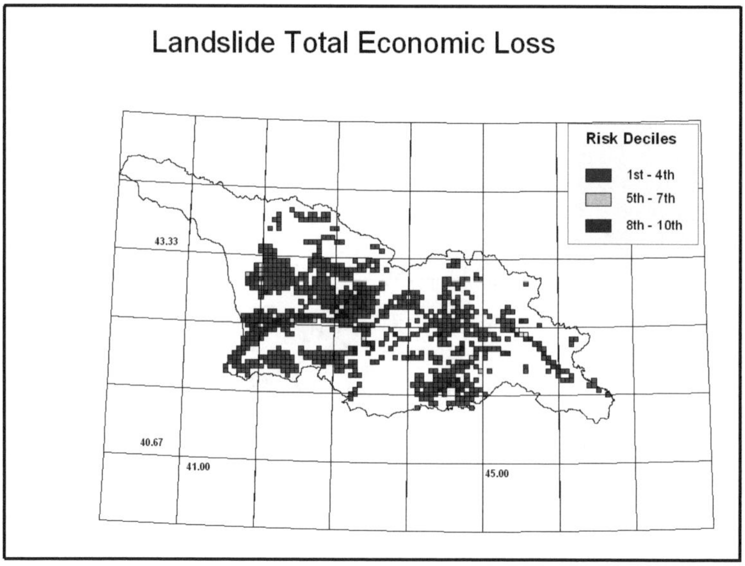

Figure 8. Landslide total economic loss.

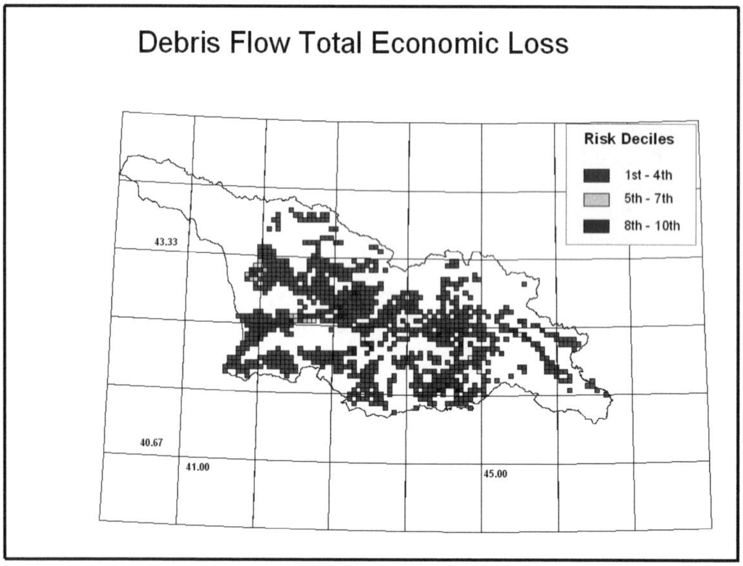

Figure 9. Debris flow total economic loss.

the Northern Caucasus is subject to geophysical and hydro-hazards. Geophysical hazards include earthquakes, volcanoes and landslides. If it can be accepted that the hydro-hazards for the both regions are the same, the relative assessment of geophysical hazards, namely earthquakes and landslides, the calculated risk for these two parts of Caucasus is wrong. The landslide risks for both parts of Caucasus are approximately the same and the seismic activity of Southern Caucasus is larger than in the North. The

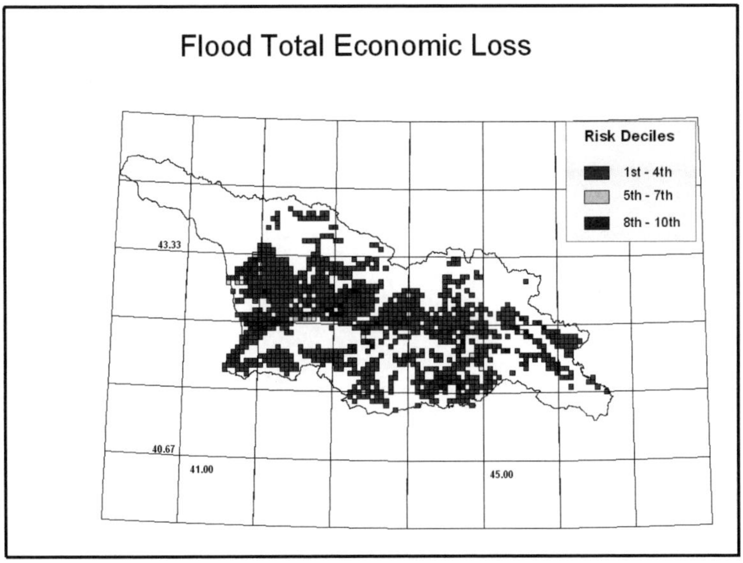

Figure 10. Flood total economic loss.

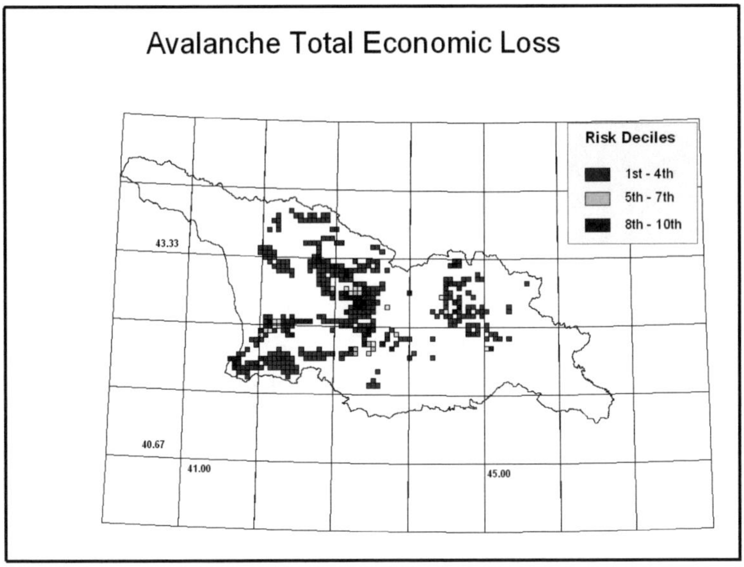

Figure 11. Avalanche total economic loss.

sources of Hotspot Map Assessments were GSHAP maps for PGA and the database of EQ of M > 4.5 occurred in 1976–2002 from the Advanced National System EQ Catalogue. It is easy to see that the GSHAP map gives for the PGA in the North mainly in the range 0.2–0.3 g and for the South – in the range 0.2–0.4 g. The number of EQ of M > 4.5 is three times larger in Southern compared to Northern Caucasus. Besides, recurrence times of M > 4.5 EQ-s in the North and South are approximately the same.

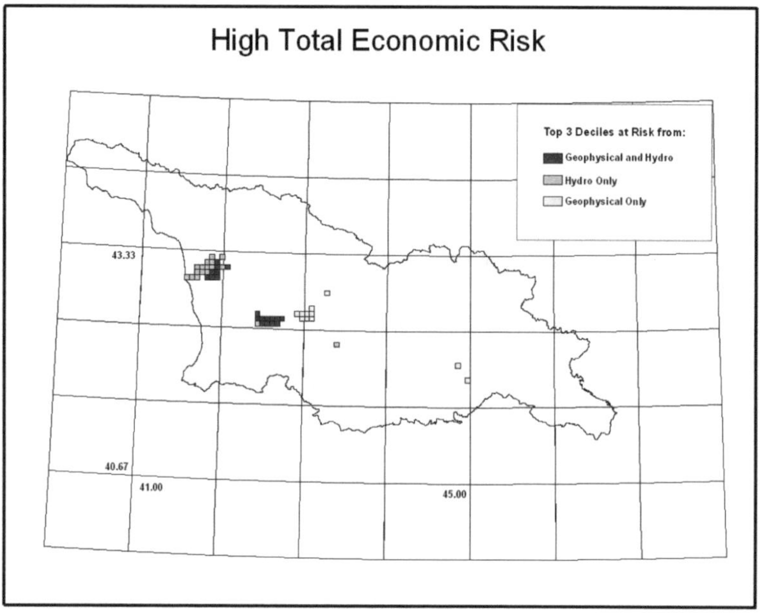

Figure 12. High total economic loss.

Thus it is concluded that during the compilation of the Hotspot map the input data were not analyzed correctly and the map needs serious revision in Caucasus region. Earthquake risks are much larger for the countries of South Caucasus, which contradicts the Hotspot Map assessments which states that this region is prone only to hydrological risks. These discrepancies may cause serious difficulties for investors and insurance companies. It is concluded that it is of major importance to refine hazard and risk assessments for South Caucasus using detailed local data.

References

Editor: Chelidze, T., 2006–2007. ATLAS of GIS-based maps of natural disaster hazards for the Southern Caucasus (earthquakes, landslides, debris flows, avalanches and flash-floods).

van Westen, C., van Asch, T. and Soeters, R., 2006 Landslide Hazard and Risk Zonation – why is it still so difficult? Bull. Eng. Geol. Env., 65; 167–184.

Dilley, M., Chen, R., Deichmann, U., Lerner-Lem, A., Arnold, M. et al., 2005. Natural Disasters Hotspots: a Global Risk Analysis. http://www.ldeo.columbia.edu/chrr/research/hotspots/.

Poverty mapping in Georgia. 2003. UNDP Country Office in Georgia.

World Map of Natural Hazards. http://www.munichre.

Optimisation of Disaster Forecasting and Prevention Measures in the Context of Human and Social Dynamics
I. Apostol et al. (Eds.)

doi:10.3233/978-1-58603-948-6-23

A Model of Sustainable Management for Forests: Prediction and Prevention of Natural and Man-Made Disasters

Valentin POPA[a,*] and Ovidiu TOMA[b]
[a]*Forest Engineer, Executive Director – Association of Landowners "Asociaţia Obştilor Vrâncene", Năruja, 627 220, Vrancea/Romania*
Tel. (+40 237)677 017; Fax (40 237)677 017
[b]*Professor PhD., Faculty of Biology, Alexandru Ioan Cuza University of Iasi, Bd. Carol I, no. 20 A, 700505 Iasi/Romania*
Tel. (+40 232) 201630; Fax (+40 232) 201472;
http://www.bio.uaic.ro; http://www.bio.uaic.ro/content/view/46/43/
E-mail: otoma@uaic.ro

Abstract. The sustainable management of forests is the key to preventing disasters, both natural and man-made. A healthy community activity fulfils, by means of education and responsibility, the concept of sustainable development.

Keywords. Sustainable management, forest, prediction, prevention, manmade disasters

Introduction

The mountain area of Vrancea County in Romania is known as being an area with a high potential for disasters. The region which represents the epicentre of the highest intensity earthquakes in Romania actually overlaps this area. The large hydrographical basins cause the collection of huge quantities of pluvial water which can lead to extensive floods in the plain areas of the county (Photo 1).

The erosion phenomena are facilitated by the friable, unstable layer of stones. The distribution of the geological layers facilitates landfalls over extensive areas, followed by the shaping of temporary and unstable reservoirs. The channels are permanently affected by these landfalls and by the flash floods during the summer.

1. General Considerations

Global climate changes increase the possibility of disasters in the area [1–4]. Uncontrolled deforestation of large areas may lead to instability in the forest medium, risking the safety of the people living in the surrounding communities (Photo 2 and 3).

* Corresponding Author: E-mail: valipopa64@gmail.com.

Photo 1. Maximal level of the floods.

Photo 2. Deforestation.

The intense downpours, with more than 110 l/m^2 over a period of 2–3 hours, can cause the destruction of more than 60% of communication routes. The large volume of infiltrated water determines its penetration into the surface layer of the soil. The land-

Photo 3. Effects of deforestation.

falls on surfaces of 20–100 ha determine the formation on the river bed of barrages of silt slime which give birth to reservoirs of more than 300,000 m^3.

An earthquake having an intensity of 5–6 on the Richter scale can lead to these barrages becoming fluid and, in consequence, the outpouring of the accumulated water and affecting downstream sites.

To view things even better, lightning during thunder storms can determine the emergence of a chain of fires in the woods which can by no means be controlled. The combustible materials, of which the resinous wood forms the majority, make it impossible for any terrestrial intervention whatsoever.

Extreme phenomena:

- Earthquakes of high intensity: 1944, 1977, and 1990;
- Heavy rainfall: 2005 – more than 110 l/m^2; 100 km of forestry roads and 40 km of public roads were damaged with more than 20 communities being isolated for several months; and
- Landfalls: Zabala 30 ha forming a reservoir of more than 300,000 m^3; the landfall still continues: Naruja Palcau 20 ha.

2. Materials and Methods

An area of more than 52,000 ha belongs to the local communities, established in associations based on an archaic organization known as "obsti". The property form is a private and not a public one but, as a peculiarity, the form of manifestation of the property right is a common one. This form of manifestation of the property has remained unchanged for several thousands of years. There are similar forms all over the Alpine-Carpathian mountain chain.

The resistance to change made it possible for this archaic form of organization to endure. One has to know that this organization based on association does not overlap the state administrative forms of organization. The only period when the administration of the property was not possible was 1948–2001, although, a legal act of nationalization of the forestry properties was never published.

During 2000–2003 some of the communities were given assistance by means of a programme initiated by the University of Auburn, Alabama, United States of America. The programme was run with the help of United States Department of Agriculture (USDA) and World Learning, and was intended to create a responsible attitude towards the process of restoring the forestry properties and towards the assessment of the impact it had on the sustainable management of the forests.

The fact that at present there is a strong Association of Landowners called "Asociatia Obstilor Vrancene", which stands for 14 rural communities having more than 15,000 inhabitants, is genuine proof of the high interest manifested by the beneficiaries of the programme.

3. Results and Discussions

Assuming management of the forests came naturally with the principle of protecting nature for the benefit of the communities. A responsible act of respecting the principles of continuity for the vegetation of the forests, which correlated with the idea of satisfying the needs of a community development, in close connection with the idea of preserving the biodiversity, led to the development of certain connections with main actors in this field of activity. These included World Wildlife Fund (WWF) Romania, Regional Research Consortium for Environment Monitoring and Protection/Faculty of Biology – "Alexandru Ioan Cuza" University from Iasi, and the Faculties of Forestry from Suceava and Brasov, all from Romania. The actions have achieved a good result by means of thematic exhibitions, by the assessment of the inventory of the natural capital, by assuming the management of the natural reservations included in "Nature 2000" network, and, finally, by obtaining the certification of the forests in accordance with the Forest Stewardship Council (FSC) model.

One might think about an important rupture between private property and environmental protection, considering the realities of the post-revolutionary Romania. No anomalies were registered where the training process for the landowners was developed systematically and responsibly. Whatever problems occurred was because some special areas of protection on the private properties were created without consulting the landowners and without paying them any compensation. The system of consulting and monitoring the phenomena in the relationship, private landowner-environmental protection on a national level is created only by means of the control function, without the feed-back function.

The educational component was accomplished mainly with the help of the NGOs which developed short and accurate programmes. There are no programmes in schools to make children aware of the importance of preserving biodiversity, or of the relations between humankind and nature, or about the healthy cohabitation between the two. Even if Vrancea is considered a seismic area, the level of training of its inhabitants is still very low. Any potential disaster here can result in huge damage and there is no educational system in the schools to train people how to reduce the effects of possible calamities.

Photo 4. Example of rebuilt road.

Responsible decentralisation began in 2003 when the local communities, known as "obsti", decided they could take the management of the forestry properties in their own hands by means of forestry districts. Their five years' existence proved to be both challenging and profitable; starting with rebuilding the whole local infrastructure, then moving on with establishing formal institutional systems and centres of culture, and ultimately assuming the responsibility for making every route accessible, in case of urgency, in the forest area. They assumed total responsibility and after two years of calamities the road system is 60% rebuilt with a particular focus on the gateways (Photo 4).

The financial efforts are significant but, nonetheless, are disregarded by the authorities. The process of sustainable management moves on. Every new year comes with new goals and targets and each and every effort is directed towards carrying them out.

For the next four years the educational component will play the leading part as it is considered a priority. Special programmes are to be expected in this respect, programmes which will be run by the "obşti" in close co-operation with the schools.

4. Conclusions

The certification of forests and the application of the "Leader" programme are two forms of self-consciously assuming sustainable development. It is easy then for anyone to observe that we are permanently living with imminent disasters which could bring enormous damage to the rural communities. One cannot separate the continuity of the rural communities from the sustainable management of the resources and of the actions of preventing disasters.

The rural community in the mountain area of Vrancea county proves that through the stability of the organisational forms and by responsibility in the sustainable management of the properties, they are not likely to repeat the mistakes they made during 1900–1925 when, after signing generous contracts with companies of forestry operation, they caused an ecological lack of balance.

The future role of the Local Action Group (LAG) – with the generic name of "The Country of Vrancea" – associated with the first forestry property certified by the Forest Stewardship Council (FSC), leads us to the conclusion of using the best actions in taking a hold of the good practices of managing in a sustainable manner the existing resources, actions which have to be correlated with models of efficiency in case of possible major natural and man-made disasters.

References

[1] Toma Ovidiu, Alexandru Ioan Cuza University of Iasi, Romania, Consortium Regional de Recherche Moldova – pour la Monitorisation et Protection d'Environnement – pour une meilleure gestion de la biodiversité, Conférence internationale, sous le haut patronage de Monsieur Jacques Chirac, Président de la République française, et de Monsieur Koïchiro Matsuura, Directeur général de l'UNESCO, "Biodiversite: science et gouvernance", 2005, 1, UNESCO, Paris, France.

[2] Toma Ovidiu, Alexandru Ioan Cuza University of Iasi, Romania, The University Regional Consortium (Moldavia) for Environment Monitoring and Protection – as a premise for the optimisation of living conditions and life, the World Conference on Ecological Restoration "Ecological Restoration – A Global Challenge", 2005, 1, Zaragoza, Spain.

[3] Toma Ovidiu, Alexandru Ioan Cuza University of Iasi, Romania, Recherche régionale pour la monitorisation et protection de l'environnement, gestion biodiversité, journees scientifiques "Recherche et développement durable: approches, méthodologies, stratégies d'action et de formation", Centre de Recherches et de Transferts Technologies de l'Université Abdelhamid IBN BADIS-Chemin des Cretes-Mostaganem, 2006, 1, Mostaganem, Algeria.

[4] Toma Ovidiu, Alexandru Ioan Cuza University of Iasi, Romania, The University Regional Research Consortium (Moldavia) for Environment Monitoring and Protection – as a premise for the optimisation of living conditions because of the prevention of natural and human ecological catastrophes, NATO Security through Science Book, IOS Press, 2007, 1, Amsterdam, Holland.

Optimisation of Disaster Forecasting and Prevention Measures in the Context of Human and Social Dynamics
I. Apostol et al. (Eds.)
IOS Press, 2009.
doi:10.3233/978-1-58603-948-6-29

GIS Application for the Assessment of Seismic Damage to Buildings

Anton ZAICENCO and Vasile ALKAZ
Institute of Geology and Seismology, 2028, str. Academiei 3, Kishinev, Moldova
E-mail: Anton.Z@softhome.net

Abstract. Geographical Information Systems (GIS) have been found to be very useful in seismic hazard and risk assessment studies. GIS can be used to integrate vast amounts of data geographically, take the spatial distribution of phenomena into consideration and communicate the results graphically, performing analysis of complex mathematical models. The central part of Chisinau (6.3 km^2 site), capital of the Republic of Moldova, has been the case study in a project aimed at the assessment of vulnerability of buildings to seismic impact. The city is exposed to Vrancea earthquakes experiencing $PGA \cong 300$ cm/s^2 for recurrence interval $T = 475$ yr [6]. Collection, classification and digitization into ArcView GIS format of the main characteristics of the subsoil, such as mean shear wave velocity, natural period of vibration and amplification factor, compilation of database for the existing structures, as well as construction of Digital Terrain Model (DTM), were performed. The final product is the GIS database and software module for purposes of evaluation of seismic damage to buildings. The incorporation of the amplification capacity of the soil through the direct utilization of the transfer function constructed on the base of geotechnical data, allows fast assessment of scenario seismic events and mapping of parameters of the ground motion (PGA, EPA, etc.). The existence of moderate-magnitude and blast records for the studied site, as well as databases of building damage, allow validation of the accepted techniques and methodologies for ground motion and damage simulations.

Keywords. Seismic damage, GIS, Vrancea earthquakes

Introduction

Vulnerability of the building stock to seismic impact is determined mainly by two factors: (i) *demand* – in terms of structural loads expected on the given site and (ii) *capacity* of the structure to withstand this level of shaking with the accepted damage. Structural seismic loads are represented by the site-response spectra and their parameters, which are influenced by such factors as: seismic source mechanism, geological structure of the region, local soft soil conditions, topography, etc. The presence of the geotechnical databases, including measured shear wave velocities for the subsoil of Chisinau, allows detailed investigation of the influence of soft soils on spectral and amplitude level of free-field ground motion. The corresponding structural damage could be obtained either from correlations with the ground motion parameters, or from damage functions [3]. GIS proved to be the adequate tool for storing, processing and mapping of the spatial information [2], such playing the main core in seismic zoning studies.

The extreme usefulness of GIS utilization is emphasized, which provides a powerful tool for performing spatial analysis of the data on building damage, soil parameters, seismic records, etc., which are geographical data by their nature. In GIS, mapping and

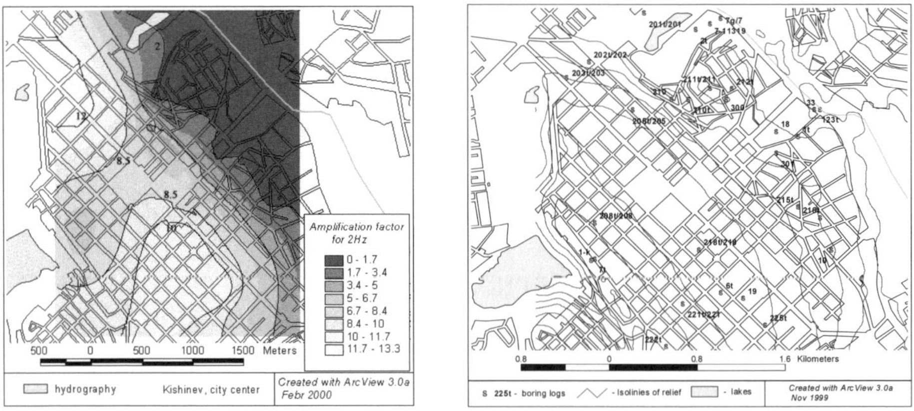

(a) (b)

Figure 1. Chisinau city centre: (a) map of values of spectral amplification function at frequency 2 Hz, (b) location of 28 digitised borehole logs with geotechnical measurements.

analysis are intimately linked, combining visual display with numerical summaries [10].

1. Geotechnical Data

The necessities of accounting for peculiarities of soil conditions are recognized by numerous guidelines for seismic microzoning [11] and usually include such parameters as natural period of soil vibration, T_o, and amplification ratios, A, of the surface response with respect to that of the free-field motion of the outcrop.

In the frame of the pilot project, 28 borehole logs with detailed geotechnical information including measured in-situ shear wave velocities v_s, were digitized and introduced into GIS for the central part of Chisinau [9]. For each of the borehole log the amplification function for horizontal ground motion was constructed on the base of 1-D wave propagation model (SHAKE [5]), thus providing values of spectral amplification. Created GIS-compatible software for processing geotechnical data, allows interpolation of selected parameters for each grid point within the studied zone. In this way, maps of spectral amplification for different frequencies could be constructed (Fig. 1).

In addition, maps of natural periods of soil vibration, influenced strongly by the depth to the bedrock ($v_s \geq 750$ m/s), were obtained using the compiled geotechnical database. The map of the bedrock surface was developed with the *TIN* model, which partitions a surface into a set of contiguous, non-overlapping triangles.

2. Database of Structural Damage

European Macroseismic Scale EMS-92 [4] was taken as a tool for building damage classification of the compiled databases for the existing building stock in the city of Chisinau. The total amount of records was 1,870, while vulnerability class *B* (masonry) buildings, constituting ≈45% of the total number of structures, were chosen as the sample space that provided the most reliable information both from a spatial distribution point of view as well as structural uniformity.

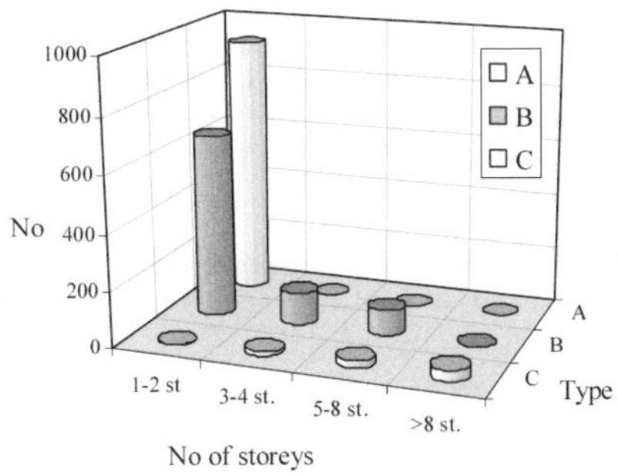

Figure 2. Statistics of building stock in the city (according to EMS-92 scale) with the recorded damage after August 30, 1986 Vrancea earthquake (magnitude $M_w = 7.2$).

Afterwards, the buildings with 5–6 stories were selected for the statistical analysis, these having approximately the same natural period of vibration T_o ($T_o \approx 2.5$ Hz, [12] Part 1–2, Annex C) and being designed according to similar building code (base shear force coefficient: 13.5–15% for $T_o < 0.4$ s). For this type of buildings the observed damage degree after 1986 earthquake was 0–3.

3. Simulation of Ground Motion Parameters

The assessment of structural damage from earthquake impact is performed on the basis of the seismic hazard assessment of the Vrancea seismic zone as well as vulnerability of the existing facilities designed according to certain building code.

The expected response spectrum for the medium soil conditions is derived on the basis of the attenuation relationship of Peak Ground Accelerations (PGA) from the Vrancea source for the sector containing Moldova [6], using Joyner–Boore model:

$$\ln PGA = c1 + c2\, M_w + c3\, \ln R + c4\, h + \varepsilon \tag{1}$$

where: PGA = the peak ground acceleration at the site, M_w = the moment magnitude, R = the hypo-central distance to the site, h = the focal depth, $c1$, $c2$, $c3$, $c4$ – data dependent coefficients and ε = random variable with zero mean, and standard deviation $\sigma_\varepsilon = \sigma_{lnEPA}$.

Normalized response spectrum shape (5% damping) compatible with Eurocode-8 format [6] is obtained on the basis of the statistical analysis of 20 components of seismic records from Moldova and neighbouring Romanian territory.

The influence of local soft soil conditions is considered by direct utilization of the soil transfer function. The corresponding database of these functions is compiled for the test zone.

The procedure of calculating the free-field acceleration damped response spectra, taking into account the influence of soft soil, is the following:

1. The normalized response spectrum for the test site is multiplied by the corresponding value of the PGA from the attenuation curve, yielding a spectrum for medium soil conditions at a given hypocentral distance and defined earthquake magnitude. The calculated spectrum is scaled afterwards to obtain the spectrum on the bedrock;

2. The damped response spectrum is divided by the peak factor, r_{sp}, at a set of target frequencies ω. Thus, the *σ-spectrum* is obtained [7]:

$$r_{s,p} = \sqrt{2 \cdot \ln(2n)} \qquad (2)$$

where: $n = [-\ln(p)]^{-1}\ \Omega_y \cdot s/(2\pi)$ – average number of cycles of response motion;

3. The power spectral density function, *PSD*, is derived from *σ-spectrum*. By this the *PSD* function, $G(\omega)$, compatible with damped response spectrum is received, for a given duration of the motion and assigned probability of exceedance [7]:

$$\sigma_a^2 = G(\omega_n) \cdot \omega_n (\frac{\pi}{4\xi} - 1) + \int_0^{\omega_n} G(\omega) d\omega \qquad (3)$$

where: ξ – structural damping;

4. The *PSD* function is convolved with the transfer function of the soft soil in frequency domain, which yields *PSD* function of the free-field motion:

$$G_{top}(\omega) = G(\omega) \cdot |H_{soil}(\omega)|^2 \qquad (4)$$

5. σ-spectrum is derived back from *PSD* function and converted into a damped response spectrum:

$$SA(\omega) = \sigma_{top}(\omega) \cdot r_{s,p}(\omega) \qquad (5)$$

Yet, the applied procedure does not take into consideration the non-linearity of the soil behaviour. The assumption of linearity could be accepted with certain confidence for the medium hypocentral distance of Chisinau in respect to Vrancea seismic source, when no non-linear effects in soil behaviour were observed during strong earthquakes.

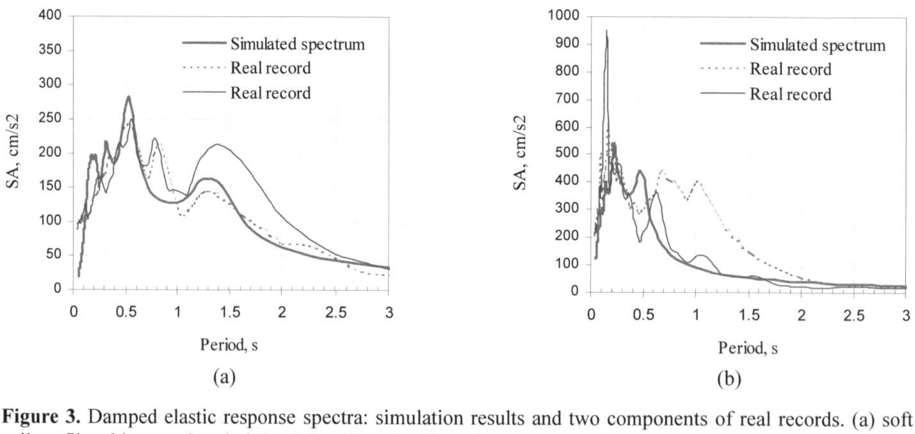

Figure 3. Damped elastic response spectra: simulation results and two components of real records. (a) soft soil profile with natural period T = 1.5 s; (b) soft soil profile with natural period T = 0.5 s.

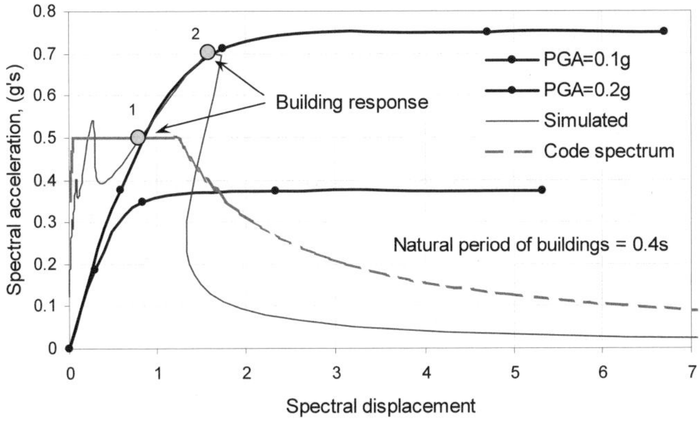

Figure 4. Evaluating building seismic damage degree (capacity spectrum method).

The degree of structural damage was studied in the context of the level of ground shaking expressed in terms of effective peak accelerations, *EPA*, and values of response spectra, SA_{2Hz}, for the natural period of building's vibration 2.5 Hz (Fig. 4).

Increased building response was observed in case of simulated response spectrum (intersection at point "2" in comparison with building code spectrum (point "1") due to consideration of local soft soil conditions influence (Fig. 3a).

4. Mapping of Ground Shaking Parameters

The accuracy of the applied GIS models depends strongly on the adopted interpolation methods, which should be properly chosen in connection with mapped physical parameters. Kriging interpolation method was employed, being in wide use in soil science and geology. It is an advanced interpolation procedure that generates an estimated surface from a scattered set of points with *z* values under hypothesis of spatial homogeneity. The digital elevation model (DEM), as well as other interpolated contours and surfaces, are created on the basis of the Kriging method.

Figure 5 provides a 3-D scene generated with ArcView GIS for Chisinau city centre, which includes terrain surface, major streets, hydrographical zones and existing structures. Due to available geotechnical databases, and taking into account results of hazard assessments for the Vrancea zone, the simulation of free-field damped response spectra for scenario earthquakes was performed, resulting in mapping of shaking parameters and building damage degree, d_i, calculations, thus providing a vulnerability map for the constructed zone.

5. Recorded Damage and Results of Simulated Ground Motion

Within the framework of the seismic risk study, the specific task of establishing the correlation between ground motion parameters and degrees of damage to the existing structures, was also investigated. This issue is tackled in the context of the European Macroseismic Scale (EMS-92, p. 25) [4] which implies that "while it is undeniable that the effects observed from which intensity values are deduced are a product of real

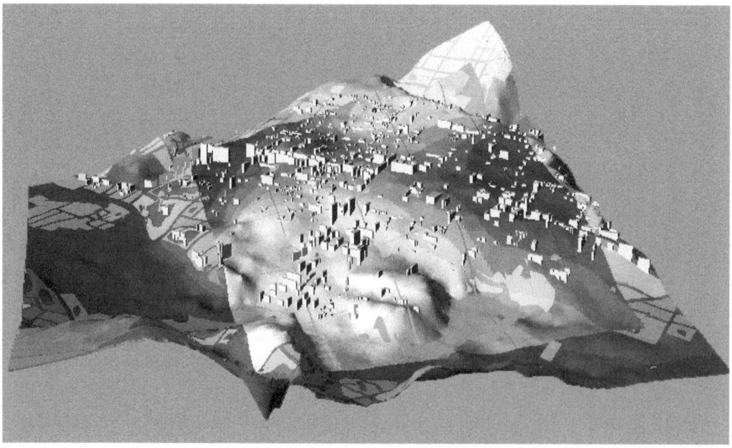

Figure 5. Chisinau centre: 3-D scene using ArcView (view from North). Ground motion simulation.

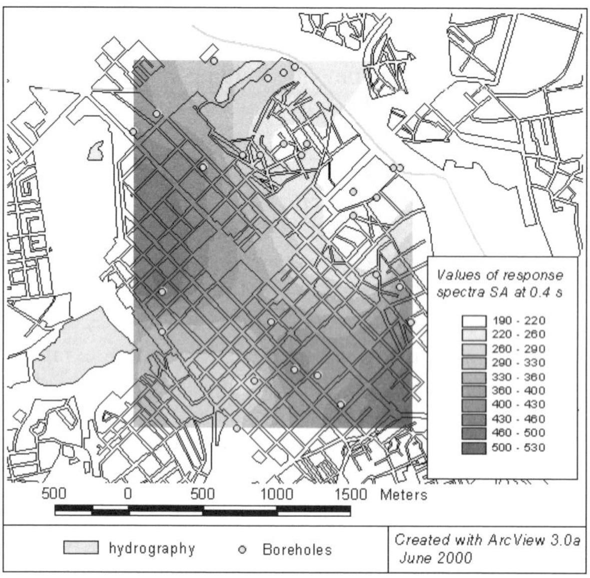

Figure 6. Map of simulated values of response spectra at T = 0.4 s for the central part of Chisinau.

ground motion parameters, the relationship between them is complex and not amenable to simple correlations".

The correlation was investigated for such parameters of the simulated ground motion as *EPA,* values of response spectra *SA* at target frequency 2.5 Hz (Fig. 6) and damage degree of the selected structures.

The better correlation with structural damage is observed for the values of spectral accelerations *SA*(*0.4s*), showing a coefficient of correlation ρ = 0.47, in comparison with the effective peak accelerations (*EPA*) of ρ = 0.18. Yet, the narrow range of the accelerations within the studied zone, depending on differences of soft soil thickness,

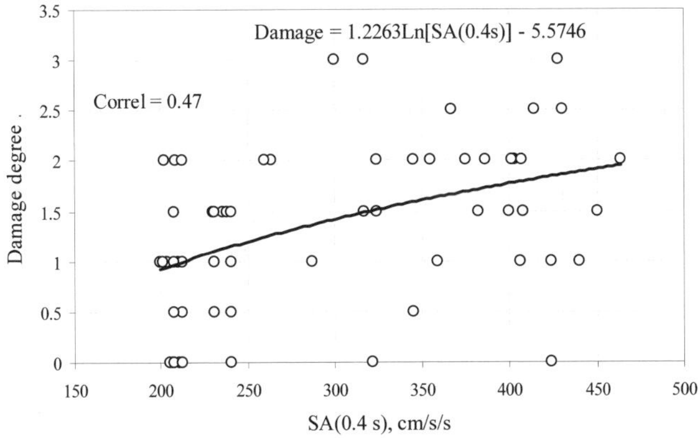

Figure 7. Bedrock *PGA* = 0.1 g at test site. Simulation results: damage degree (126 buildings) vs. values of response spectra SA (cm/s^2) at $T = 0.4$ s.

does not allow the more detailed investigation into correlation of structural damage and parameters of ground motion. Also, the non-homogeneity of masonry structures within the test site in terms of quality of materials and methods of construction makes a contribution to the larger scatter of the degree of seismic damage.

6. Conclusion

The usefulness of GIS for seismic damage assessment studies is emphasized due to its capacity to store, process and visualize spatial information. The following databases have already been compiled for the test site: geotechnical, building stock, records of blasts, microseisms and strong motion. GIS, which is capable of providing a powerful tool for the analysis of geographically distributed data, was used as a core instrument in the research work. The technique for structural damage assessment has been worked out and tested.

Results of the simulated ground motion parameters are compared with the observed damage degrees of buildings, which are better described by values of spectral accelerations *SA* at target frequencies than effective accelerations *EPA*.

Acknowledgments

All the data were kindly provided by Institute of Geology and Seismology, Moldavian Academy of Sciences, which monitors seismic activity of the Vrancea zone in the territory of the Republic of Moldova and concentrates information related to seismic hazard and risk.

The material presented in this paper is the result of the joint research based on collaboration between several institutions, performed in the context of the seismic microzoning of Chisinau.

References

[1] Alkaz, V. and Zaicenco, A., (1999). Spatial Correlation Between Level of Water Table and Damage of Buildings After 1986 Vrancea Earthquake in Kishinev using GIS, DACH-Tagung Conference, Berlin, Nov 24-25. pp. 19-25.

[2] Andy Mitchell, (1999). The ESRI Guide to GIS Analysis. Volume 1: Geographic Patterns & Relationships. ESRI Press, Redlands, California

[3] Charles A. Kircher, Aladdin A. Nassar, Onder Kustu and William T. Holmes, (1997). Development of Building Damage Functions for Earthquake Loss Estimation, Earthquake Spectra, Vol. 13, November 4, p. 663.

[4] Grunthal, G., et al. (1993). European Macroseismic Scale 1992.

[5] Idriss, I.M. and Sun, J.I., (1992). User's Manual for SHAKE91, Department of Civil & Environmental Engineering, University of California, Davis.

[6] Lungu, D., Cornea, T., Aldea, A. and Zaicenco A., (1997). Basic representation of seismic action. In: Design of structures in seismic zones: Eurocode 8 – Worked examples. TEMPUS PHARE CM Project 01198: Implementing of structural Eurocodes in Romanian civil engineering standards. Edited by D. Lungu, F. Mazzolani and S. Savidis. Bridgeman Ltd., Timisoara, p. 1-60.

[7] Vanmarcke, E.H., (1974). Structural Response to Earthquakes, MIT, Cambridge, Mass., USA.

[8] Vucetic, M., Doroudian, M. and Martin, G.R., (1998). Development of Geotechnical Data Base of Southern California for Seismic Microzonation, 3rd Annual Caltrans Seismic Research Workshop, Sacramento.

[9] Zaicenco, A. and Alkaz, V., (2000). Development of 3-D Geotechnical GIS-oriented Database for Seismic Microzonation Studies. Proceedings of the 3rd Japan–Turkey Workshop on EQ engineering, Istanbul, Feb. 21-25, pp. 159-165.

[10] ArcView Spatial Analyst and ArcView 3D Analyst, (1996-97). Manuals for utilization of ArcView GIS software, Environmental Systems Research Institute, Inc. (ESRI).

[11] Guidelines for Seismic Microzonation Studies, (1995). AFPS.

[12] EUROCODE 8. Earthquake Resistant Design of Structures.

Optimisation of Disaster Forecasting and Prevention Measures in the Context of Human and Social Dynamics
I. Apostol et al. (Eds.)
IOS Press, 2009.
doi:10.3233/978-1-58603-948-6-37

Actuarial Risk Management Through Geological Risk-Geoinformation Systems (RiskGIS)

T. RUDOLPH
De Perponcherstraat 79, 2518 SP Den Haag, The Netherlands
tobias.rudolph@uni-muenster.de

Abstract. Actuarial geo-information systems are mainly used for the spatial analysis of surface data sets. But geological, hydrogeological as well as hydrological aspects should be considered in the calculation of premiums in the insurance industry. A first deployment method for these geo-scientific subsurface information is shown in this paper.

Keywords. Insurance, Geo-information systems, GIS, Geology, Hydrogeology, Hydrology

Introduction

The application of geo-information systems for geological, hydrological and hydrogeological aspects in risk management is a further development of the geographical underwriting of the insurance industry. The knowledge of the geology, hydrology and hydrogeology is fundamental for the understanding and spatial analysis of an insured object before and during loss-events, for example with contamination of the aquifer. Furthermore possible loss scenarios could be prevented or minimized if the subsurface geology and hydrogeology is already known and integrated in the initial insurance appraisal. The modelling and classification of the geo-scientific knowledge also enables the definition of Action Zones, which allow a better appraisal and assessment of the insurable objects. This leads to an optimized and transparent premium calculation for both the insurer and the policyholder. The visualization of the geo-scientific subsurface information in geo-information systems is simple and economically feasible, it is quick to analyze, it can be combined with additional information and gives important insights into the subsurface structures.

1. Set-Up of a Risk-Geoinformation System

A Risk-Geoinformation system (RiskGIS) contains a digital geo-information system which is populated with geographical, geological, hydrological and hydrogeological datasets. With the combination of these datasets specific actuarial conclusions can be derived. In this paper the tool ArcGIS 9.1, ESRI, was used with the extensions of the Spatial Analyst and 3D Analyst.

1.1. Data Fundamentals

For a RiskGIS exercise is it important to first complete preliminary work such as a review of the existing datasets. Datasets about topography, geography as well as geology, hydrology and hydrogeology have to be obtained and digitized to get imported to the RiskGIS. A quality control on these datasets should be applied and, if needed, the dataset versions should be updated. But it is also considered that historical and old datasets contain important useful information. The distribution of these datasets is carried out by local, state and governmental authorities and other technical authorities as well as by private facilities [2,14]. Private important datasets are, for example, subsurface reports. Important datasets which should be used are [4,11]:

- Basic information like topographical maps, satellite and aerial pictures, coordinate systems and altitude information;
- Recent and historical wells for subsurface investigations, groundwater wells and other exploration wells;
- Hydrological information about rivers and lakes with flow direction, flow rates, protected areas along water works and water catchment areas;
- Hydrogeological information about groundwater, groundwater contours, groundwater flow directions, flow rates;
- Geological-hydrogeological information like rock- and sediment-formations, lithologies, porosities, permeabilities and soil retention;
- Topographic information about important traffic and transportation routes (roads, railroads and waterways);
- Residential areas with important buildings, land use and important object information like storage tanks, bonding depth of buildings in the groundwater;
- Vulnerability maps with contaminated areas; and
- Maps with possible subsurface background contamination.

For the usage of wells it is important to consider that the well itself does not provide only information about the subsurface geology and hydrogeology but the position of the well also shows possible contamination of the subsurface structure. As an example reference can be made to the explosion of the oil depot in Bruncefield, Great Britain on December 11, 2005 where infiltrating hydrocarbons along non-abandoned wells contaminated the aquifers [9].

For the initial interpretation of objects, different resolutions could be applied; the coarser the resolution of a model the coarser is the conclusion. Thus, for a small-scale area, where distances of less than a few metres can decide between a loss-event and non loss-event, preferably fine scales should be used [13]. Therefore, small-scale as well as large scale topographical maps should be used in the RiskGIS. With fewer work-steps, information about altitude and slope dipping is deducible from these maps. Simultaneously these topographical maps give insights in the surrounding area like nature reserves or water protected areas, which are possibly not visible during the initial site inspection of the object but are relevant during a loss event.

Additional to this basic information more detailed information, like pipe and cable plans, should be used. Along these pipes and cables routes exist high ground permeabilities and pathways which enable contamination of the groundwater. These pipes and cables have also an influence on operations to minimize damage.

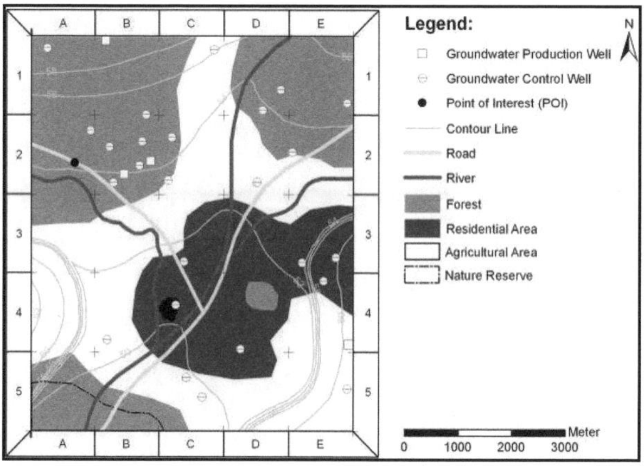

Figure 1. Map of the model.

Only with hydrological and hydrogeological datasets is it possible to appraise uncertainties in the subsurface, because this information also minimizes the loss scenarios and contamination. The geological subsurface information could be used in the classical way as for a geological map. Sometimes geological maps in higher resolution are also available and additional geological special maps like Hydrological Maps or Engineering Maps give important insights. But independent from the scale, geological maps only give information about the geology at the surface. However, datasets about the subsurface are important to understand the extension and connectivity of geological factors. This information must be gathered with geological cross-sections, wells or with special maps. Additional information of subsurface models could be implemented in the RiskGIS by special interfaces with subsurface models [15]. Overall the subsurface data must be principally combined, summarized and simplified depending on the desired conclusion. An additional perspective and/or three-dimensional visualizations of the datasets simplify the overview and therefore the interpretation.

1.2. Model of the RiskGIS

To show the functionality of a RiskGIS for an actuarial risk management the model of Coldewey and Schütz [5] was used, improved and enhanced. The presented model, with its visualizations and workflows, are idealistic and the points of interest used are imaginary.

In a first step topographic information, including an altitude model, is displayed on a map (Fig. 1).

For a better visualization of the possibilities of a RiskGIS two example scenarios (or, points of interest) are integrated in the model, which are, on the one hand, a single object in quadrant A2 and, on the other hand, cumulated objects in quadrant C4. The visualization was chosen this way for different reasons, namely the single object could be a moving object like a vehicle on the street or a fixed object like a gas station or a building with an oil tank in the cellar. An example could be a truck with dangerous goods and materials which has an accident [16]. For additional loss scenarios with dangerous goods reference is made to the Transport-Accident-Information and Support-System (Transport-Unfall-Informations- und Hilfeleistungs-System TUIS) of the Asso-

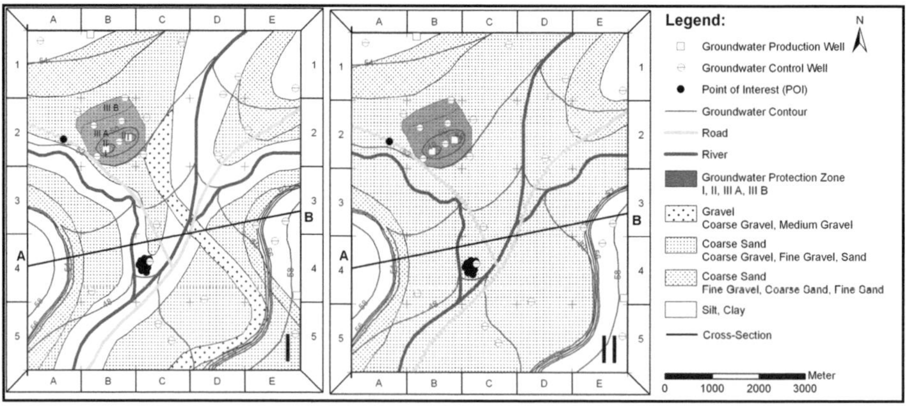

Figure 2. Map of the model with geological, hydrological and hydrogeological information for i) a depth of 0 m to 2 m, and ii) a depth of 2 m to 5 m.

ciation of the Chemical Industry and the Ministry of the Interior [17]. The cumulative objects are an accumulation of buildings on an industrial site close to two rivers. As an example the loss scenario of Sandoz in 1986 is mentioned where fire water, which was polluted with mercury and phosphoric pesticides, flowed into the Rhine, and caused severe pollution of the river water. To show the overall sensitivity of the system more realistic and significant supplemental environmental parameters are added. Therefore, in the vicinity of the single object in the quadrant A2 a water works with active wells is located. Along the course of the river are located residential areas and further downstream agricultural areas and nature-protected areas.

In a second step for the RiskGIS the geological and hydrogeological datasets have to be incorporated (Fig. 2). Furthermore, by the interpolation of point data like groundwater measurements groundwater, contours must be generated.

The groundwater flow direction, which is perpendicular to the contours, could also be developed, plotted as arrows on the contours and used to describe the flow of the pollutants. For the overview in the model area these arrows are not displayed but the general groundwater flow is to the southwest. On the basis of the surface understanding, subsurface information could be extracted from the geological map or, like in this case, from the wells to create geological maps for deeper structures (Fig. 2) as well as a hydrogeological cross-section (Fig. 3).

Through a comparison of the surface geology with the subsurface geology a change is visible. The reason for this is that the model area represents a river valley where, especially at the surface, recent river deposits are located and which are completely different from the deeper subsurface geology. The next interpretation step is the generation of the map with distances between the aquifer and surface (Fig. 4). The shorter the distances between the aquifer and the surface, the quicker the possible contamination of the groundwater. The map shows that in this model the distances are in the most cases less than three metres. In the quadrant B2 an increase of the distance is visible due to active producing groundwater wells with a drawdown of the groundwater table.

To show the impact of the distance of the aquifer and the surface, additional parameters have to be incorporated in the model. By using the permeability it is possible to understand the rate of contamination of the underground regime whose permeability

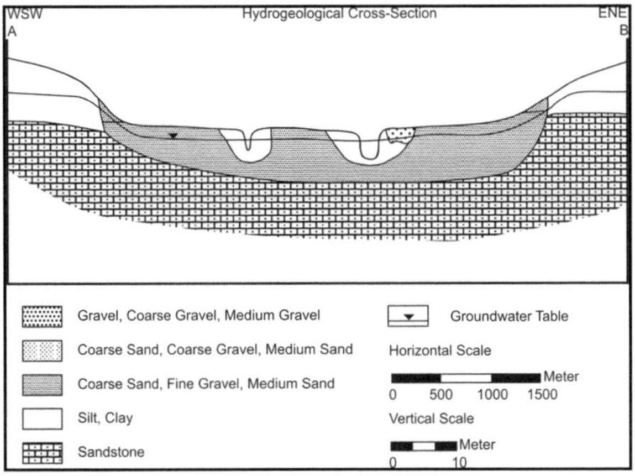

Figure 3. Hydrogeological Cross-Section of the model.

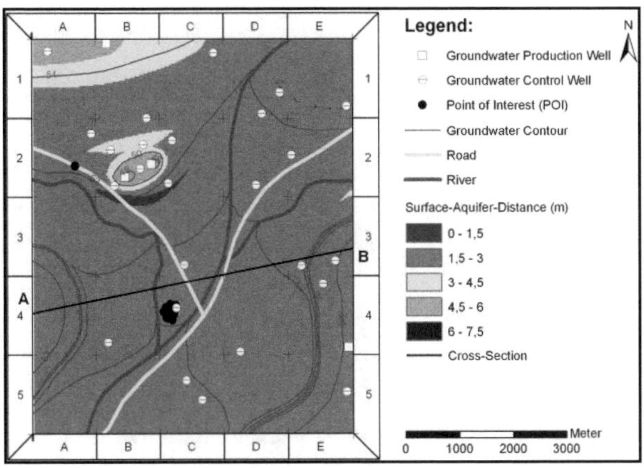

Figure 4. Map of the distances between the aquifer and the surface (Classification after [3]).

is a measure of the ability of rocks and sediments to transmit (ground)-water [8]. The lower the permeability value the lower the propagation velocity of substances in the underground. Typical permeabilities are shown in Table 1.

Based on DIN 18130-1 a classification of permeability values is possible (Table 2) [6]. These permeabilities will be allocated to the geological units of the geological map and in a second step using the DIN 18130-1 classification are coloured in five classes and plotted as an additional map (Fig. 5).

The permeability values in the active riverbed are very low and in the old riverbed very high because of the depositional history of the rivers. The comparison of both maps shows clearly the more homogenous and permeable sediments at a depth more than two metres. Additional analysis of the deeper geological situation, using the (hydro-) geological cross-section, shows possible high permeable rocks and sediments with permeabilities of $k_f = 10^{-4}$ m/s to $k_f = 10^{-5}$ m/s (Fig. 3). On this basis of classification of permeabilities, Zones of Exposure could be identified. In the already mentioned example of the Sandoz loss scenario, the distance to the Rhine was only a couple of

Table 1. Permeability (Durchlässigkeitsbeiwerte) k_f [8]

Description	Permeability (Durchlässigkeitsbeiwert) k_f
Gravel	10^{-1} m/s -10^{-2} m/s
Coarse grained Sand	≈ 10^{-3} m/s
Medium grained Sand	10^{-3} m/s - 10^{-4} m/s
Fine grained Sand	10^{-4} m/s - 10^{-5} m/s
Silty Sand	10^{-5} m/s - 10^{-7} m/s
Clayey Silt	10^{-7} m/s - 10^{-9} m/s
Clay	< 10^{-9} m/s

Table 2. Classification of Permeabilities (Durchlässigkeitsbeiwert k_f) in five classes (DIN 18130-1) and colour code

Permeability (Durchlässigkeitsbeiwert) k_f (DIN 18130-1)	Class of Permeability	Colour code (RGB colour palette)
> 10^{-2} m/s	Very high permeable	255,0,0
10^{-2} m/s - 10^{-4} m/s	High permeable	255,102,0
10^{-4} m/s - 10^{-6} m/s	Permeable	255,255,0
10^{-6} m/s - 10^{-8} m/s	Low permeable	0,255,255
< 10^{-8} m/s	Very low permeable	0,0,255

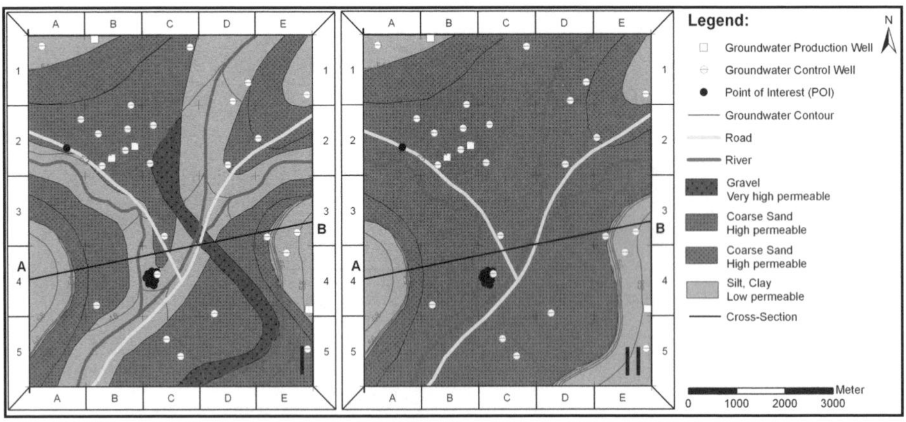

Figure 5. Permeabilities for i) a depth of 0 m to 2 m, and ii) a depth of 2 m to 5 m.

hundred metres. To validate the Zones of Exposure the established classification of the protected zones for water works, which show the sedimentary aquifers, the following, are the recommended values [7]:

- Protected zone I: The upstream flow distance for groundwater wells should be two to one kilometres;
- Protected zone II: The upstream flow distance for groundwater wells should be 50 days (retention period) and not less than 100 metres; and
- Protected zone III: The upstream flow distance for groundwater wells should be more than 20 metres.

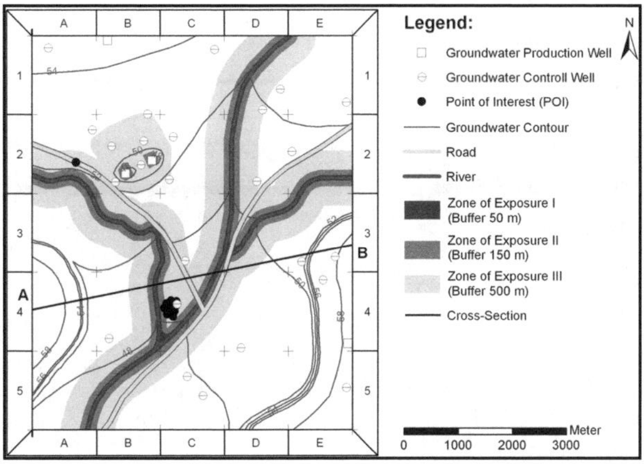

Figure 6. Model with the Zones of Exposure.

The classification of the buffers along the rivers should also incorporate local parameters like width, depth of rivers and flow velocity, as well as usage of the water for water supply. Therefore the size of the buffer zones vary. In the model buffers with a width of 50 m, 150 m and 500 m are used which result in three Zones of Exposure (Fig. 6):

- Zone of Exposure I with a width of 100 m;
- Zone of Exposure II with a width of 500 m; and
- Zone of Exposure III with a width of 1,000 m.

In the presented model the protected zones for water works are similar to the Zones of Exposure. The final step of the workflow to define Action Zones is the spatial analysis of the permeability classes with the distances of the aquifer from the surface and with the Zones of Exposure. For this all the different classes are combined and multiplied. The possible 75 combinations are sorted and ordered in Action Zones (Table 3). An Action Zone is defined in this model as a zone where an action/workflow has to take place because of the geological, hydrological and hydrological situation. On this basis five Action Zones plus one are established:

- Action Zone A (Combination 1 to combination 43);
- Action Zone B (Combination 44 to combination 63);
- Action Zone C (Combination 64 to combination 70);
- Action Zone D (Combination 71 to combination 74);
- Action Zone E (Combination 75); and
- Action Zone F.

The generation of these Action Zones is effected using the Rastercalculator in the RiskGIS by multiplying the values of the different maps. Therefore a very high permeability soil with a small distance between groundwater and surface in the Zone of Exposure I results in Action Zone A. The Action Zone F is established for areas where no area-wide information coverage exists for permeability values, aquifer surface distances or Zones of Exposure. On basis of the two permeability maps, the spatial distribution of the Action Zones are visualized for the different depths (Fig. 7).

Table 3. Part of the decision tree as a spatial analysis of the permeability classes, distance between the aquifer and surface and the Zones of Exposure

Permeability	Distance between Aquifer and Surface	Zone of Exposure	Action Zone	No.
Very high permeable	1	I	A	1
		II	A	2
		III	A	3
	2	I	A	4
		II	A	5
		III	A	6
	3	I	A	7
		II	A	8
		III	A	9
	4	I	A	10
		II	A	11
		III	A	12
	5	I	A	13
		II	A	14
		III	A	15
High permeable	1	I	A	16
		II	A	17
		III	A	18
	2	I	A	19
		II	A	20
		III	A	21
	3	I	A	22
		II	A	23
		III	B	44
	4	I	A	24
		II	B	45
		III	B	46
	5	I	A	25
		II	B	47
		III	B	48
Permeable	1	I	A	26
		II	A	27
		III	A	28
	2	I	A	29
		II	A	30
		III	B	49
	3	I	A	31
		II	B	50
		III	B	51
	4	I	A	32
		II	B	52
		III	C	64
	5	I	A	33
		II	B	53
		III	C	65
Low permeable	1	I	A	34
		II	A	35
		III	A	36
	2	I	A	37
		II	B	54
		III	B	55
	3	I	A	38
		II	B	56
		III	C	66
	4	I	B	57
		II	C	67
		III	D	71
	5	I	B	58
		II	C	68
		III	D	72
Very low permeable	1	I	A	39
		II	A	40
		III	A	41
	2	I	A	42
		II	B	59
		III	B	60
	3	I	A	43
		II	B	61
		III	C	69
	4	I	B	62
		II	C	70
		III	D	73
	5	I	B	63
		II	D	74
		III	E	75

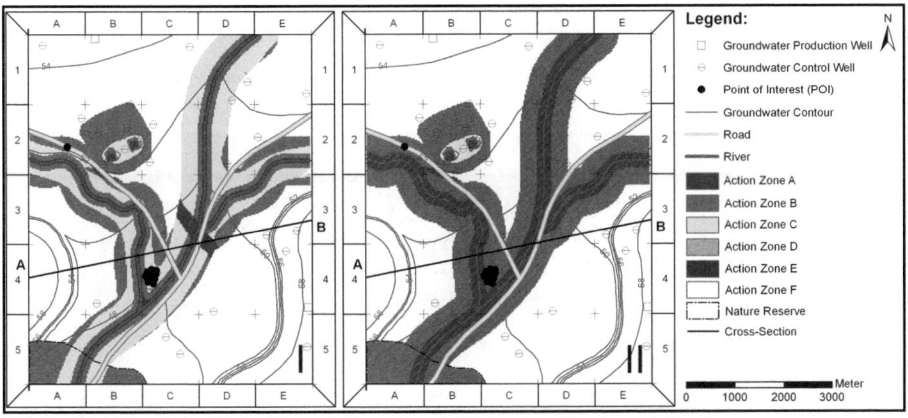

Figure 7. Action Zones for i) a depth of 0 m to 2 m, and ii) a depth of 2 m to 5 m.

Table 4. Description of the premium levels for insurances on basis of the Action Zones

Action Zone	Description	Premium Level
A	A contamination of the subsurface must be always avoided. In every case of a contamination a direct action must prevent a dispersion of the pollutants.	During a loss scenario with a dispersion of pollutants exists a very high exposure of the subsurface structures. Special safety systems and control mechanisms must exist.
Urgent need for action		Highest premium level or not insurable!
B	A contamination of the subsurface must still be avoided. In every case of a contamination immediate action must prevent a dispersion of the pollutants.	During a loss scenario with a dispersion of pollutants exists a high exposure of the subsurface structures.
Higher need for action		Higher premium level!
C	Special attention must be give to the distances to the endangered object, to the permeabilities and to the distances between aquifer and surface. Often are permeabilities low and an action to avoid a dispersion of the pollutants is needed	During a loss scenario with a dispersion of pollutants exists an exposure of the subsurface structures.
High need for action		High premium level!
D	The exposure is lower because the distances between the surface and the aquifer are bigger and the permeabilities are lower.	During a loss scenario with a dispersion of pollutants exists an exposure of the subsurface structures.
Increased need for action		Increased premium level!
E	In case of an accident it is important to consider the deeper subsurface structures to avoid a dispersion of pollutants into the deeper structures.	During a loss scenario with a dispersion of pollutants exists an exposure of the subsurface structures.
Moderate need for action		Moderate premium level!
F	A need of action is based on the assessment of the local permeabilities and the distances between the surface and the aquifer. Therefore the need of action could vary.	Local appraisal of the geological, hydrological and hydrogeological parameters important to assess possible loss scenarios.
Observation status		Normal premium level

For a better visualization, Action Zone B for the nature reserve is assigned to the quadrants A5 to C5. During the visualization of the Action Zones, information about a depth trend has also to be acquired so as to understand the change of an Action Zone with depth. In the model the shaded zone along the river displays an increase of the action zone from A to B caused by the increase of permeability values with depth (Fig. 5). Table 4 with the six Action Zones is a further development after MELCHERS, C.,

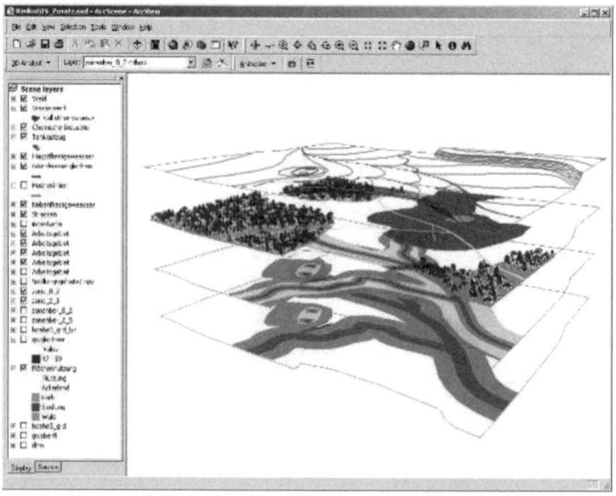

Figure 8. Screenshot of ArcSCENE with a perspective view on the model.

GÖBEL, P. & SCHÄFER [10]. The table shows in the second column the action that is needed to prevent contamination being dispersed. Together with the first column it clearly indicates also the timing of required actions. The last column highlights the implications for insurances and the impact on the calculation of the premiums.

2. Interpretation of the Data

For a summarized interpretation of the data it is useful to display the model in a perspective, or three-dimensional, view because the spatial relationship of the datasets is more apparent. In this case the visualization tool ArcSCENE is recommended (Fig. 8). All the layers of the RiskGIS are displayed as separate layer stacks:

- First layer: Hydrology and the groundwater contours;
- Second layer: Topography;
- Third layer: Action Zones for a depth between 0 m and 2 m; and
- Fourth layer: Action Zones for a depth between 2 m and 5 m.

The spatial junctions, especially of the hydrological and hydrogeological datasets with the geological datasets of the deeper subsurface, result in important information which, if separately analyzed, is not obvious. An example is the depth trend and the change of the Action Zone in the riverbed with an increase of permeability. Another example is the analysis of the industrial areas in the vicinity of the two rivers. Often industrial areas are built in the river plains because only there are enough and wide areas available. But the possible interaction of the industrial area with the river and the influence to the subsurface are often not considered. Therefore, pollutants infiltrate the subsurface structure easily during loss-events because of the high permeability.

The need for action increases if the pollutants infiltrate deeper structures because the permeability values there are even higher. Furthermore, an analysis of the slopes of the elevation model shows that the pollutants during a loss-event run off the surface into the river, which will then be transported towards the nature reserve. The example of the loss scenario in the Northwest of the model is an object, which is lying on the edge of an Action Zone. Only with a spatial analysis of the position of the object it is possible to detect in the vicinity an active water works and a river (Fig. 2). The increased exposure will be visible only by the interpretation of the two geological maps. These maps show a minimum five metre thickness of coarse sands which are highly permeable. During loss-events contamination infiltrates the subsurface structure and is transported with the aquifer because the distance between aquifer and surface is also only between 1.5 m and 3 m (Fig. 5). Already the combination of these parameters would result in an Action Zone B. Overall the increase in Action Zones shows clearly the increased need to prevent the dispersion of contamination and the bigger impact on the calculation of insurance premiums.

The presented RiskGIS, and the presented model with examples of possible loss scenarios, show how spatial workflows can generate new and important information which have an impact on the assessment of objects. This information reduces the consequences and expenditure on loss scenarios and therefore has an impact on the premium calculation. The installation of safety systems and control mechanisms, like adequate collecting systems for tanks, reduces even more the risk and therefore the premium. On the other hand, an extreme case leads to a non-insurable object.

3. Conclusions

Recent investigations by the author have shown that no Risk-Geoinformation systems are available which work with geological, hydrological and hydrogeological datasets and are used by insurance companies to assess insurable objects and loss scenarios [1]. The presented RiskGIS is a further development of a normal Geoinformation system and Geoscience Based Fire Decision Systems (Geobasierte Feuerwehr-Entscheidungshilfe-Systeme). The presented RiskGIS is even more enhanced by keeping the compatibility with other GIS tools. Although the workflow of the RiskGIS is not described in detail it is possible to implement the general set-up in other systems. Important for the setting-up of a RiskGIS is the collection of the necessary geological, hydrological and hydrogeological information. As presented, different ways and opportunities could be used. Only with the overall knowledge of the geology, hydrology and hydrogeology it is possible to understand the subsurface structures and how the subsurface is linked to the surface. All these datasets are generally available area-wide in urban areas. Often additional information is where a very detailed scale is accessible. The datasets have to be combined with the RiskGIS and must be interpreted for the actuarial problem. The disadvantage of the non-availability of these datasets, especially in rural areas, should be compensated through a spatial analysis of the existing datasets with a combination of the information from an initial site inspection. But already the analysis of the existing datasets is an essential work-step before the site inspection to address possible problems.

The combination of the knowledge of the site inspection with the RiskGIS shows the exposure and the necessity for an action during a loss-event. After the initial investment to build a RiskGIS, this system is cheap and could easily be improved as well as enhanced by incorporating new and additional datasets.

In summary, the setting-up of a RiskGIS and an initial assessment are much cheaper than the expenditure for a loss-scenario. Even if a loss-scenario occurs is it possible with an updated RiskGIS to assess quickly the subsurface structures and define feasible decontamination methods and procedures. Therefore, the opportunity for risk management with a RiskGIS tool provides insurers and policyholders with an optimized, economical tool.

References

[1] BUSINESS GEOMATICS (2007): Ökologischer Schaden – SPACE analysiert das ortsabhängige Risiko für Umweltschäden. – [online in the Internet: http://www.business-geomatics.com/archiv/507/42.html, Status October 9, 2007].

[2] COLDEWEY, W.G. (1993): Archivmaterial. – In: Weber, H.H. & Neumaier, H. [Hrsg.]: “Altlasten. Erkennen, Bewerten, Sanieren”, 2. Aufl., S. 44–73, Berlin (Springer).

[3] COLDEWEY, W.G. & GÖBEL, P. (2001): Hydrogeologisches Geländepraktikum [“Hydrogeologische Kartierung”]. – 38 S., 6 Abb., 6 Tab.; [unveröffentlicht].

[4] COLDEWEY, W.G. & KRAHN, L. (1991): Leitfaden zur Grundwasseruntersuchung im Festgestein bei Altablagerungen und Altstandorten. – 124 S., 34 Abb., Anh.; Düsseldorf.

[5] COLDEWEY, W.G. & SCHÜTZ, H.G. (1990): Untersuchungen der hydrogeologischen und hydrochemischen Verhältnisse kontaminierter Standorte.- Müll und Abfall, 1, 12 S., 10 Abb.; Berlin (ESV).

[6] DIN 18130-1 (1998): Baugrund; Untersuchung von Bodenproben; Bestimmung des Wasserdurchlässigkeitsbeiwertes – Teil 1: Laborversuche. Berlin (Beuth).

[7] DVGW ARBEITSBLATT W101 (2006): Richtlinien für Trinkwasserschutzgebiete; Teil I: Schutzgebiete für Grundwasser. – Bonn (Eschborn).

[8] HÖLTING, B. & COLDEWEY, W.G. (2005): Hydrogeologie – Einführung in die Allgemeine und Angewandte Hydrogeologie. – 326 S., 118 Abb., 68 Tab.; München (Elsevier).

[9] KASTL U. (2006): Verheerende Explosion in einem Öldepot. – MÜNCHENER RÜCK (2006): Schadenspiegel 2/2006 – Themenheft Risiko Feuer. – 49,2, 2-5., 3 Abb.; München.

[10] MELCHERS, C., GÖBEL, P. & SCHÄFER (2003): Entwicklung eines Konzeptes zur Bewertung der Umweltgefährdung während des Feuerwehreinsatzes aus hydrologischer Sicht. – VFDB; 3, 143-148, 5 Abb.; Stuttgart (Kohlhammer).

[11] MELCHERS, C., RUDOLPH, T. & COLDEWEY, W.G. (2005): Geologische Aspekte der angewandten Risikobewertung. – Münster. Forsch. Geol. Paläont.; 100, 8 S., 7 Abb.; Münster.

[12] MÜNCHENER RÜCK (2002): Topics 2002. – 49 S., 33 Abb.; München.

[13] MÜNCHENER RÜCK (2003): Topics geo 2002. – 53 S., 33 Abb.; München.

[14] PÄLCHEN, W. (2006): Bei der Geologie gespart?. – Geowissenschaftliche Mitteilungen; 25, 20-21; Bonn.

[15] RUDOLPH, T., ELFERS, H., JUCH, D., LINDER, B. & THOMSEN A. (2006): Untergrundmodelle in Nordrhein-Westfalen – Möglichkeiten der Zusammenführung unterschiedlicher Modellansätze. – Schriftenreihe der Deutschen Gesellschaft für Geowissenschaften (SDGG) [Hrsg.]: GeoBerlin 2006 158. Jahrestagung der DGG, 50: 61; Hannover.

[16] WESTFÄLISCHE NACHRICHTEN (2007): Gefahrgut-Transporter brennt – Großeinsatz der Feuerwehr – A2 bei Essen stundenlang gesperrt. – Newspaper article of April 30, 2007.

[17] VCI (2007): Transport-Unfall-Informations- und Hilfeleistungssystem (TUIS). – [Online im Internet: http://www.vci.de/TUIS/default2~cmd~shr~docnr~114675~nd~~rub~ 741~ond~tuis~c~0.htm, Status August 21, 2007].

Theme 2

Water-Based Hazards/Risks

Optimisation of Disaster Forecasting and Prevention Measures in the Context of Human and Social Dynamics
I. Apostol et al. (Eds.)
IOS Press, 2009.
doi:10.3233/978-1-58603-948-6-51

Bulgarian Policy for Water Resources Management and Flood Protection

Plamen GRAMATIKOV*
The Neofit Rilski South-West State University, 2700 Blagoevgrad, Bulgaria

Abstract. Fresh water resources of Bulgaria and water management in the country are estimated and compared with other countries. The government policy, and organizational structure of the network for flood monitoring, forecasting and warning are presented in this paper, as are the types of activities and divisions of work with neighbouring countries in the framework of trans-border cooperation in this area of the Balkans.

Keywords. Fresh water resources, organizational structure, flood monitoring and forecasting

Introduction

The big security issues of 21st century are energy, water and climate. New security challenges require new approaches. World water resources are very important because all living creatures depend on them. The world ocean covers 74% of the Earth and saline and mineral waters are about 97.5% of all water reserves. The potential resources of fresh water, excluding glacial water, are 4.2 million km^3 or 0.3% of all hydrosphere reserves. There is no living creature or plant that can live without water. Water constitutes 14–16% of the seeds' content and up to 90–95% of the fruits' content. Almost 2/3 of the human body consists of water and we need about 2 litres of water intake daily. One litre is supplied by drinking liquids and another one is supplied by food. The great importance of water is connected with its amazing properties, which are not typical of other substances. Water is eternal because it is constantly renewed by water circumrotating and it is the only substance that has the three aggregate conditions – liquid, solid and vapour. Besides, it is a universal solvent – depending on temperature, pressure and other factors, it can dissolve almost all chemical elements.

Throughout history water has confronted humanity with some of its greatest challenges. Water is a source of life and a natural resource that sustains our environments and supports livelihoods – but it is also a source of risk and vulnerability. In the early 21st Century, prospects for human development are threatened by a deepening global water crisis. In a world of unprecedented wealth, almost 2 million children die each year for want of a glass of clean water and adequate sanitation. And water-borne infectious diseases are holding back poverty reduction and economic growth in some of the world's poorest countries.

* Correponding Author: Department of Physics, South-West University "Neofit Rilski", 66 Ivan Mihailov Blvd., 2700 Blagoevgrad, Bulgaria; E-mail: psgramat@yahoo.com.

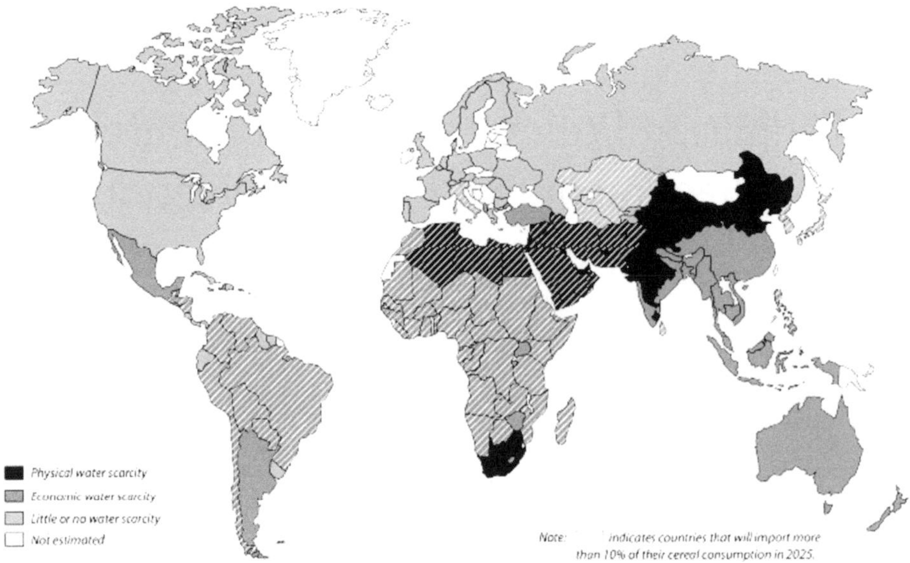

Figure 1. Projected Water Scarcity in 2025.

A clean and safe source of drinking water is regarded by the United Nations as a fundamental human right. Many European countries depend on groundwater for drinking water supplies. As this water becomes increasingly polluted, they are faced with two options:

- Develop increasingly complex and expensive methods of cleaning the water, or
- Risk the consequences to human health of drinking polluted water.

Groundwater pollution is, of course, also of concern in environmental terms. Most of the groundwater participates in the hydrological cycle although the residence time may vary from months to centuries.

On the other hand natural disasters including floods were always part of the environment and humans were combating them. The ability of human race to successfully mitigate them is a criterion for its development.

1. Integrated Water Resources Development and Management

1.1. Global Water Shortage in the New Century

About 80 countries now have water shortages that threaten health and economies while 40% of the world (more than 2 billion people) has no access to clean water or sanitation (Fig. 1). In this context, we cannot expect water conflicts to always be amenably resolved by the European environmental standards.

More than a dozen nations receive most of their water from rivers that cross borders of neighbouring up-stream countries which are often viewed as hostile. These include Botswana, Bulgaria, Cambodia, Congo, Gambia, Sudan and Syria, all of whom receive 75% or more of their fresh water from the river flow from such neighbours. In

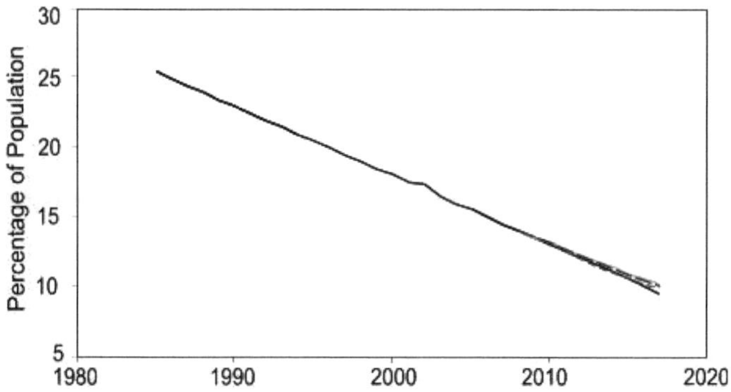

Figure 2. Population Lacking Access to Improved Water Sources (percentage of population).
Source: World Health Organization and UNICEF, Meeting the MDG Drinking Water and Sanitation Target, and Millennium Project estimate.

the Middle East, a region marked by hostility between nations, obtaining adequate water supplies is a high political priority. For example, water has been a contentious issue in recent negotiations between Israel and Syria. In recent years, Iraq, Syria and Turkey have exchanged verbal threats over their use of shared rivers.

Almost 14% of the European Union (EU) population has been affected by water scarcity. Over 80% of the original floodplain area along the Danube and its main tributaries has been lost as a result of dams, pollution, and climate change. The Belgian government recognizes water as a human right, and its development aid will focus on water. Water utilities in Germany pay farmers to switch to organic operations because it costs less than removing farm chemicals from water supplies.

Global water problems are attracting increasing attention. A prime cause of the global water concern is the ever-increasing world population. As populations grow, industrial, agricultural and individual water demands escalate. World-wide demand for water is doubling every 21 years, more in some regions. Water supply cannot remotely keep pace with demand, as populations soar and cities explode (Fig. 2).

Population growth alone does not account for increased water demand. Since 1900, there has been a six-fold increase in water use for only a two-fold increase in population size. This reflects greater water usage associated with rising standards of living (e.g., diets containing less grain and more meat). It also reflects potentially unsustainable levels of irrigated agriculture. World population has recently reached six billion and United Nation's projections indicate nine billion by 2050.

Meanwhile many countries suffer accelerating desertification. Water quality is deteriorating in many areas of the developing world as population increases and salinity caused by industrial farming and over-extraction rises. About 95% of the world's cities still dump raw sewage into their waters.

1.2. Water Exploitation Index

Over the last 10–15 years the Water Exploitation Index (WEI) decreased in 18 European countries, representing a considerable decrease in total water abstraction (about 9% of total abstractions corresponding to 23,081 million cubic metres decrease of water) [1].

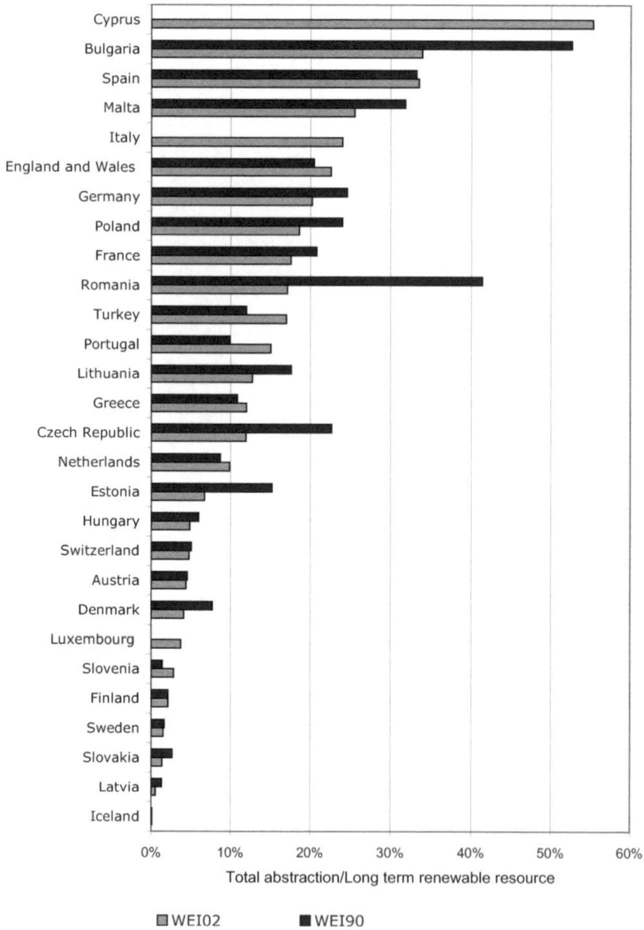

Figure 3. Water exploitation index. Total water abstraction per year as percentage of long-term freshwater resources in 1990 and 2002.

Data source: EEA-ETC/WTR based on data from Eurostat data tables: Renewable water resources (million m^3/year), LTAA & annual water abstraction by source and by sector (million m^3/year) – Total freshwater abstraction (surface + groundwater).

The warning threshold for the water exploitation index (WEI), which distinguishes a non-stressed from a stressed region, is around 20% (Fig. 3).

Severe water stress can occur where the WEI exceeds 40%, indicating unsustainable water use. But nearly 44% of Europe's population still lives in water-stressed countries (approx. 255 million inhabitants).

In Europe there are eight countries that can be considered water-stressed based on the Eurostat data available for the period 1997–2005 (Germany, Cyprus, Spain, Bulgaria, Italy, UK, Malta and the FYROM), representing about 44% or almost half of Europe's population. Based on the 2005 available data Cyprus (60%) and Bulgaria (>35%) have the highest WEI. However, it is necessary to take into account the high water abstraction for non-consumptive uses (cooling water) in Germany, England and Wales, Bulgaria and Belgium. Most of the water abstracted in the remaining four water-stressed countries (Italy, Spain, Cyprus and Malta) is for consumptive uses (especially irrigation) and there is therefore higher pressure on water resources in these four countries.

The WEI decreased in 18 countries over the last 10–15 years, representing a decrease of about 9% in total water abstraction (in absolute number there is a 23 km^3 of abstracted water reduction as compared to the 268 km^3 total abstraction in 1990). Most of the decrease occurred in the new EU Member States as a result of the decline in abstraction in most economic sectors. This trend was the result of institutional and economic changes. However, seven countries (The Netherlands, the UK, Greece, Portugal, Slovenia, Spain and Turkey) increased their WEI during the period 1990 to 2005 because of the increase in total water abstraction. The WEI has also increased in Cyprus from 1998 to 2005 (lack of data do not allow comparison to the pre-1997 period).

1.3. Bulgarian Water Strategy and Policy

Environmental policy in Bulgaria has evolved with overall political and economic changes in the last 17 years. The Strategy and the Environmental Action Plan for the period up to 2000 introduced new approaches to environmental management and set Bulgarian environment policy on track with modern environmental policy-making – addressing environmental issues in their inter-sectoral complexity, thus providing the initial framework for integrating environmental, economic and social issues as a basis for the country's sustainable development. Particularly intensive in terms of legislative changes and new policy implementation was the period after 1997 when Bulgaria signed an association agreement with the European Union (EU) thus formally undertaking an obligation to meet European environmental standards. This process continues after the acceptance of Bulgaria as a member of the European Union (EU) at the beginning of 2007.

In 1997 the Government adopted a Strategy for Integrated Water Management in the Republic of Bulgaria [2]. This document identifies the main objectives towards achieving sustainable water management:

- meeting different water-use needs (drinking water supply, recreation, industrial and agricultural use) while conserving water resources; protecting the environment and the aquatic ecosystems;
- limiting potential impacts of floods and drought.

The Strategy also takes into account new challenges arising from market economy development. It introduces the approach of a government regulated water-use balance as a way of ensuring the social function of water supply and the protection of the environment. The Strategy identifies economic aspects of water use and water management structures and mechanisms to be established.

Recognising the new stage of environmental policy development the Ministry of Environment and Water (MoEW) initiated the elaboration of a National Environmental Strategy and Action Plan 2000–2006. The strategy prepared with the valuable contribution of Bulgarian governmental institutions, scientific organisations, non-governmental organisations (NGOs) and societies and with international support was approved by the Government on 31 May 2001 [3]. This strategic document has a primary focus on end results, i.e. on activities and measures for practical implementation of existing legislation with the aim to improve the quality of life and to protect the environment.

The strategic objectives for the country for this period go beyond mere environmental concerns and integrate sustainable development considerations in two aspects:

1. Preserving and expanding the large clean territories in the country and protecting Bulgaria's rich nature in conditions of economic growth and improved social welfare;
2. Overcoming existing local environmental problems, thus improving quality of life.

In line with overall environment and economic policy development, water management priorities have also been set.

A few years ago a working group chaired by the MoEW, and including representatives of other organisations, was set up to draft a National Strategy for Water Management and Development [4]. The new strategy will also reflect international commitments, particularly those related to EU accession and the trans-boundary water management implications it has. The strategy will identify measures and mechanisms for adequate water supply to all citizens and other users in the country while guaranteeing social acceptability of water services and integrating other sector priorities set out in strategic documents such as national economic and regional development plans, district and municipal social-economic development strategies, municipal programmes, etc.

1.4. Legislation

Based on the key issues and priority actions of the National Environmental Strategy and Action Plan 2000–2006, the Environmental Protection Act (EPA) of 2002 [5] provides a comprehensive legal framework for environmental policy. It ensures a common approach in all environmental sectors and at the same time provides a basis for integration of environment into other policy sectors. The EPA also provides for wider opportunities for public involvement in decision-making and policy implementation.

While the EPA provides the framework for environmental policy, sectoral legislation (laws and regulations) details the particular requirements, and specifies enforcement and control mechanisms, and deadlines. Sectoral legislation ensures that EU accession related commitments and other international obligations are observed. At the same time it gives stakeholders clear guidance on compliance with particular requirements as well as establishing mechanisms for public participation in the decision-making process.

The general principles of the EU policy in the water sector are introduced in the national legislation through the Water Act [6]. The Water Act introduces the principle of integrated water management on the basis of river basins. The development of river basin management plans, and programmes for water bodies' pollution reduction and elimination, is regulated. The main rules for the operation of the national water monitoring system are specified. Internal monitoring combined with periodical inspections by the state institutions is projected for the big enterprises. A permitting regime for water use and use of water bodies is introduced, including discharges of wastewater from urban collection systems as main tools for regulating water resources use and protection of water from pollution.

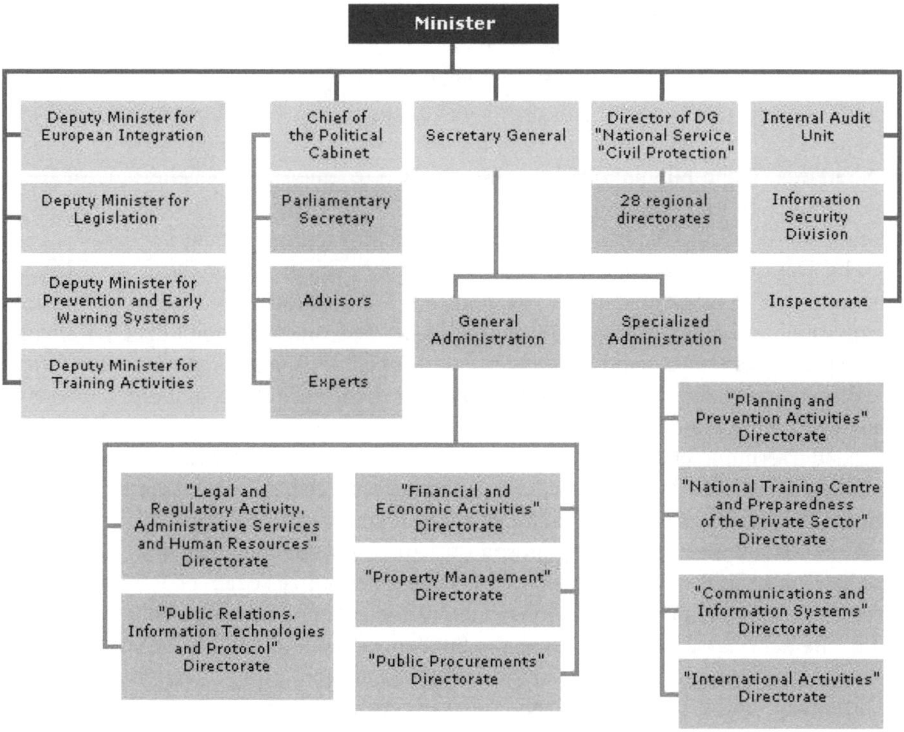

Figure 4. Organisational structure of Bulgarian MSPDA.

2. Crises Protection Organisation

2.1. Management of Crises Prevention and Protection in Bulgaria

The Ministry of State Policy for Disasters and Accidents (MSPDA) unites the existing agencies responsible for prevention, response, management and recovery in case of crises (Fig. 4).

The main aim of MSPDA is to establish a working and efficient, suitably technologically and materially equipped integrated system for the prevention, preparation, response and recovery in case of crises, meeting the real needs of Bulgarian citizens in such cases. The MSPDA policy will aim to establish a unified model for action in emergency situations, efficient crisis management communication, strengthening transparency of the administration work in crisis management.

The Ministry's policy aims to enhance the skills and improve the training of state bodies, legal entities and citizens in the country in these situations. The MSPDA along with the Ministry of Education and Science takes actions to improve the training at secondary schools and universities in the field of civil protection. Along with insurance companies, they work out policies for the prevention and improvement of the population's insurance culture. The Ministry will also introduce standards for preparation for crises-specific criteria that all public administrations will have to meet.

The state policy concerning disasters and emergencies is implemented both with funds allocated under the state budget of the Republic of Bulgaria and funds under the

PHARE Programme and international donor organisations working in the field of civil crisis management.

One of the major priorities of the MSPDA is to ensure preparedness of the population for actions in the event of disasters and accidents. In the aftermath of the floods that hit the country two years ago some heavy problems came to the surface, the poor preparedness of the population to act in response to emergencies being one of them. And it is namely the prompt and adequate conduct in crisis situations that is the major prerequisite for saving the life of people, as well as for reducing material losses.

The National Training Centre Directorate has prioritised for the time being the preparation of management bodies responsible for the population protection against disasters and accidents on a local level. The Directorate, together with the Civil Protection National Service Directorate General, developed a training programme for district governors, deputy district governors and mayors of municipalities and districts in Bulgaria. The topic of the training course is "Strengthening the Management Capacity of District Governors, Deputy District Governors and Mayors of Municipalities in the Republic of Bulgaria to Carry out Events Concerning Population Protection and Protection of the National Economy in the event of Disasters and Accidents".

The objectives of the training programme are to improve the quality of the work of local and municipal management bodies in the event of disasters and accidents concerning the duly and prompt decisions with a view to population protection and protection of the national economy; to enhance the theoretical and practical preparedness on planning, organisation, management, governance and monitoring of activities related to the protection and elimination of the consequences in the occurrence of disasters and accidents.

The National Training Centre Directorate, in cooperation with the Ministry of Education and Science, has drawn up several projects in implementation of the policy and priorities of the Ministry of State Policy for Disasters and Accidents regarding the training and preparation of the population within the system of education and science.

A team of experts and specialists is particularly working on:

- Developing a syllabus and timetable of a training course in response to disasters and accidents for trainers;
- Drawing up a syllabus of a course in response in the event of disasters and accidents for pupils in Grade I–XII;
- Preparing textbooks and training aids for children at kindergartens;
- Developing textbooks and training means for pupils at primary and secondary schools, as well as methodological handbooks for teachers;
- Preparing study resources for university students; and
- Creating e-textbooks and an educational web site.

2.2. International Activity

MSPDA's major objectives in its international activity are:

- Implementation of the foreign policy priorities of the Republic of Bulgaria concerning fulfilment of the requirements for NATO and EU membership;
- Cooperation with the United Nations, the Council of Europe, and other international organisations;

- Strengthened cooperation with the countries in South East Europe and the Black Sea Region and establishment of our country as a guarantee for stability and security in the region;
- Enhanced bilateral international cooperation;
- Improved organisation and coordination in the preparation and implementation of international humanitarian operations and trainings.

These major objectives underpinning the international activity of the Ministry are carried out through a range of measures focused primarily on:

- Planning, coordination and management of bilateral, regional and multilateral activities related to collaboration in the state policy for disasters and accidents;
- Preparation, management and coordination of the activities on drafting and harmonising bilateral and multilateral international treaties, agreements and other international laws relevant to the state policy for disasters and accidents:
- Fulfilment of the commitments under international agreements in the field of state policy for disasters and accidents to which the Republic of Bulgaria is a party;
- Full participation of the Republic of Bulgaria in international structures and organisations connected to the state policy for disasters and accidents and security;
- Organisation and coordination of activities related to international humanitarian operations;
- Participation in the coordination and management of the planning and implementation of initiatives relevant to Bulgaria's NATO membership and of activities aimed at catching up with EU Member States in the field of civil emergency planning and civil protection;
- Establishment of international cooperation and advance of bilateral and multilateral relations with the aim of guaranteeing alignment with European standards and good practices in prevention, response, management, consequences mitigation and recovery from the damages inflicted by disasters and accidents;
- Establishment of beneficial cooperation and assistance in introducing systems for monitoring, early forecast and warning, a single information system for disaster and accident management, a single emergency call system '112', providing up-to-date and highly effective equipment for response in the event of disasters and accidents and others;
- Creation of the necessary administrative capacity to effectively use the pre-accession financial instruments, as well as the EU Structural and Cohesion Funds in order to ensure maximum population protection and protection of the economy against disasters and accidents.

The Ministry's international activity is implemented by officials of the International Relations Directorate in cooperation with the Minister's Political Cabinet, the Ministry's Administration and other Ministries and agencies. In the fulfilment of its functions the International Relations Directorate works in cooperation with other bodies and institutions, as well as civil society structures. In pursuit of its objectives the Ministry, through its Directorate, drafts, presents for funding and implements programmes and projects, participates in inter-agency working groups, ensures maximum transparency and publicity of its international activities.

2.3. Flood Risk Assessment

Floods are natural phenomena which cannot be prevented. However, some human activities and climate change contribute to an increase in the likelihood and adverse impacts of flood events. It is feasible and desirable to reduce the risk of adverse consequences associated with floods.

Floods have the potential to cause fatalities, displacement of people and damage to the environment, to severely compromise economic development and to undermine the economic activities of the Community.

Between January 2006 and October 2007, the institutions of the European Union negotiated the text of the Flood Risk Directive 2007/60/EC (FRD). The FRD was formally adopted by European Parliament and by the Council of the European Union (council of ministers) on October 23rd, 2007. The directive was published in the Official Journal of the European Union on November 6th, 2007 and will enter into force on November 26th [7]. Member States now have two years to implement the Directive's requirements into their respective national legislations.

According to the Article 2 of the FRD, *flood* is "temporary covering by water of land not normally covered by water" and *flood risk* is combination of:

- the probability of a flood event
- the potential adverse consequences.

Often used definition: Risk = probability x consequences:

$$\text{Flood risk} = \int_0^{\infty} D(h).P(h).dh \qquad (1)$$

where: D(h) = damage associated with a particular flood event. P(h) = probability of that flood event occurring (0......1).

2.4. Flood Protection Activities in Bulgaria

The basic characteristics of Systems for Emergency/Disaster Aid are:

- Rapid deployment by air, sea, land transport;
- Medium water production volume (main priority is drinking water), small decentralised units;

Table 1. Total flood risk [8]

Flood level (h)	P(h)	x	D(h)	=	P(h).D(h)
low	0.8		€ 150,000		€ 120,000
medium	0.5		€ 2,500,000		€ 1,250,000
high	0.1		€ 10,000,000		€ 1,000,000
			Total flood risk:		€ 2,370,000

- Proven function and reliability;
- Operability in remote locations with insecure supply chains (scarce fuel, no grid, and destroyed infrastructure); and
- Deployment, operation and maintenance usually carried out by donor organizations.

The heavy rainfall and floods that occurred in Bulgaria in the period May-August 2005 left substantial devastating consequences for the infrastructure and the regional and local economies. In addition, many human lives have been jeopardized, 17 persons lost their lives, more than 2,000,000 persons directly suffered from the devastating power of the high water. The rainfall caused extensive flooding and material damage and the potential danger of new and larger floods has not yet been overcome.

Preliminary data and analyses show that most of these areas have received over 300 $l/m^2/day$. A crisis situation was officially declared by municipal and district authorities in 24 municipalities within Shoumen, Stara Zagora, Targovishte, Veliko Tarnovo, Lovech, Pleven, Pernik, Vratza, Pazardjik, Plovdiv, Smolian and Sofia regions. The population affected by the disasters is around 3.2 million persons. As a result of the floods a great majority of the above-mentioned municipalities are currently deprived of electricity, adequate water supply and lack communications. The analysis carried out shows that more than 14,500 private and public buildings have been flooded and 1,292 of them have been partially or completely destroyed, part of the local population – about 14,000 persons have been evacuated and left without proper housing conditions and amenities, 125 road and railway river bridges have been damaged, 5,736 km of roads and highways were affected and 124 km of railway tracks have been destroyed, including also several railway stations. 52 km of water protection dykes were destroyed. 42 hospitals and health-care establishments flooded and 435 schools, kindergartens and other educational facilities were damaged.

Aside from the clear need to restore accessibility and the ecological balance in the most affected areas, the seriousness of the damage to public health and educational institutions cannot be underestimated. The schools receive funding through the municipalities, which in crisis situations leads to major funding delays or even cuts. The health system comprises primarily integrated health care complexes, which, especially in small municipalities, provides both inpatient and outpatient treatment.

The Permanent Commission for Civil Protection from Natural Disasters, Calamities and Catastrophes (PCCPNDCC) and the Ministry of State Policy for Disasters and Accidents, also the Permanent Municipality Committees, have organized evacuations and provided emergency assistance to the population in these areas. Preliminary calculations have estimated that the total damages caused to the affected areas amount to 435.7 MEUR.

The purpose of the proposed investment (grant scheme) is to reconstruct the local transport, environmental and public health and educational infrastructures damaged by floods, thereby contributing to the economic development of the areas worst affected by the flooding. It should help to ensure that Bulgaria's process of preparation for accession is not slowed down by this natural disaster and reduce the adverse effects it has on the budgetary situation of the country. Based on the damage assessment findings, well-targeted and urgent investment measures will be implemented in order to rehabilitate the transport infrastructure, revitalize the environment and restore the public health and educational facilities in the areas concerned.

2.5. Cross Border Cooperation for Flood Protection with Neighbouring Countries

Most European rivers are shared by two or more countries. Management of water resources is therefore an important issue in border regions. Since floods are basin-wide phenomena, they do not respect borders, whether national, regional, local or institutional.

People react differently to the flooding of land by sea or rivers. It may be accepted as being caused by nature, which means it cannot be influenced. Flooding may also be seen as a process that produces fertile agricultural land and develops nature, and is therefore very welcome in some situations. Or it may be considered as a threat because of economic damage and danger to life, and is therefore something that should be prevented.

These different perceptions of floods lead to discussions about how to manage floods. The cross-border effects of floods make it more difficult to solve flood problems. In addition, preventive measures were not possible in all situations because of differences in legal systems and culture, and because of a lack of understanding or even the lack of the right contacts.

Real time monitoring and collection of hydrological data is implemented by the National Institute of Meteorology and Hydrology. Most of the river cross-sections are monitored by observers through foot gauges and the reporting of water levels via telephone or telegraph. From the existing 210-river level measuring stations, 44 are reporting at real or semi-real time. Daily data collection is arranged for 12 of those 44 stations, while for the rest a weekly cycle of daily values for the previous week is arranged. The stations and the frequency of data collection for the Danube region are given in the Table 2 below.

Similar sets of real-time data are being made available by Bulgaria to the Romanian partner.

The Maritza/Evros/Meric basin, including Arda, Tundja and Ergene tributaries, is one of the major river systems located in the eastern Balkans, with a total length of 550 km and a total catchment's area of 39,000 km^2. About 66% belongs to Bulgaria, 28% to Turkey and 6% to Greece. About 218 km of the river are located in Greece, with 203 km of the river forming the borderline with Turkey. Although Maritza/Evros/Meric River, shared by

Table 2. Operational data used for flood forecasting services and received operationally from the Romanian side

River	Cross-section	Frequency	Data type
Danube	Corabia	Daily	levels/discharges
Danube	Tr. Magurele	Daily	levels/discharges
Danube	Giurdjiu	Daily	levels/discharges
Danube	Oltenita	Daily	levels/discharges
Iron Gate 1	Orsova	Daily	Levels
Iron Gates 1 & 2	n/a	daily evacuated discharges	
Iron Gates 1 & 2	n/a	3 days forecast of daily evacuated disch.	
Iron Gates 1 & 2	n/a	3 days forecast of daily incoming disch.	
Jiu	Podari	Daily	levels/discharges
Arges	Budesti	Daily	levels/discharges

Greece, Bulgaria and Turkey, is the second longest river after the Danube in the Balkans, this river and its tributary Arda (shared by Greece and Bulgaria) are lacking considerable recent bilateral or trilateral agreements. This situation is mainly due to the past non-trusted political relationships between the three countries. Parts of the Evros/Meric River bed serve as a state border between Greece and Turkey. Thus, both Evros and Ardas rivers are located in a military controlled area. A special permit from military authorities is needed for all scientific or other activities near the rivers. Its delta is an important bird area protected by the Ramsar Convention and the Bern Convention on special species of flora and fauna. It is also cited in the list of regions of special protection according to the EU Directive 79/409/EEC and the national Greek legislation 66/81.

Maritza River, which originates in Bulgaria, joins with Arda and Tundja Rivers near Odrin, which are flowing through Bulgaria and Western Thrace respectively. Tundja establishes a 61 km length of Turkey-Bulgaria border until joins to Maritza River. Ergene River, which also joins to Maritza River, is 280 km long and it totally flows in Turkey.

Two major tributaries of the Maritza have transboundary sub-catchments themselves:

- Arda river flows eastward from the Eastern Rhodope mountains (240 km and 5,200 km^2 in south-eastern Bulgaria; only 30 km and 345 km^2 in Greece) including Kardjali (60,000 inhabitants) and includes various big reservoirs.
- Tundja river (350 km length and 7,982 km^2 in Bulgaria). Main cities are Kazanlak, Sliven (136,000 inhabitants) and Yambol (110,000).

The tributaries Ergene (from Easter Thrace/Turkey) and Arda (Bulgaria and Greece) may induce severe floods and cause a lot of damage to downstream areas.

The lower Maritza River regions suffer from floods on Turkish, Bulgarian and Greece territory. Recent years' floods frequency and magnitude are getting higher and higher. In the past few years the floods occurred at a scale which was not seen in the past twenty years. Besides the floods, decreasing the channel capacity is another polemic part of the region. It becomes clear that improvements in measures for flood prevention and diminishing of flood hazardous effects, could be achieved only through co-operation and use of common information sources.

Turkey and Bulgaria have developed three projects – one for information and real-time data exchanging, and two for flood forecasting and warning. These projects are the first common projects which are applied in the region and in the hydrology area.

Regarding Greece (GR) and Bulgaria (BG), bilateral cooperation in the use of water dates back to 1964. Both countries ratified the Helsinki Convention for protection and use of trans-boundary watercourses (1992; in Greece it has been in force since 1996) and the Espoo Convention. Since the implementation of Helsinki Convention, Greece and Bulgaria have been cooperating by a joint monitoring in the three common river basins, i.e. Struma, Mesta (including the tributary Dospat) and Maritza (including the tributary Arda). In the following years, bilateral agreements on the use of other trans-boundary rivers waters were signed. Also, cooperation in scientific and technical field for the best management of water resources is established.

The main agreements on the protection and use of trans-boundary watercourses are:

- GR-BG agreement on co-operation for the use of watercourses flowing through the two countries (Legislative Decree 4393/1964).
- Second Protocol of the GR & BG agreement about the regulation of economic questions and development of the economic co-operation (Legislative Decree

4393/1964) agreement between GR & BG concerning the formation of a joint committee for the cooperation in the field of electric energy and the use of cross-border river waters (Sofia, 1971).
- Agreement between GR & BG on scientific and technical cooperation (Athens, 1973/1976).
- Protocol for the Joint GR-BG Technical Working Group and Environment Group (approved 1990).
- Protocol for the co-operation of GR-BG Experts for flood control of Strymonas River (approved on 1980); The Agreement from 1964 on flood protection refers to the section downstream of a series of reservoirs in Bulgaria. It operates between local authorities (when the BG reservoir gates release excess water upstream, they send a warning to the GR local authorities).
- Protocol of the Meeting of the Joint GR-BG Committee of Experts for the preparation of a common proposal to the EU for the joint monitoring and control of water quality and quantity of the transboundary rivers Maritsa/Evros, Mesta/Nestos and Struma/Strymonas. (1991).

At national level in Greece, the Ministry of Environment is responsible for integrated water management. For trans-boundary rivers the Ministry of Foreign Affairs is also involved. Next are the Ministries of Economics, of Agriculture and of Defence to some extend. Concerning planning of water quality, parts of the Evros and Ardas catchments on Greek side are designated as NATURA 2000 sites. For these, planning and decision-making is carried out according to the provisions of the relevant national and EU legislation.

The State Hydraulic Works (DSI) is responsible for all surface and sub-surface water resources in Turkey (monitoring and planning, design, construction, and operational activities). Decisions are shared between the Ministry of Energy and Natural Resources and Ministry of Environmental and Forestry, and local communities. The State Hydraulic Works (DSI) does the planning and has several irrigation projects. The Ministry of Environment and Forestry has carried out several wastewater treatment projects in the basin.

3. Conclusions

As a result of the issues discussed above the following conclusions can be made:

1. The Mesta/Nestos River (Greece/Bulgaria) has a number of hydroelectric power plants on the Greek side which need an adequate river flow to operate properly. At present the inflow into Greece is satisfactory. However the hydroelectric and irrigation complex in the Greek part is very vulnerable to interventions by the equipment in the Bulgarian part of the river basin. It is feared that future interventions in Bulgaria may cause shortfalls in the required water levels in Greece.
2. Common legislation or conventions do not apply to the management of cross-border rivers in Greece like the Evros and Nestos. Greece and Bulgaria are EU members (obliged to comply with the WFD), whereas Turkey is a non-member, although accession negotiations are currently in progress with the EU.

3. It is important to realise that the amplitude, frequency, duration and impact of floods depend on natural characteristics and man-induced changes within the entire river basin area.
4. In many cases, good cross-border cooperation between local and regional flood management authorities can improve the effectiveness of flood management services in these regions. This will ultimately result in better protection of citizens and the environment and reduction of damage.
5. Climate change appears to increase the chances of flooding, while human intervention and activities appear to reduce the resilience of water systems and their environment.
6. The ongoing occupation of flood plains has not only increased the risk of potential damage, but has also resulted in a loss of ecological, economic and social benefits of wetlands. Simultaneously, the increasing investments in safety have reduced the public awareness of flood risks.
7. The need to develop appropriate strategies, policies and programmes to adapt to the changing circumstances, to reduce the negative impact of flooding and to protect the dynamic function of ecosystems, is now widely recognised.
8. Strategy, policy and measures on the prevention, mitigation and protection of floods should be based on a holistic approach. Achieving this requires cooperation between authorities on a river basin scale, as well as integration of spatial planning and water management, integration of the various functions and uses of water, joint disaster management and increased cross-border public awareness.

References

EEA-IMS Indicators: Use of freshwater resources (CSI 018), Assessment Draft: http://ims.eionet.europa.eu/IMS/ISpecs/ISpecification20041007131848/IAssessment.

Strategy for Integrated Water Management in the Republic of Bulgaria, Council of Ministers, Sofia, 1997.

National Environmental Strategy and Action Plan 2000–2006, Republic of Bulgaria, Council of Ministers, MoEW, Sofia, 2001.

Freshwater Country Profile – Bulgaria, Decision-Making Programmes and Projects, Sofia, 2003.

Environmental Protection Act, Republic of Bulgaria, National Assembly, State Gazette No. 91/25.09.2002, Corrected, SG No. 96/2002.

Water Act, Republic of Bulgaria, National Assembly, State Gazette No. 67/1999, enforced on 28.01.2000, amended in SG No 87/2000.

Flood Risk Directive 2007/60/EC, OJEU/6.11.2007.

Jan Verkade, A brief introduction to the Flood (Risk) Directive, National flood conference, Sofia, November 2007.

I. Apostol et al. (Eds.)
IOS Press, 2009.
doi:10.3233/978-1-58603-948-6-66

Operation of Automatic Water Monitoring Systems for Emergency Planning

Stephan ANKE, Werner BLOHM and Michael LECHELT
Institute for Sanitation and Environment, Hamburg,
Division: Water Studies, Water Quality Measuring Network,
Marckmannstr. 129b, 20539 Hamburg, Germany
E-mail: stephan.anke@hu.hamburg.de

Abstract. Time and again, shipping accidents and major incidents at industrial establishments have demonstrated how quickly serious water pollution can occur, with effects such as fish mortality and other harmful impacts on the aquatic habitat. In order to minimise the consequences of such incidents, continuous water monitoring is indispensable in the interests of early identification and timely countermeasures. This is all the more essential in an industrial conurbation like Hamburg. Here the water quality measuring network, with a current total of ten measuring stations, has been operating on all important bodies of water since 1988. In addition to averting danger, continuous water monitoring makes a contribution to prevention (detection of illegal discharges) and to observing short-term and long-term changes in water quality.

Keywords. Automated surface water quality measuring stations, continuous water monitoring, water surveillance network Hamburg

Introduction

Inputs of hazardous substances into flowing waters as a result of accidents or illegal discharges can give rise to substantial risks and cause harm to man and the environment. In Europe, with its many countries, such incidents frequently assume transboundary proportions. For this reason a number of arrangements and agreements exist to protect man and the environment from industrial accidents at national, river basin, EU and European UN level. They include technical requirements for operating facilities, liability issues, the preparation of warning and emergency plans, and mutual assistance. Timely and speedy action is always necessary to ensure the success of measures to avert danger and minimise damage. The first step is to register the incident, ascertain its scale and then alert the competent authorities.

The warning and emergency plans of the commissions of Europe's major river basins are largely based on a reporting system that requires the author (pollutant emitter) to notify an incident to the competent authorities immediately after it occurs, supplying all relevant data. In this form the warning and emergency plans are "emission oriented". If the notification is not made – whether deliberately or as a result of ignorance – there remains a risk of major irreparable harm to man and the environment. This danger could be considerably reduced by means of a networked automatic system for identifying incidents along the watercourse and for raising the alert in the notification system of a warning and emergency plan.

This task could be performed by measuring stations sensibly distributed along the river if they were equipped with a technology that enabled them, by means of suitable automatic measurements in the water – i.e. on an "immission-oriented" basis – to first detect "unusual events", then identify them as "natural" or "incident-induced", and finally take an alarm decision based on an assessment of their "relevance".

1. General Aspects of the Location and Tasks of Automated Surface Water Measuring Stations

The quality of stagnant waters and streams is monitored regularly as part of regional, national and international measurement programmes. For this purpose, measuring stations are set up at suitable locations. Measuring stations should be set up in such a way that all major streams can be recorded; that causal relationships can be detected across the country; and that anthropogenic and geogenic impacts can be measured. The issue of trans-boundary effects may also be of importance to the setting up of measuring stations.

Possible/suitable locations for measuring stations should relate to one or more of the following:

- upstream from the place where streams (relevant to water management) empty into lakes or coastal waters;
- at important trans-boundary waters near the border;
- upstream and downstream from conurbations and larger industrial settlements;
- within important river sections of larger bodies of water;
- at crucial tributaries immediately upstream from the point of confluence; or
- at anthropogenically unpolluted river sections ("zero measuring points", reference measuring points, measuring points of background pollution).

Tasks of measuring programmes should include:

- Collection, evaluation and assessment of data on water quality (substance concentration with respect to water, suspended matter, sediments and biota; determination of the biological-ecological state of quality and the ecological structure) as a basis for describing the quality of water all across the country;
- Long-term recording of the quality of streams including background pollution (quality of anthropogenically unpolluted streams) to allow for longer-term and larger-scale trends to be recognised (level of pollution, trends);
- Compliance with international and national obligations under statutory provisions (e.g. EC directives, state regulations, agreements between states such as ICPDR); monitoring of compliance with predetermined requirements as to water quality (targets/quality standards, for example, under directives 76/464/EEC and 2000/60/EC);
- Detection and monitoring of critical water conditions, as well as securing of evidence in the case of unforeseen events (e.g. accidents, incidents, and fish mortality);
- Analysis of substance transport and substance load; and
- Monitoring of the impact of water use.

Figure 1. Measuring station “Seemannshöft” on the Elbe in Hamburg/Germany.

The goal of the investigation is to create a database of sufficient size that allows for

- the evaluation of the ecological state of bodies of water (chemical and biological);
- the determination and assessment of water pollution; and
- the determination of loads and “substance flows” (with estimates of the selective and diffuse portions).

Depending on the water quality-related importance and dynamic (temporal variability of data), a decision is taken on whether the measuring point is sufficient or whether a measuring station has to be set up.

2. Automated Surface Water Quality Measuring Stations

Measuring stations are always fixed facilities (e.g. buildings) near bodies of water – it is here that the measured variables are recorded continuously (Fig. 1). Furthermore, these mainly deal with measuring stations that are used for detecting acute waves of pollutants resulting from accidents, incidents, or unintended or illegal inputs.

The measuring stations, which are automated and working continuously, create the framework needed for continuous water-quality monitoring, which is not possible with “normal” investigative programmes by taking individual or random samples. The continuous measurements taken at those stations enable scientists to record parameters over a period of time that are characterised by a high degree of variability (mostly strongly dependent on seasonal and meteorological factors). They are also crucial parameters in assessing the results gleaned from laboratory tests (through random samples).

These stations are not only equipped with measuring devices, but also with data-collecting devices. Depending on the task at hand, these data-collecting devices can be

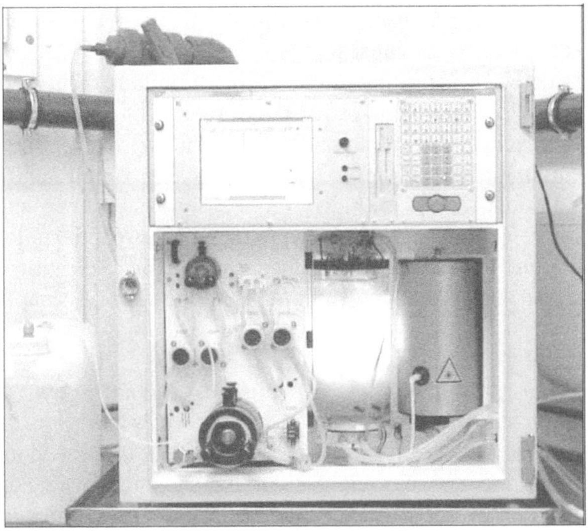

Figure 2. Algae toximeter (*Manufactured by bbe*).

of varying sophistication. The simplest form is the combination of a measuring device and data logger. In this case, data are read manually and evaluated at a later point in time. The most complex – and thus the most effective – form is a PC-based measuring-station control system. Such a system can collect, evaluate and assess data, automatically trigger sample-taking and alarms, and make available the data online (in real time) to any competent agency. Several measuring stations can be linked with each other to create a measuring network.

Short-term changes in waters, typical of accidents, can only be recorded by having continuously working measuring stations. The various measuring devices at the stations allow for a continuous and complete monitoring of water quality. Early detection through continuous water monitoring is an essential element in limiting the effects of accidents and unauthorised input.

In view of the large number of potential pollutants, it is not possible to detect in real time all individual substances as part of regular quality monitoring efforts for streams or by means of continuous physical/chemical monitoring. Besides monitoring chemical/physical parameters, there is also continuous active bio-monitoring as a possible solution to close that gap. In bio-monitoring, "standardised biological material" is exposed under defined conditions in the test facilities of the measuring stations. This bio-monitoring may take the form of pollutant monitoring that serves to monitor bioavailability and bioaccumulation of water pollutants, in particular for the purpose of detecting long-term effects. Or it can be effect-monitoring to show the effects of pollutants – especially in the case of acute pollutant waves, for example, as a result of peak loads caused by accidents – on organisms of various trophic stages of the food web. (see example of algae toximeter in Fig. 2).

The objective behind using continuous bio-monitors is the ability to detect toxic effects even before the ecosystem of the body of water suffers conspicuous damage. This allows for a quick response to pollutant inputs and leads to information on the polluter. The combination of continuous monitoring of chemical/physical parameters and biological-effects monitoring, promises to be more effective in uncovering acute water damage and helps to identify the causes.

By using automated measuring stations with continuous measurements – and this applies to both biological and chemical/physical monitoring – it is possible to isolate the best time for an event-triggered sample-taking, thus allowing for a targeted chemical analysis. In continuous bio-monitoring, the detection of pollutant effects on water organisms is, in addition, always an indication that the water bio-coenosis is at risk. A triggered alarm, therefore, may also give rise to more in-depth biological testing. In addition, bio-testing methods applied to known inputs can yield estimates regarding the risk to the water. Measuring stations equipped with automated sample-takers provide for a variety of ways of automated sample-taking. This may involve time and alarm-triggered single or composite samples.

Water-quality measuring stations with continuous chemical/physical and biological monitoring have been shown to be valuable tools for water monitoring in Germany and other countries over the past few years. Bio-monitors in Germany along the Rhine and Elbe rivers and their tributaries, as well as along the Danube, have made great contributions to the detection and assessment of unauthorised and/or unintended inputs of pollutants. Another positive effect of measuring stations is that they serve as a deterrent to illegal polluters – an aspect that should not be underestimated. Therefore, they also serve as a preventive measure to ward off accidents caused wilfully or by gross negligence.

3. Equipment of Automated Measuring Stations

The equipment of measuring stations and measuring networks depends on the objectives of analysis and local circumstances, including personnel and financial resources. As well as collecting basic chemical-physical data to track water-quality parameters over time, detecting incidents and accidents is efficiently possible. This is why a graduated, modular-style equipment concept has been developed in the EASE project (www.ease.hamburg.de) that gives an overview of the range of instruments in detecting accidents as well as the efficiency and costs of each stage.

The equipment components listed in the equipment concept for measuring stations (see Chapter 2.2.1 – www.ease.hamburg.de – *Equipment of measuring stations*) are intended as examples or suggestions and so only serve as guidelines. In no way does it imply preference for, or judgement on individual manufacturers or certain products.

4. Demands on Observation Network

Measuring networks are combinations of several measuring stations; ideally, a measuring network would cover an entire river basin. Various states in Germany operate measuring networks for continuous water-quality monitoring. With data of all measuring stations along a river being concentrated in one place, it is possible to take targeted, manageable action to control the tasks in a way that is affordable. Operating a measuring network is costly – in terms of organisation and finance – but the number of stations pooled in a network is of relatively little importance. The more that stations are operated effectively within a network, the lower the cost will be for each station, because this system helps to avoid redundancies. In Hamburg, one control centre currently manages a network of 10 measuring stations; even if this number were to double, it still would not pose a major technical or organisational problem for the control centre.

At this point, equipment and control programmes for measuring control centres are not yet available as standard products. This helps to explain the high costs attached to setting up a measuring network control centre.

Apart from data management, other essential components of a control centre include quality assurance, evaluation and export to graphics and lists and export of such results to files for further use. The largest expenditure in data management is for the transmission of data from stations to the control centre (i.e. data alignment) and for the maintenance and archiving of data in a central database. One approach that has proved useful is to store data in professional relational databases in stations and at the control centre. This ensures reliable operations and secures data. Databases with such a structure are more expensive to buy and maintain than, say, Freeware databases or spreadsheets, but they are still the best way of managing larger data volumes. The data volume in Hamburg involves about 4 million data sets per year; the control centre has been up and running since 1988 and contains approx. 80 million data sets. In order to execute different tasks and address various problems, these data must be accessible at all times.

The evaluation and export functionalities of a measuring network are the channels through which data reach their target groups. These channels are crucial for passing on data directly and swiftly to the requesting offices and for processing inquiries speedily.

Data management in measuring network control centres offers a number of advantages. Given that the data of all connected stations are available at the centre, analysis and evaluation will be much simpler. The same is true of data maintenance and all other administrative tasks related to maintenance such as archiving, storage media, etc. The service, maintenance and calibration will be easier to control and manage and less expensive. For cost reasons, the number of measuring network controls should be kept as small as possible. It does seem sensible, though, to operate joint control centres in river-basin communities. Technically speaking, this is very easy to do without any problems – via the Internet.

The connection for transferring data between the centre and stations can be implemented through dial-up connections. Dedicated lines are, as a rule, too expensive and unnecessary. In future, we will surely see a substantial increase in the speed of data transfers once the new phone networks are in place (e.g. ADSL).

5. Demands on Automated Measuring Stations

The measuring tasks are often correlated with the use of the water bodies which can be used in many different ways, such as:

- Habitats for plants and animals (protection of aquatic communities);
- Fisheries;
- Recreation;
- Drinking water; and
- Commercial and industrial purposes (like shipping, generation of electricity, cooling water, waste-water disposal, or agriculture).

These different uses may create conflict which, traditionally, consisted of the production of drinking water and the input of municipal waste water. Uses with high "protection targets", such as subjects needing protection ("aquatic communities" or "drinking water"), can be secured only if the relevant targets (quality standards/"limit values") are met. In order to estimate pollution, systematic water-quality tests are carried out. Automatic measuring stations with continuous quality monitoring were supposed to play a central role in this respect. The apparently bad condition of some bodies of water that are not polluted by what is known as "spots", but by "diffuse inputs", has also led to the setting-up of measuring stations in order to investigate the exact progression of such pollution over time.

Measuring stations also fulfil tasks of immediate protection. In Germany, more and more measuring stations have been set up or expanded since the 1980s as a result of major chemical spills (e.g. near Basle in 1986). These stations were to detect pollution early in order to trigger counter-measures. Drinking-water supply and eco-systems were to be protected in this, targeted, way.

In addition to water-quality assessment, accident recognition and sample-taking within local, regional, national and international measuring programmes, measuring stations can perform other tasks too. Such additional tasks may include programmes for monitoring nuclear power stations, redevelopment of bodies of water (such as weirs or re-naturation measures) or hydrological measurements such as water levels or run-offs. Frequently, measuring stations for bodies of water also record meteorological parameters crucial to the water's quality (precipitation, direction and speed of wind, global radiation, air pressure and temperature).

6. Locations of Measuring Stations Along Bodies of Water

The site should essentially be representative of a certain section of a body of water and the river cross-section, but it can also be used for specific emission-monitoring, in which case it would deviate from this principle. In selecting a site, one must take into account the measuring task at hand as well as the logistics of the location. For measuring purposes, preference should be given to measuring stations set up in the middle of a body of water, but this is rarely possible. The station, however, should be as close to the channel line as possible (and, in terms of the flow, that would be the centre of a river) in order to achieve more representative measurements and samples. The easiest place to achieve this is at an undercut bank. If the station is to observe certain "plumes", special attention should be paid to this.

Preliminary tests should be done during the planning phase of a measuring station, using longitudinal, cross and depth profiles. For example, this can be done by means of mobile measurements. The results of laboratory tests can also provide useful insights into selecting the right location.

In selecting a location, also consider any interference that may result from the catchment area itself. Flotsam, for example, may damage water-sampling facilities or the entire station; smaller flotsam may clog pipes or impair the functioning of pumps and measuring systems. Water levels that fluctuate severely, floods or floating ice often cause interruptions at stations. Such problems and others can be minimised if the location is chosen wisely.

Figure 3. Measuring station "Bunthaus" on the Elbe in Hamburg/Germany.

7. Different Styles of Measuring Stations

The style of a measuring station is primarily determined by its location and its measuring tasks. The location defines the type of building, while the measuring task defines the scope and space required for equipment.

All this requires a certain level of logistics. For example, when planning a station, one should allow sufficient room for all measuring systems and supply lines/connections (like electricity, phones and water). Maintenance/repair work on measuring systems and station facilities also require room and some measuring systems need a special climatic environment (e.g. air conditioning for complex measuring systems, such as GC-MS). Also, precautions need to be taken against theft and vandalism.

The most common type of measuring station, surely, is the land-locked form. Land-locked stations are generally less expensive to set up and allow for easier access. They are a particularly good choice if there is a chance of using existing buildings or adapting existing former commercial facilities close to the water. Flooding, however, may be a problem, because such a set-up would often require long pipelines, and frost protection for intake facilities would be particularly complicated. Ensuring that sample water has been drawn for all measuring systems and samplers can be quite difficult and thus more expensive than for floating stations. As a rule, pumps for such stations have to be bigger because of higher lifts (and thus more expensive to buy).

In areas of rivers affected by the tide, measuring stations are often designed as floating facilities (Fig. 3). This type of station is surely a complex form of measuring station, but it also has its advantages, for example in the way that pipelines are designed. With relatively short transport routes, even if water levels fluctuate, the lifts of pumps are always constant.

It is quite conceivable that floating measuring stations can be designed in such way that they can do without any pipelines whatsoever if the measuring sensors in the measuring station are placed directly inside the water. Protection against flooding can also be a decisive advantage. For cost reasons, when setting up a station, one should always try to use existing facilities whenever possible. Problems that may arise with floating stations include difficult access by means of bridges or boats or the issue of securing the station (e.g. by means of piles). The installation of supply and drainage lines can be difficult and it might also be difficult to obtain a licence if the river is used for commercial shipping.

Measuring buoys and offshore stations are special types of floating stations. Measuring buoys may be an inexpensive alternative to measuring stations if only few measuring systems (low-maintenance) are used and if the location is not accessible to the public (thus reducing vandalism and theft risks).

8. The Hamburg Automatic River Surveillance System

The Water Surveillance Network in Hamburg is provided by the Department of Environmental Analyses and Assessment and is the automatic river surveillance system for Hamburg. The department has a total staff of 122 (including 25 scientists and 19 engineers), and the yearly budget is €10 million. The quality management is according to DIN EN ISO 17025.

The department tasks relate to chemical, biological and physical laboratories, automatic continual surveillance of water and air quality, development for analytical methods (especially for legislation and standardization), admission of laboratories for legally required analyses, and scientific expert advice for the state administration.

Time and again, shipping accidents and major incidents at industrial establishments have demonstrated how quickly serious water pollution can occur, with effects such as fish mortality and other harmful impacts on the aquatic habitat.

In order to minimise the consequences of such incidents, continuous water monitoring is indispensable in the interests of early identification and timely countermeasures. This is all the more essential in an industrial conurbation like Hamburg. Here the water quality measuring network with a current total of eleven measuring stations has been operating on all important bodies of water since 1988.

In addition to averting dangers, continuous water monitoring makes a contribution to prevention (such as the detection of illegal discharges) and to observing short-term and long-term changes in water quality.

The data recorded provide a basis for decisions on water management measures. Another aspect is protection of drinking water extraction areas. The automatic continual surveillance of water (WGMN) provides currently ten monitoring stations in the area (Fig. 4).

9. Equipment of the Measuring Stations

All stations perform automatic, continuous round-the-clock recording of the following chemo-physical parameters: conductivity, oxygen content, pH, temperature and turbidity.

The specially important Elbe-stations "Bunthaus" and "Seemannshöft", the "Fischerhof" station at the river Bille and the "Wandsbeker Allee" station at the river Wandse also operate a biological early warning system that detects the presence of toxic substances in the water. These stations are equipped with automatic samplers, thus ensuring immediate availability of samples for detailed laboratory analysis if an

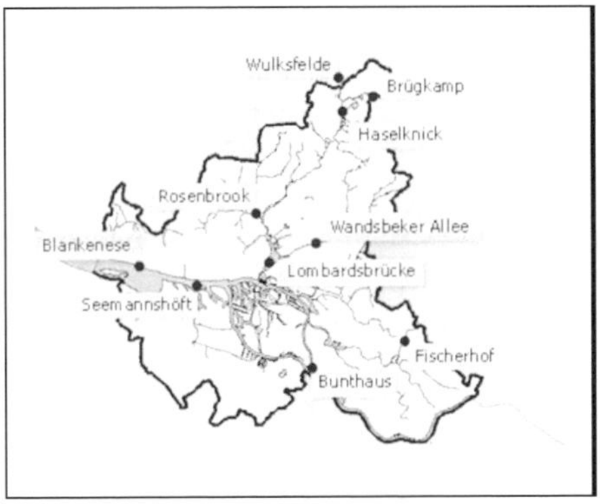

Figure 4. Monitoring stations in the Hamburg area.

incident occurs. Some of the stations also have devices for oil detection and for measuring UV absorption (for identification of organic contamination).

The measuring stations record the water data at 10-minute intervals. The large quantities of data recorded are temporarily stored on station computers before being transmitted, complete with any alarm reports, by ISDN to the central computer. The entire measuring network including the "biological early warning system" forms the "water quality measuring network".

10. Biotest Devices – The Biological Early Warning System

In continuous monitoring of bodies of water over a long period, the necessary investment in measurement technology severely restricts the number of parameters that can be recorded. More than 50,000 different chemicals are produced in Germany alone. It simply is not possible to test water for such a vast range of individual substances. These substances may find their way into waters in various ways, for example as a result of accidents or leaks, during transhipment in ports, during use in the agricultural sector. Nevertheless, in order to obtain as complete a picture as possible of pollutant levels in water, it makes sense to use methods with biological-effect monitoring, which reflect acute toxic effects in summary form.

For this reason automatically operating test systems with water fleas (Daphnia magna) and green algae (Chlorella vulgaris) are employed in four of the Hamburg measuring stations. The Daphnia toximeter monitors the movements of the water fleas with the aid of a camera. Any significant changes in behaviour can be presumed to be due to acute water pollution. In the case of the algal toximeter, harmful effects on the algae are registered in the form of an inhibition of photosynthesis activity.

11. Alarms – What Happens in an Alarm Situation

If the measuring systems supply readings that are outside the limits of statistical fluctuation, this is reported to the headquarters of the water quality measuring network. If unusual events are registered for several parameters at the same time, the alarm is raised.

In an alarm situation it is important to assess the damage and the origin and nature of the pollutants as fast as possible. If the central computer receives an alarm, it automatically forwards it by e-mail or SMS to the responsible staff so that they can take the necessary action. At the same time the station initiates a special alarm sampling programme. The samples obtained then undergo thorough chemical analysis to determine the type of contamination. From this analysis it may be possible to draw direct conclusions about the identity of the polluter.

In this way the water quality measuring network with a biological early warning system ensures that sudden occurrences of toxic water pollution are recognised at an early stage and timely countermeasures can be taken. For example, if an alarm occurs at a station, discharge of water from a river to a drinking water extraction area can be stopped.

References

Institute for Sanitation and Environment Hamburg (2004): Development of Alarm Criteria and Detection of Major Incidents in Measuring Stations in the Elbe Catchment Area for International Emergency Planning (EASE) – Final report to the German Federal Environmental Agency project FKZ -200 48 314/02-Subproject 2, Hamburg.

2000/60/EG: Richtlinie des Europäischen Parlaments und des Rates vom 23. Oktober 2000 zur Schaffung eines Ordnungsrahmens für Maßnahmen der Gemeinschaft im Bereich der Wasserpolitik, Amtsblatt Nr. L 327 vom 22/12/2000 S. 0001 – 0073.

LAWA (Länderarbeitsgemeinschaft Wasser): Fließgewässer in der Bundesrepublik Deutschland – Empfehlung für die regelmäßige Untersuchung der Beschaffenheit der Fließgewässer in den Ländern der Bundesrepublik Deutschland-LAWA-Untersuchungsprogramm in den Ländern der Bundesrepublik Deutschland – (1997).

LAWA (Länderarbeitsgemeinschaft Wasser): Einsatzmöglichkeiten des Biomonito-rings zur Überwachung von Langzeit-Wirkungen in Gewässern (2000).

LAWA (Länderarbeitsgemeinschaft Wasser): Empfehlungen zum Einsatz von konti-nuierlichen Biotestverfahren für die Gewässerüberwachung (1996).

Design and Installation of Automatic River Surveil-lance Systems with Automatic Alarming by P. Friesel, W. Blohm, M. Lechelt.

Optimisation of Disaster Forecasting and Prevention Measures in the Context of Human and Social Dynamics
I. Apostol et al. (Eds.)
IOS Press, 2009.
doi:10.3233/978-1-58603-948-6-77

Cost-benefit in Water Hazard Management

Marcel Fälsch
Institute for Infrastructure and Resources Management,
University of Leipzig, Germany

Abstract. The protection of water against pollution from hazardous substances of industrial plants is normally implemented by a combination of technical and organisational safety measures within strategic hazard management. The demand for hazard prevention results from several motivations that can be allocated to the public as well as to the plant operators. But rarely is economic benefit the reason why preventive measures are seen as mandatory. This paper focuses on prevention activities to detect kinds of costs and benefits in water hazard management. Furthermore, the first basic approaches are shown for using economic perspectives in the fields of hazard prevention for a practical method of improving efficiency.

Keywords: Water hazardous substances, Safety Management, Cost-benefit, Cost-effectiveness, Prevention

Introduction

The operation of technical installations normally causes a variety of risks to the environment that can adversely affect its condition in different ways. Especially when dealing with hazardous substances, the prevention of water pollution is of particular importance and has to be considered by the safety approach of a company.

To rate as a water hazard, a substance needs to have the ability to reduce the quality of water and limit its human and natural usability. For the assessment of the chemical inventory of an industrial plant it is necessary to analyse the special properties of the involved substances. Water hazardous substances are at least (very) toxic, corrosive, harmful to health, environmentally hazardous, harmful to aquatic organisms or accompanied by long-term harmful effects to water bodies. If a substance or a mixture shows one or more of these properties, measures for hazard prevention will be required. [1]

But what are the particular motivations of the involved actors or parties to implement effective hazard prevention?

The initial point to start considerations about implementing preventive security measures is to fulfil the social and ecological liability for saving human health und protecting the environment. This general need for safety does not necessarily entail the proactive interest of the operator, not at least because there can be other factors involved with no immediate effect on the scope of the operator's activities. Therefore, safety needs are clarified through legal requirements to induce the concerned actors to implement appropriate measures to reduce existing risks. From this safety-oriented perspective, preventive hazard management first of all results from administrative regulations and is due to the greater demand.

In the European Union these requirements are reflected within various legal regulations, found both in water-related and plant-related environmental legislation. Both of them share the aim to prevent and/or to mitigate the effects of accidental pollution incidents and have to be implemented by the several member states. [2]-[4]

However, the legal part is not sufficient to explain the whole scope of motivation to implement measures of hazard management. Furthermore, safety can be seen as a status resulting from operators' interest to optimize technical processes and increase the efficiency of resource utilization. Measures of hazard management are conducive to these aspects to a substantial extent.

It can be concluded that in addition to social and ecological arguments, an economic benefit also has to be supposed. Among economic conditions the operator will create a far higher motivation from an efficiency perspective. This point of view so far doesn't really play a reasonable role in the current development of hazard prevention. This paper describes, in respect to this problem, the basic relationships and outlines approaches to possible solutions for increasing effectiveness in water hazard management. Figure 1 shows the connection between the several motives to implement hazard prevention measures.

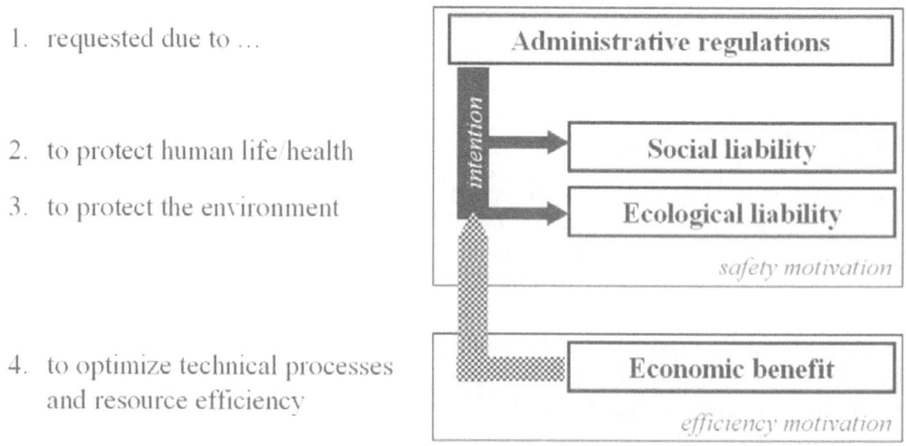

Figure 1. Motivation for implementation of hazard prevention

1. Implementation of Hazard Prevention

Preventive measures play a particular role in water protection. Compared with typical technologies of pollution control that try to reduce continuous and known arising emissions, measures of precaution are applied to work against pollution, which cannot be allocated to the proper operation. Therefore, it is not the effective emission of a process that is in focus, but the possible emission resulting from a connected and not foreseeable factor. Hazard management pursues some functional and organizational approaches to consider this special character.

1.1. Classification of Preventive Measures in Water Protection

Hazard management is closely based on the precautionary principle. Possible pollution should be prevented before it causes adverse effects on the water status. Effective emissions in connection with the adopted measures do not exist in the normal operation of a plant.

The initiation of effective hazard management was due to a few serious hazardous incidents in the past, which first caused an intensified risk awareness among the public and between the involved actors, and subsequently the increased claim to protect from such events in the future or to mitigate their effects. With the use of non-natural and hazardous materials, but also of organic substances whose impact on the environment is not known at all, the precautionary principle demands the isolation on the natural environment, especially the water cycle. Alternatives that allow the substitution of hazardous with more harmless substances should receive priority.

The narrow interpretation of the precautionary principle causes the application of appropriate measures to reduce the release of hazardous substances as far as possible. Both in usual and unusual operations, zero-emission is the ideal intention but against this argument acts the principle of proportionality. According to this, safety requirements for an installation can be selected in relation to the existing hazard. The assessment is based on the probability of occurrence and the appreciable extent of damage, which basically depends on the amount of hazardous substance. Out of this, differential demands result in the appropriate and proportional safety level for the concerned installation. Simply, it may be assumed that the lower the probability of negative effects, the lower the prevention level to be applied, also because of the comparatively low apprehension. Due to this relationship, demands for graduation of technical safety arise that are based on a subjective risk assessment and hazard perception.

1.2. Multiple Barrier Approach

The precautionary principle can be realised by the implementation of a *Multiple Barrier Approach* to segregate hazardous substances from environmental circles. Two barriers (illustrated in Figure 2) are mandatory to meet the criteria of the principle. [5]

The first barrier is focused on the usual operation and aims at the direct control of a hazardous substance to prevent an uncontrolled release as the main trigger for an accidental event. Examples of measures used in the first barrier are:

- technical safety devices;
- thresholds for substances;
- safety-orientated maintenance;
- monitoring; and
- overfill safety.

Nevertheless, the case of process failures and damage caused by accidents cannot be entirely excluded. As well, improper handling can also contribute to unwanted substance releases. In those situations, measures for the second barrier take effect and released substances are retaining from further distribution. Possible measures for the second barrier can be:

- retention volume;
- sealing surface and basin;

- fire prevention strategies; and
- internal warning and alert systems.

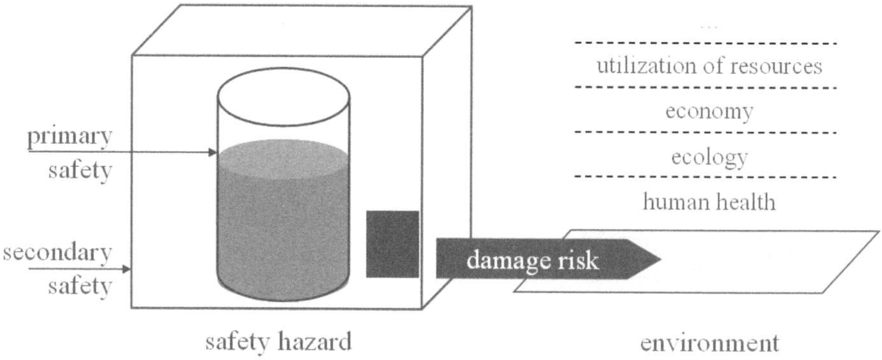

Figure 2. Multiple barrier approach: first and second safety barriers

In addition, further barrier systems in the form of specific or additional measures can be necessary, if it is required by the specifications or the hazard scale and risk potential of a plant. This advanced safety approach is especially subjected to the relationship between precaution and proportionality. It requires, next to the technical approach especially, an expanded organizational consideration. Within the first and second barriers, safety measures concentrate on physical or constructive opportunities. On the advanced level, functional, symbolic or incorporeal measures can also be considered. [6]

1.3. Management Approach

Hazard management, however, should be more than a combination of preventive measures to assume a reduction of probability on unintended releases and negative effects for human health and the environment. Rather, it is to consider as an interaction between an existing situation, which is intensively analysed and assessed, and the adequate reaction on the specific requirements resulting from that analysis. Note that this procedure should not be interpreted as a static process. Both the initial situation and the options of effective (and efficient) prevention are subject to a continuous change and enhancement. Only active and detailed realisations of this process will permanently achieve a decrease of emissions from technical incidents, regardless of whether the process is evaluated in a economic sense or not.

Hazard management can be seen as a strategic element of the complex organisational structure of an establishment and so it is only one part of the overall management system. The special function is the implementation of safety strategies and operational tasks to decrease the risk of an incident. Appropriate instruments like defined procedures, applications and improved strategies are used to identify, evaluate and control safety hazards through specified adoption due to the individual risk situation. So the approach is integrated in the overall concept of the industrial enterprise. [7] That is also the reason why it is closely oriented towards the approach of a typical information system as shown in Figure 3, that can be visualised by the management cycle.

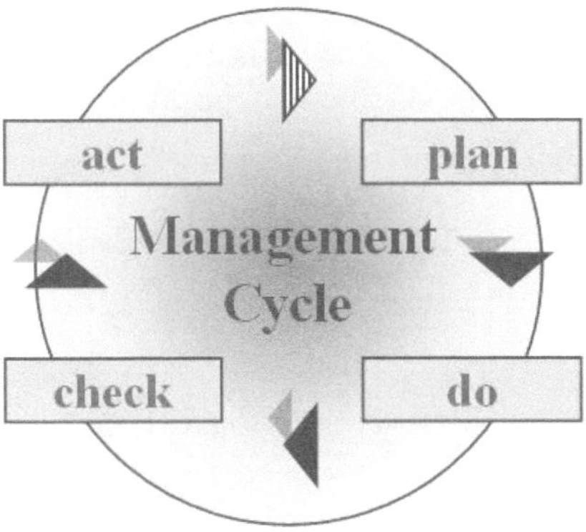

Figure 3. Management cycle

The cycle can be divided into four sections. The first section is the planning phase determining the enterprise safety policy and objectives where the necessary structures and responsibilities have to be identified and allocated. Accordingly, the focus concentrates on the question of what is to be done to achieve the planned objectives. The first step is the accomplishment of hazard identification and evaluation to identify the existing safety hazards and assess the inter-relationships between the likelihood and severity of possible events. On the basis of the investigated information, appropriate measures, like those referred to earlier, have to be implemented within the operational procedures. Possible adaption of the installation or process may need to be considered at an early stage. After the implementation is completed, monitoring procedures are essential for checking the performance of the hazard management process and investigating any occurred events and near-miss situations. The collected information and experience will be evaluated at regular intervals to review and audit the effectiveness of the management system. As a result, adjustments and improvements for single elements of not only the safety chain but also for the overall safety system have to be integrated into the steps of the cycle. So the whole management procedure restarts.

2. Basic Approach to Cost and Benefit Assessment in Hazard Management

As mentioned before, the implementation of safety requirements within the scope of hazard management results in costs which relate to those activities. The hazard management doesn't only answer the purpose of duteous realization of legal restraints but also relates to the economic benefit for the operator, who implements the measures, and for the public, whose security demands are represented by the public authority. Consequently the question arises as to the costs of hazard management and what is the expected benefit? Moreover, is it possible to oppose and compare both parameters (see Figure 4)?

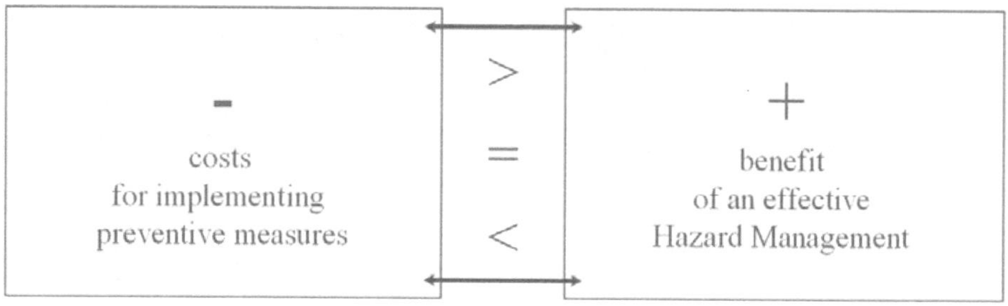

Figure 4. Quantity comparison of costs and benefits

The objective of this mirror-effect may be to compare the efficiency of several safety concepts and, furthermore, it may encourage the discussion about an adequate level of prevention in a constructive way. Therefore, it has to succeed in making visible the relationship between costs and benefits in hazard management. To achieve this objective, first of all several basic considerations are unavoidable.

2.1. Identification of Costs

The costs for hazard management are basically borne by the involved operators and the public authorities. The task for the public authority is the application of various instruments (as shown in Figure 5) to motivate the operator to realize hazard prevention measures and, if necessary, to undertake some additional territorial measures by itself. According to that, the effort which has to be calculated arises especially in the following areas:

- Information (information systems / communication): Public authorities have to give information about what is the precautionary demand of the public and how the operator could achieve it;
- Legal implementation: Includes the creation of legal basics and the adaption to changing requirements, as well as the allocation of responsibilities and the provision of necessary facilities;
- Permission: Includes the granting of permissions for relevant plants or the verification of safety concepts, both in the case of new construction and changing of existing components; and

- Control: Control of information submitted by the operator, including on-site reviews and logging by official experts, as well as the investigation of incidents.

To realize the legal requirements, the operators have to find out what are the appropriate precautionary measures necessary for achieving them. Inside the concept for hazard management there can be chosen some different kinds of measures, which are equally coupled with different kinds of costs:

- Identification of safety hazards: Operator's activity to identify internal and external factors that cause safety hazards (risk analysis);
- Planning: Depending on the types of measure, various planning processes are required to integrate them into the organisational and technical structure of the installation;
- Investments: Substantive measures such as architectural designs or technical components generate investment costs;
- Maintenance: Includes maintenance, adoption or change of elements to preserve and improve the safety level and requires special qualified personnel or external know-how;
- Training and qualification: Activities to train and improve safety relevant knowledge of personnel and increase the risk awareness;
- Additional personnel: Safety measures can possibly cause a demand for additional personnel for monitoring or supporting processes; and
- Revision and redevelopment: Costs incurred for audits, expert reports or review of the effectiveness of the hazard management concept.

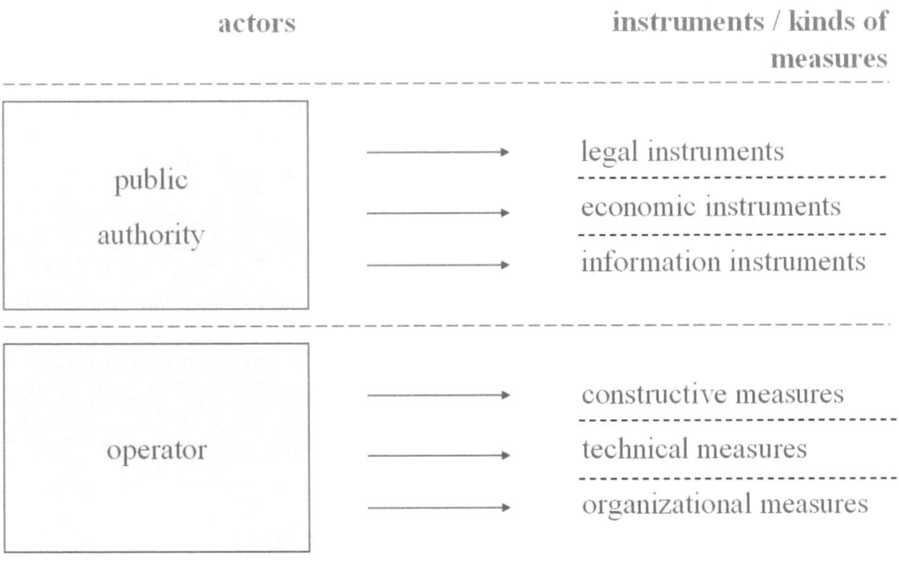

Figure 5. Cost units and activities

2.2. Identification of Benefits

Direct benefits, as a result of effective hazard management, can be expected on the part of the plant operator. Due to the resulting effects the operator benefits partially by saving costs, but also a part of hazard management is the indirect benefit. Both can be divided into the following categories:

- Reduction of substance losses: The focal point of hazard management is the prevention of events with hazardous substance releases. These are not necessarily always caused by accidents, but can also result from improper handling of installations. With the right safety measures, such losses can be reduced crucially, together with the effect of increased resource efficiency;
- Continuous functionality: Hazard management can achieve a more stable situation because, due to a reduced failure rate, the efficiency of production increases;
- Protection of installation components. Due to an active prevention of incidents, system components are protected, the physical life of installations is extended and replacement costs decrease;
- Protection of personnel: The absence from work of personnel (for example, through disease or injury) leads to additional costs for the operator. A better safety level is accompanied by less personnel absence; and
- Image-effects: Publicity generating events can cause negative consequences for the image of a company. Prevention can protect against such image-damage and the related financial losses. A more effective public monitoring for better event-detection can intensify this effect equally.

The public primarily benefits from the preservation of the environment as a result of reduced pollutant emissions. In this case the fields of benefit can especially be found in the distribution of resources, the sustainable usage of resources and, of course, the correspondingly increased levels of safety. The following (social and economic) benefits are a result of the implementation of safety demands:

- Preserving resource quality;
- Ensuring drinking water supply;
- Preserving ecological diversity;
- Providing diverse economic usage; and
- Resource preservation for coming generations.

While the demand for protection often is a social interest, economic interests aren't brought to mind until consequences due to critical events occur. So a profit in the economic sense can also be generated by the elimination of unwanted occurrences which would affect the existing economic and cultural uses by humans in a negative way.

Figure 6 illustrates the fields of benefit for the involved actors or parties.

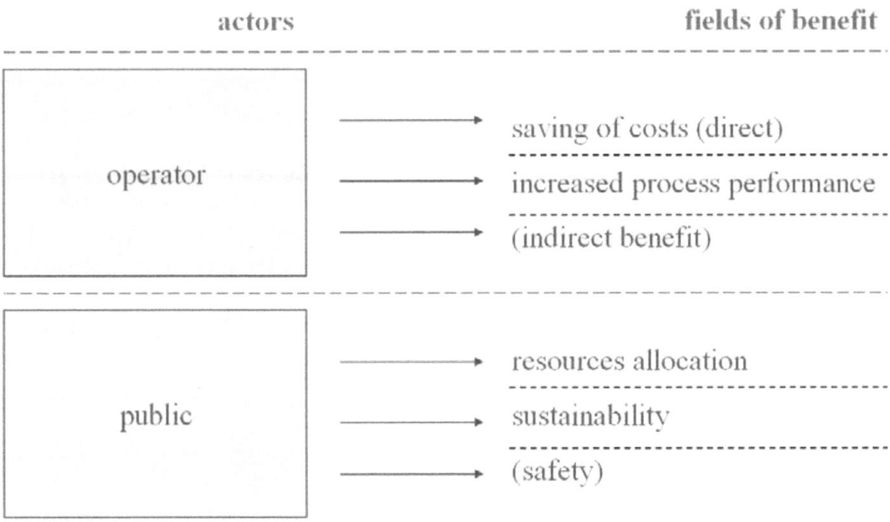

Figure 6. Fields of benefit for involved parties

3. Integration of Economic Instruments into Hazard Management

So far, the following can be concluded: the reason for hazard prevention is based in the social and ecological liability. But with the implementation of safety measures, an economic benefit is also associated. So why could this consideration of the economic perspective be relevant?

Indirectly, in this approach can be seen an instrument that could be conducive to the implementation and development of hazard management. If it succeeds in explaining to the operator that there is an economic incentive for a particular behaviour, then a stronger initiative to improve effectiveness of the related parameters can be assumed. To achieve this, first of all the economic benefit must be more transparent. If this objective can be achieved, the economic incentive will produce a wider scope, as is only to be expected from social and environmental claims. But note that, also, the non-economic (non-profit) benefit can determine financial factors, although these are much more difficult to evaluate. Figure 7 gives an idea of the influence of economic benefit.

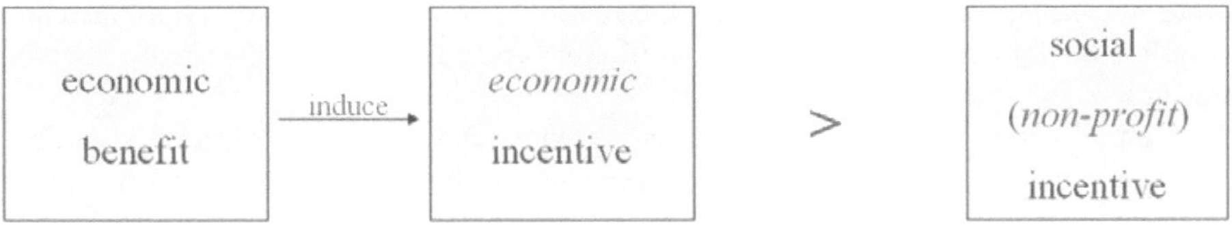

Figure 7. Influence of economic benefit

3.1. Cost Analysis

In order to establish or to improve transparency in hazard management, the first step has to be an explicit analysis of the costs incurred. For this purpose two factors have to be considered more exactly.

First, it will be interesting to analyze how the different types of costs are distributed within a hazard management concept because of the combination with regard to different measures. The next step is then to compare different concepts in regard to types of measures and related costs. Additionally, the total costs of an activity in comparison with the investment costs become more important. The considered types of costs can be divided as follows, for example:

- Investment costs;
- Operating costs;
- Personnel costs;
- Planning costs; and
- Assessment costs

It isn't possible to derive a reliable statement about the costs due to safety measures through the calculation of the costs alone because of the distribution of costs into different functional fields within the establishment. Usually the costs of a measure have to be distributed to different functions; for example, if an employee has tasks in the security sector and also in the production sector. The distribution of the costs is imaginable in the following categories, for example:

- Hazard management (only safety related costs);
- Production process; and
- General organisation.

In this regard, the expected difference between complex management systems and safety hazards, which only require basic measures, will be interesting to analyze.

3.2. Effectiveness of Measures and Cost-effectiveness Analysis

The second step is to analyze how the effectiveness of safety measures can be established. Therefore, it has to be clarified in terms of which factor the effectiveness should be expressed.

As a visible impact of the prevention measures, decreases of negative health effects for humans and a decline in environmental pollution are to be expected. However, these parameters will be hard to predict and to determine afterwards because it isn't sure that general facts about pollution and negative health effects (without precautions) are available.

Consequently there has to be chosen another reference that makes it possible to illustrate the efficiency. The level of precaution is also oriented towards hazard potential in connection with the risk of an arising incident. So a possible solution would be the analysis of risk reduction due to safety implementation.

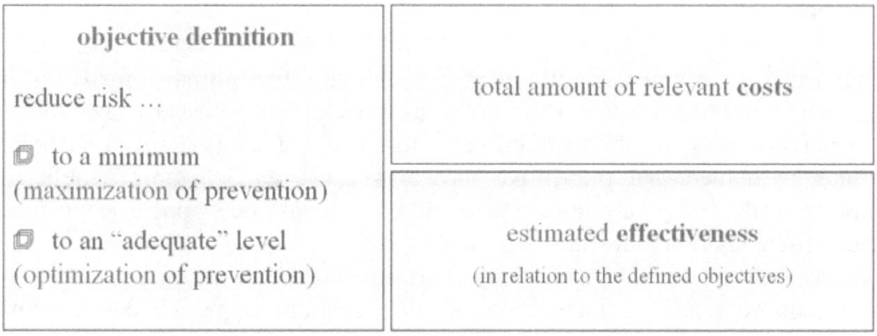

Figure 8. Elements of cost-effectiveness analysis

As a result of a successful cost analysis and the estimated effectiveness, the next objective has to be to perform a cost-effectiveness analysis, using the elements shown in Figure 8. In this case, the target is the desired safety level as a result of the reduced risk. There are two approaches for defining the decisive objective:

- Maximization: Prevention to reduce risk to a minimum level; and
- Optimization: Prevention to reduce risk to an "adequate" level

With the application of the cost-effectiveness analysis accrues the possibility to compare different hazard management concepts also in terms of economic aspects. The expected effect is to find lower-cost solutions to give priority, without reducing the safety effectiveness of the chosen measures or combinations of them.

3.3. Assessment of Total Benefit

The final step has to be the benefit assessment resulting from the hazard management. As already mentioned the total benefit is distributed to several scopes. While the detection of the economic benefit through cost savings or avoidance of adverse effects still appears as corporeal, the (monetary) evaluation of social benefits is probably more difficult to prove. Nevertheless this is part of the total benefit and therefore a part of what the accruing costs have to be set against. Figure 9 tries to illustrate this circumstance.

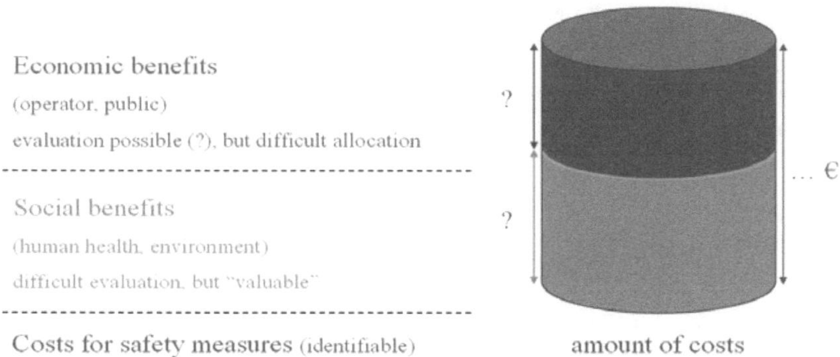

Figure 9. Structure and elements of total benefit

Conclusions

The solution of technical requirements is actually the primary focus in hazard management activities. To date, the view from an economic perspective is rarely shown and financial aspects are often subordinated to the social safety demand. Principally it could not be denied that preventive measures cause an economic benefit for the operator as well as for the public. These effects, for instance, appear as an increased resource efficiency or a more steady process.

Nevertheless, without any on-going investigation it will not be possible to clarify whether a more intensive integration of the economic approach can improve the efficiency of hazard prevention and generate a feasible instrument to encourage an incentive for safety measures to implement. But what is to be expected, approximately, from a more extensive research activity in this topic?

First of all it will be possible to clarify what kind of measure causes what kind of costs, and how the allocation of costs is effected inside different management concepts? If this consideration is extended to the assessment of the particular effectiveness of the combination of measures, the collected data allow an investigation of the cost-reducing potential without compromising the effectiveness of the precautions.

If it succeeds in realising these deliberations it would be a consequential basis for the evaluation of the economic (as well as social and environmental) benefits. All of these elementary steps can contribute to the integration of economic incentives to encourage transparency in water hazard management and enforce operator initiatives.

References

[1] Federal Environmental Agency (2006): Checklists for surveying and assessing industrial plant handling materials and substances which are hazardous to water. No. 1 – substances. available online: http://www.umweltbundesamt.de/anlagen/Checklistenmethode/homeen.html.

[2] Seveso II Directive: Council 96/82/EC of 9 December 1996 on the control of major accident hazards involving dangerous substances.

[3] IPPC Directive: Council Directive 96/61/EC of 24 September 1996 concerning integrated pollution prevention and control.

[4] Water Framework Directive: Directive 2000/60/EC of the European Parliament and the Council of 23 October 2000 establishing a framework for Community action in the field of water policy.

[5] Lühr, H.-P. (2006): Vorschläge zur Weiterentwicklung des anlagenbezogenen Umgangs mit wassergefährdenden Stoffen. KA Abwasser, Abfall, vol. 53, (5), p. 503-513

[6] Hollnagel, E. (2008): Risk + barriers = safety? Safety Science, vol. 46, p. 221-229.

[7] Harms-Ringdahl, L. (2004): Relationships between accident investigations, risk analysis, and safety management. Journal of Hazardous Materials, vol. 111, p. 13-19.

Optimisation of Disaster Forecasting and Prevention Measures in the Context of Human and Social Dynamics
I. Apostol et al. (Eds.)
IOS Press, 2009.
doi:10.3233/978-1-58603-948-6-89

The Environmental Benefits from the Treatment of Waste Water and Slime Derived from Crude Water pre-conditioning at S.C. CET Iaşi S.A.

Mugurel Rotariu [1], Dorin Ivana [2], Lavinia Tofan [3], Monica Rotariu [4], Ovidiu Toma [5*]

[1] *Associate professor, Faculty of Electric Engineering, Department of Energetic*
"Gh. Asachi" Technical University of Iaşi,
Bd. D. Mangeron, nr. 51-53, 700505 IASI / ROMANIA
Tel (0040 232) 278680, Fax (0040 232) 236283 E-mail: vrotariu@ee.tuiasi.ro

[2] *Engineer S.C. CET Iasi, Bd. Calea Chisinaului, nr 25, IASI / ROMANIA*

[3] *Associate professor, Faculty of Chemical Engineering*
"Gh. Asachi" Technical University of Iaşi,
Bd. D. Mangeron, nr. 71 A, IASI/ ROMANIA
Tel (0040 232) 278680

[4] *Assistant, Faculty of Electric Engineering, Department of Energetic*
"Gh. Asachi" Technical University of Iaşi,
Bd. D. Mangeron, nr. 51-53, 700505 IASI / ROMANIA
Tel (0040 232) 278680, Fax (0040 232) 236283 E-mail: vrotariu@ee.tuiasi.ro

[5] *Professor, Faculty of Biology*
Alexandru Ioan Cuza University of Iasi
Bd. Carol I , no. 20 A , 700505 IASI / ROMANIA
Tel. (+40 232) 201630 ; Fax (+40 232) 201472;
http://www.bio.uaic.ro; http://www.bio.uaic.ro/content/view/46/43/
** Corresponding author, E-mail: otoma@uaic.ro*

Abstract. Pre-conditioning of crude water at S.C. CET Iasi S.A. is affected with lime and $FeSO_4$, the result consisting of a semi-liquid substance (slime). The slime is temporarily stored in tanks and is transported by truck to the CET Iasi II Holboca slag and ash storage centre. The transportation and storage conditions, as well as the need for environment protection, led to a new approach being adopted. A new installation for slime transport from the temporary storage tanks to the drying installation was created; the result consists of a solid product that is easy to transport and store.

Keywords: slime, pre-conditioning water, slag, environmental impact

1. Introduction

S.C.CET Iaşi S.A. is one of the medium capacity plants for electrical energy generation in Romania. The production is made in a CHP format (i.e. a co-generation plant), and water is one of the principal flows. Any thermal power plant needs essential volumes of industrial water consumption and so water can be considered as one of the major utilities, that is apart from the fuels. The water is used, for example, in boilers and turbines, cooling and in district heating flows. Used industrial water does not have the necessary chemical qualities to be used in the process and so it needs to be improved through chemical pre-conditioning treatment. This pre-conditioning is done in order to reduce water salinity by using, for example, a lime solution and ferrous (iron) sulphate [1-7].

2. General Considerations

CET Iasi uses almost 1.1 million m^3/year of industrial water and to treat this quantity of water there is a need for about 250 tonnes of lime and 50 tonnes of iron sulphate. As a result, in addition to the treated water, there also appears a new product, namely a slurry/slime. This is in a sludge state and it is considered by Romanian environmental law as a "toxic and dangerous substance". In the Romanian specialized terminology it is called "shovelled mud", giving it a high risk level in the loading and storage processes. The resulting slime quantity is around 300 tonnes per year. To avoid accidental pollution with slime (like entering the city sewerage system) it is temporarily stocked in four tanks and, periodically, it is loaded in special trucks and carried to the slag and ash CET Iasi II storage centre.

The slime circuit from production to final storage is very complex and involves important costs including both its temporary storage and, what is more important, the transportation to the slag and ash storage centre which is almost 18km away. In a single year, the transportation costs (fuel price) rose to 2,000 RON and, in addition, there are staff costs, and exploitation and maintenance costs, which also rose to 25,000 RON.

In Romania the environmental legislation (Order no. 118/2004; the Order 211/2004; Order no. 2/2004; Procedure from 05/01/2004; Procedure from 06/02/2004; and Procedure from 02/03/2004) introduced strict rules regarding the transport of toxic and dangerous substances, one of which is slime.

3. Materials and Methods

In order to comply with the new regulations of Romanian and European environmental laws a real solution to simplify the process of slime treatment was to be found in a new method to dehydrate the slime. Integrated equipment for slime dehydration, of the FILTER PRESA type, was developed. The equipment operates to a high level of slime dehydration, reducing the moisture content from 86-90% down to 40%, and in this way reduces or minimizes the impact on the environment, and also reduces the possibility of some accidental pollution event in the city sewerage system.

The FILTER PRESA equipment assures the continuous separation of solid suspensions from liquids and total re-utilization of the waste water, a fact that gives it a technological advantage. In addition, this equipment needs only some simple maintenance operations, thus giving it a financial advantage too. The filters have automatic washing systems, and another advantage is that the equipment operation is automatic and so the staff costs are significantly reduced.

4. Pre-conditioning Slime Plant

The pre-conditioning equipment was modernized and it is represented in Figure 1. A plant to dehydrate the slime coming from untreated water pre-conditioning was planned to be build next to the pre-conditioning plant equipment. Before the construction of this plant, the mixture of slime and waste water was transported in special trucks to the slag and ash of CET Iasi II storing centre (18km from Iasi) at high cost.

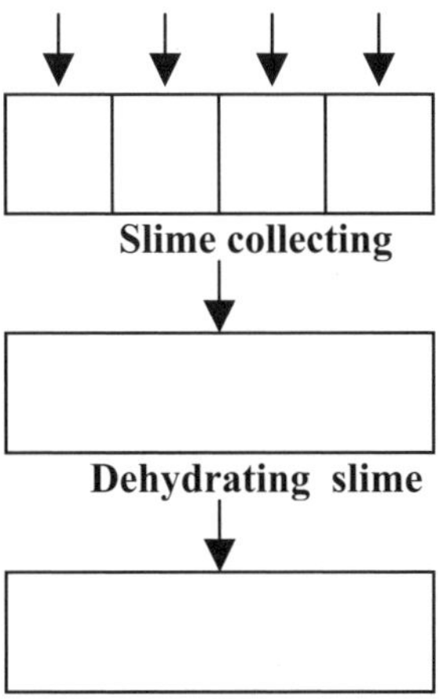

Fig. 1 Storage plant for dehydrated slime

The classical management method for slime coming from the crude water pre-conditioning (with lime and $FeSO_4$) includes:

- collecting the slime in tanks;
- slime dehydration; and
- dehydrated slime storage and transportation to the slag and ash CET Iasi II storing centre.

Entry parameters for the slime are:

- flow: 5-25m^3/day;
- moisture content: 85-96%;
- pH: 9-10;

The technological scheme contains:
a) The slime collecting equipment;
b) Dehydrating slime equipment (Press filter K800); and
c) Transport and storage of pressed slime called ‘pat’.

5. Results and Discussions

The slime collecting equipment consists of 4 concrete tanks having 200m^3 capacity each. The slime is extracted from the basin by pumping it directly into the tanks of the dehydrating equipment. The slime basins have level indicators. On each tank there are vertical pumps which suck the slime to the dehydrating plant. The scheme of the new slime collecting equipment is shown in Figure 2.

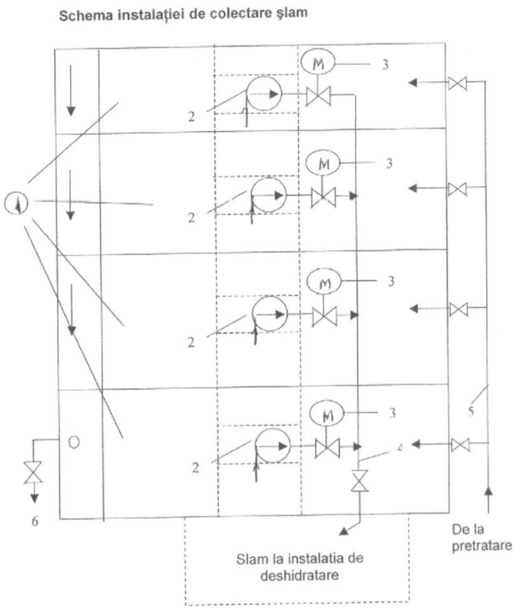

Fig. 2 Scheme of slime collecting equipment. 1. Tanks for slime collection; 2. Vertical pumps; 3. Electrical driven valve and knife type actuator (SIBAR); 4. Ascending collector; 5. Tanks supplying collector of slime coming from pre-conditioned water; 6. Discharge to sewerage system of clean water

Slime dehydrating plant

The plant comprises:
- 2 conditioning accumulators tanks (each of 20m^3);
- pumping and electrolyte solution-preparation group, with a 20m^3 tank;
- electrolyte solution pumps group having a 1,200 litre accumulator tank;
- pumping equipment for completed slime (the mix consists of slime and electrolyte solution);
- press filter ; and

– dehydrated and pressed slime band transporters.

From the slime tanks, using vertical pumps, the slime is sent into the conditioning tanks. Here it is mixed with electrolyte solution pumped from an electrolyte tank. It is mixed with water to get the solution and this water flow is controlled by a batching device in order to have a constant solution concentration. The process computer commands the batching-preparation of the solution equipment and the solution is sent through a pump from the reservoir to the conditioning tanks.

The press filter K800 is created in order to filter the liquids and suspensions under pressure and to obtain the maximum possible dry substance in the material that remains in the filtering rooms, the so called 'pats' (Figure 3).

Fig. 3: Filter press K800

When the equipment is closed between the plates covered by the filtering canvas, spaces appear where the material requiring filtering is inserted through an adapter positioned on the support plate. A pump fills the press filter and the solid particles are retained by the filter canvas. The filtered fluid which passed through the canvas gets into the bleeder system of the filtering plate and then into the collecting pipe. The plate pressing/depressing is achieved with a double-action hydraulic cylinder, with a maximum work pressure to close (i.e. press) the pack being 400 bar. To open the filtered plates pack a 50bar-pressure is needed.

Transporters with band for dehydrated-pressed slime

The transport equipment contains two bands; a horizontal one attached under the press filter and the other one reclined (Fig. 5) which receives the 'pats' from the horizontal band and carries them outside the plant (equipment) into the transporting storage truck.

Automatic equipment conducts and supervises the dehydrating installation in all the phases as follows:

- the transport of untreated slime;
- conditioning solution preparation;
- pressing (dehydrating) installation; and
- pressed slime transport.

The control of the equipment is made at the local level from three commands points (benchboard desks) as follows:

- one for solution preparation;
- one for conditioning technological equipment and slime transport; and
- one for the press filter.

The last benchboards has a specialized computer for dehydrating equipment and this computer has dedicated software specially conceived for the dehydrating process.

The size of the slime dehydrating equipment was chosen with due consideration of important physical-chemical characteristics such as:

- the slime resulted from water pre-conditioning;
- $25m^3$/day quantity;
- overall 15% of dry substance; and
- dry substance volume.

In order to complete the dehydrating process of the slime, a polymer is added and the necessary quantities are almost 2kg/tonne of dry substance. The polymer preparation system (FLOQUIP-type) is an autonomous one, designed to dilute and activate polymer emulsions. Clean polymer solution is pumped from a recipient tank through a variable speed pump without flow pulses. The emulsion is injected in order to initiate the transforming process of the polymer, through a dynamic mixer made from non-corrosive steel. Next, the polymer solution is diluted through water addition in a static mixer. After the mix is complete, everything goes into a 1200 litre tank in order to prepare the mixing process with the crude slime in the conditioning tank. Both manual and automatic work can be done with the solution-preparation equipment. In the manual cycle; all preparation phases are independent.

The polymer is actually introduced from the mixing reservoirs in conditioning tanks through pumping; another method is by directly pumping the slime feeding pipe into the press filter. The water flow entering the mixing reservoir can be controlled by the batching device so that the polymer solution concentration is kept constant; even the pressure of the mixing water varies. The polymer batching preparation equipment is controlled by the process computer placed next to the dehydrating installation.

From the mixing tank which is supplied with an agitator, the polymer solution continuously enters the conditioning tanks – which have agitator systems – and it can be delivered continuously.

The command signals which adjust the batching polymer equipment are sent through the process computer. The batching device is controlled by the frequency converter which modifies the capacity in stages, depending on the indicated concentration by the process computer.

The working phases are presented in Figure 4.

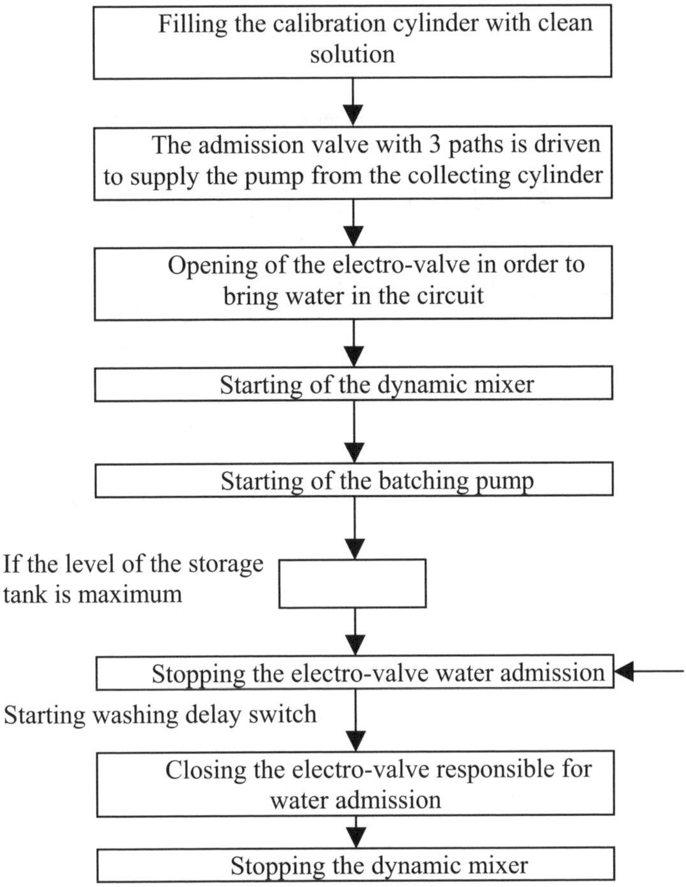

Fig. 4 Working phases

Fig. 5 Reclined band

Fig.6 shows a general view of the slime dehydrating equipment positioning.

Fig. 6 Slime dehydrating equipment – general view

6. Conclusions

The use of the slime dehydrating equipment has a direct and positive impact on the environment as well as on the thermal power plant. The results of the dehydration of slime, in terms of the environmental impact, are:

- a decrease of the slime-leaking risks during handling and transportation; and
- a decrease in accidental infiltration of the slime into the ground or the urban sewerage system.

Regarding the positive impact on the thermal power plant the following can be observed:

- a decrease in the slime tank dimensions;
- a decrease in transportation and storage costs; and
- an important economy of industrial water (i.e. over 50 %) contained by the non-dehydrated slime, water that could be reused in the industrial water circuit of the thermal plant.

References

André P., Delisle C.E., Reveret J.P., 1999 - L'évaluation des impacts sur l'environnement – Presses Internationales Polytechniques, Québec,

Ngo C, Regent A., 2004 - Déchets et pollution. Impact sur l'environnement et la santé, Dunot, Paris

Rotariu M, Ivana D., Ivas D-tru, 2002 – Producerea energiei in centrale electrice. Editura VENUS, Iaşi

Rotariu M., Tofan L., Toma O., 2007 – Preliminary study regarding local potential for developing of wood-energy way, like alternative to the energy which produce gas emission-as manmade disaster. NATO Security through Science Series Book , 1 , IOS Press., Amsterdam

Sarlos G. , Haldi P. , Verstaete P., 2003 - Systèmes énergétiques – Offre et demande d'énergie: méthodes d'analyse. Presses Polytechniques et Universitaires Romandes, Lausanne

*** , 2001 - Contract de colaborare S.C. CET Iasi - ICPIAF Cluj Napoca, S.C. ASAM Iasi

*** , 2002 - Manualul de instalații – Instalații de încălzire Vol. I, II - Editura ARTECNO, Bucureşti

I. Apostol et al. (Eds.)

doi:10.3233/978-1-58603-948-6-98

Economic and Legal Aspects Related to the Prevention and Mitigation of Flood Risks and their consequences for Tirlisua (Bistrita-Nasaud): A Case Study from Northern Romania

Ruxandra Malina PETRESCU-MAG[a], Dacinia Crina PETRESCU[b], Doina PETRI[c], Alexandru OZUNU[a]
[a]*Faculty of Environmental Sciences, Babes-Bolyai University, Cluj-Napoca, Romania,*
[b]*Faculty of Business, Babes-Bolyai University, Cluj-Napoca, Romania*
[c]*Public Health Directorate, Bistrita-Nasaud, Romania*

Abstract: Water is becoming more and more a scarce commodity in many parts of the world where they are also facing tremendous extreme events such floods. In all these areas, due to the inefficient use of water, water pollution caused by the human intervention and by changes of the human habits, the water becomes an important issue of debates, making the sustainable use of water a mandatory requirement. In the last decade, Europe suffered major floods, causing fatalities, economic losses, displacement of population and, last but not least, a huge impact on nature. The paper raises the reader's awareness of the fact that floods have always existed and will continue to exist - they are part of the nature; as United Nations and Economic Commission for Europe Guidelines on Sustainable Flood Prevention pointed out, as far as feasible human interference in the processes of nature should be reversed, compensated and prevented. The "lesson" from Tirlisua (Northern Romania) remains in the collective memory, strengthening the necessity of approaching a new paradigm of humans and nature living together.

Keywords: floods, legislation, assessment, management, risk, sustainability

Introduction

Floods are natural phenomena and represent a component of the natural hydrological cycle of the Earth; they have always existed and will continue to exist, unable to be prevented. From this perspective, the assessment and management of flood risk become essential. Risk assessment comprises the identification, assessment and interpretation of risk perception and the comparison with accepted social risks for the purpose of directing the decisions and actions in the process of risk management. The occurrence of floods cannot be avoided, but they can be managed, and their effects may be reduced through a systematic process that leads to a set of measures meant to contribute to diminishing the risks associated with these phenomena.

Floods do not have national boundaries, so that the joint actions of the affected states are mandatory, and the principle of solidarity should play a fundamental role. A holistic approach to the flood phenomenon at the level of the hydro graphic basin is necessary through the promotion of a coordinated development and integrated management of activities concerning water, land and adjacent resources, transportation, urban development and nature preservation. Flood management is a multidisciplinary and interdisciplinary activity, and comprises the management of water resources, land planning and urban development, nature protection, agricultural and forest development, and individual protection for example, with the achievement of some specific actions being assigned to each branch. The damage caused by floods may vary from one country and region of the European Community to another. The flood risk management plans should, thus, take into consideration the characteristics peculiar to the areas they cover and foresee solutions adopted according to the needs and priorities of these areas, providing at the same time relevant coordination in the hydrographical districts and promoting the achievement of environmental objectives stipulated by the community legislation.

Flood risk management (Jaap Bouma et al., 2005) means applying certain policies, procedures and practices, pursuing the objective of risk identification through analysis and assessment, treatment, monitoring and reassessment of risks, in order to reduce them so that human communities may satisfy their needs and aspirations in a sustainable physical and social environment. The main activities of flood management consist of:

1. prevention, protection and preparedness activities;
2. operational management activities performed during the development of the flood phenomenon; and
3. activities performed after the occurrence of the flood phenomenon.

A famous article (Wildavsky cited by Ionescu, 2006) had a special contribution for the understanding of the fact that not taking risk into consideration represents the greatest risk of all, and that risk management is a profitable solution, economically and socially. But the acceptance of a certain risk level is inevitable, because it is not possible to completely eliminate risk.

Between 1998 and 2002, Europe experienced more than 100 great floods, including, among others, the ones from the Danube and Elba in 2002 (Dworak & Görlac, 2006). The floods between 2005 and 2007 showed both certain weaknesses in the techniques used for flood protection, and the response capacity for the phenomenon management. Since 1998 the floods in Europe caused over 700 deaths, the migration of over half a million persons, and more than €25 billion material losses. Under these circumstances, the European Commission underlined the necessity of creating a legal framework for flood risk prevention and management. The joint efforts of the European officials were materialized in the directive 2007/60 regarding food risk assessment and management. The new directive completes the community's legal framework concerning water. Despite the fact that the 2000/60/CE directive (concerning the creation of a community-wide political framework relating to water) imposes the creation of certain management planning for the hydrographical basin for each hydrographical district, in order to achieve a good ecological and chemical state. This, in turn, contributes to the mitigation of flood effects, but the reduction of flood risk does not constitute one of the main objectives of the respective directive and does not take into consideration the future modifications of the flood risks as a consequence of climatic changes.

The purpose of the 2007/60 directive is that of establishing a framework for the assessment and management of flood risks, with the purpose of reducing the negative consequences for human health, the environment, cultural patrimony and economic activity associated with floods within the Community. Article 2 provides us with the definition of certain terms as “floods”,”flood risk” etc. The new directive will have to be transcribed into the national legislation by the middle of 2009. The implementation by the member state of the 2007/60 directive will be achieved in three stages:

a) preliminary risk assessment concerning floods that has to be completed by December 2011;
b) development of hazard maps and flood risk maps by December 2013; and
c) design of flood-risk management plans by December 2015.

The directive describes the technical content of these three stages according to certain common criteria meant to improve considerably the management of flood-risk situations at the European level. The Euro-parliamentarians asked that the directive make reference to climatic changes, a fact that will have to be taken into consideration in the preliminary assessments as well as in the report that the Commission will establish in 2018 concerning the enforcement of the directive.

1. Case Study: 20th June 2006 Floods in Tirlisua, Bistrita-Nasaud

1.1. The Importance of the Case Study

The case study intends to present a description of the floods that took place on 20th June, 2006, analyzing the manner of the event, the factors involved in this catastrophe, but also the impact on the community. The impact is appreciated by the effects on the health condition (appreciated by the raw mortality rate and the morbidity structure in relation to the state of post-traumatic shock), on the environment, but also by the economic losses. The evaluation of the environmental impact was made by analyzing the water quality that indirectly offers indications also on soil pollution which is directly connected to health condition.

The importance of the Tirlisua phenomenon is reflected in the following aspects:

- If there existed (or not) a prevention strategy of such a phenomenon, as well as how prepared the community was for such a situation: also, if the interests of the community during recent years decreased or increased the vulnerability to floods;
- If the authorities managed fluently, rapidly, and efficiently the emergency situation that occurred; and
- If the lesson Tirlisua was correctly learned and brought a post-recovery strategy that would decrease the community’s vulnerability to such phenomena.

Although since 1970, in Bistrita-Nasaud the floods occurred annually, the Tirlisua phenomenon is unique by the impact it produced. It proved, by the human losses and by material damage, that the ability to cope with nature has gaps, and the management of such an emergency situation, at least in this community, should be improved.

1.2. Description of the Event

On June 20, 2006, for around 17 hours, in the area of the Sendroaia locality under the Omusor peak on the Southern slope of Tibles, there were rainfall levels up to 125 l/m^2, (Fetea et al., 2006). The deluge lasted 20 to 30 minutes. The rains washed the slopes of the hills causing the flow of significant amounts of soil, raw and fine sediments, vegetation and wood residues, towards the river Ilisua. The exceptional flash-flood had a pluvial genesis; the hydro-meteorological context previous to the flash-flood was characterized by the amount of rainfall in the period of 1st – 20th June, namely values between 28 l/m^2 in the low areas of the basin and 53 l/m^2 in the higher areas. On 20th June, based on the intrusion of a mass of hot air accompanied by high temperatures, intense accumulations occurred, which was doubled by the high altitude existence of a nucleus of very cold air, and led to the occurrence of some very powerful downfalls in a relatively short period of time (Fetea et al., 2006).

The flood-wave on the river Ilisua was 4 to 5 metres high compared with the level of the thalweg (i.e. the lowest point of the valley) within 10 minutes of its formation, thus having a catastrophic character and causing a disaster in the area, with consequences difficult to assess and with extremely low recovery possibilities. In the Tirlisua locality, according to the information provided by the local authorities, the phenomenon began at 16.15 hours with half an hour duration when it rained quietly. From 16.45 until 17.15 it rained heavily and at 17.00 hours significant increases of water flows were noticed. The flood-wave reached Tirlisua at 17.15, 10 minutes after it was formed (Presentation of the Bistrita-Nasaud County, 2006).

The disaster is identical to the situation of hydro-technical accident in a reservoir that is destroyed and the flood-wave rapidly propagates downstream. Practically speaking, the inhabitants of Tirlisua had no chance due to the magnitude and violence of the phenomena. In the 1955 - 2006 period on the Ilisua river at the Cristestii Ciceului hydrometric station, the greatest flood-wave occurred in May 1970 when the maximum flow recorded was of 294 m^3/s, which corresponds to an occurrence probability p=1.5%. In June 2006, as the amount of water was very large, a flood-wave was created, a high wave which destroyed everything in its way: roads, houses, bridges. The water had a rich content of fine and raw sediments. By washing the slopes, the water had driven the sandy and weakly-fixed soils, and the wood residues resulting from deforestation, and these increased the impact force of the water, causing injuries to people and reducing the rescue chances for the victims. The wood residues blocked the bridge at the confluence of the two streams; the water levels increased and flooded even more the houses located in the riverbed. The impact was so powerful that it caused the death of 13 people and the disappearance of another three (Fetea et al., 2006). Many houses were destroyed and dozens of them damaged, public institutions flooded, and agricultural fields rendered inactive; the electricity supply networks, and fixed telephony networks were interrupted, a part of the county road was broken and a great part of the village road flooded, completely isolating the locality.

1.3. Description of the Site

Tirlisua village is situated north-west of Bistrita-Nasaud county in a sub-mountain area. This explains the presence of some petrographic characteristics specific to the Northern Group of Eastern Carpathians (volcanic rocks - granite, andesites, basalt, metamorphic rocks, and towards the south are sedimentary rocks-marls and sandstones). The relief of the zone is highly fragmented by the streams that separate narrow-base hills with 500-600m level differences. The most important watercourse of the area is Ilisua that is a minor tributary stream of Somesul Mare. Ilisua has as a main component Valea Izvor, the hydrographical network of the river Ilisua which is characterized by a dendritic configuration and has three or four streams that flow towards it and which during the summer periods are supplied intermittently, arriving sometimes in the drought period. The Izvor valley crosses a trajectory from Sendroia hamlet, between the forested hills or is covered by pastures toward Tirlisua at a distance of approximately 2.5km with a 300m level difference. The course of the river Ilisua in Tirlisua is low riverbed between two narrow hills.

The rain that fell on 20th June 2006 was fortuitously located and supplied only the main tributary stream of Ilisua valley, respectively only the Izvor valley. It is estimated that the flash-flood transported an amount of water 100 times larger than it was transported under low water regime conditions.

1.4. General Data of Community Characterization

Demographic Data: The annual rate of decrease in the population between 1993 and 1999 was 2.05%, a trend which also continued in the subsequent period. The population density is low, ranging between 30 and 25/km^2 since 1990 with the county average being 60.9/ km^2; The demographic dependence ratio, showing the pressure exerted by the inactive age groups and characterizing the activity potential of a community, shows a value of 55.6 % (i.e. smaller than the county value of 56.7%) and consequently a higher economic and social support pressure on the active age group.

Socio-economic Data: The percentage of the remunerated persons in 2000 was 5% which is very small when compared with other communities. The basic occupation of the residents was an agricultural one, to which forest exploitation can be added and which led to important deforestation after 1990. The forests taken into ownership were deforested without the observance of the sylvan regulations of clearance and replanting, thus increasing the gravity of the disaster in June 2006. The access to the telephone network, TV and written mass media is possible in the central village and in the sub-ordinate villages. In terms of access to infrastructure, the village locality benefited from an asphalted road, and all the localities were partially connected to electricity; in addition, there existed the possibility to use both fixed and mobile telephone systems. With regard to potable water, 93.65% the population used the water table through installations of the drilling-fountain type,; 5.62% of the population had access to a supply network coming from a catchment system based on deep sources; and 0.73 % of the inhabitants were supplied from fountains that were considered public (common). There was no organized system for waste disposal; medical care comprised one physician for 1300 inhabitants, and in the village there were 2 general practitioners, a dentist and 3 nurses. Diseases of the cardio-vascular type and hepatic cirrhosis were predominant in the mortality structure.

2. The Vulnerability and Risk Analysis

Considering vulnerability as a function that displays the degree of exposure of humans and their goods to the different types of risks, raises the issue of how vulnerable Tirlisua was. If we are to analyze the three elements of vulnerability, namely hazard exposure, economic resources and access, to services and infrastructure, we notice the following results.

2.1. Risk Exposure

- The physical factors increasing vulnerability: geographical location, mountain area, fragmented relief, pluviometric annual regime, general climatic conditions favourable to extreme events, including for Romania;
- The human factors/activities increasing vulnerability: deforestation and urban planning.

The predominant activity in the area is agriculture and sylviculture. While the agricultural *pattern* did not change very much after 1989, the logging pattern did because the forests were given to the people, who also undertook the right to exploit them. The deforestation did not preserve the organized character, thus becoming a common phenomenon, especially when the wood as a raw material represented a rapid source of income. Many wood residues are left at the logging site. It must not be forgotten that, besides these, other quantities of wood residues occurred at the logging site are added, so that the volume of the residues increases. Partly, wood is stored in the house yards of the people who use it as a heating source during winter. The flood-wave caused by the Tirlisua disaster drove significant amounts of wood residues and this increased the impact force of the water and the resultant damage as well as loss of human lives (Figure 2). Practically, the river Ilisua was blocked by wood residues driven from the upstream.

Figure 1. Wood waste carried by the high flood

The second element of human nature that increased the community vulnerability to floods was the poor planning of the locality. The easiest development of Tirlisua, as favoured by the geography of the place, was along the Ilisua riverbed. Without any precautionary measure, new houses were built in the river floodplain based upon the history of the place where the floods were considered insignificant.

2.2. The Chance of Economic Resources

Being a relatively isolated locality with good infrastructure, but with very low economic potential and an active population with high tendencies of migration, the resident population is composed especially of old people and children.

2.3. The Chance of Service Access

The village is located 45km from the centre of the county and 20km of the nearest town, and for the Somes Tisa Basin, the flood vulnerability degree was established. In this assessment, Bistrita-Nasaud County presents a high vulnerability. The Tirlisua lesson should be taken into consideration by all the villages of the county and not only Tirlisua, especially when the experts foresee possible similar phenomena for the future as well. The Tirlisua event reflected the vulnerability of a human community exposed to risk, manifested by the low capacity of processing the effects of the phenomenon.

3. Impact Assessment

The environmental, economic and health impacts were also studied. The environmental impact is analyzed through observations on the evolution of plants and the quality parameters of the subsurface, knowing that the subsurface water could be an indirect indicator of the quality of the soil, but also an important indicator of human health. The economic impact is presented as material damage and lost monetary value. The impact on the health of the population is assessed by two indicators: the raw mortality rate (RBM) and the structure of the post-impact morbidity, through the analysis of the diagnosis recorded in the first two weeks following the phenomenon.

3.1. The Environmental Impact

In most situations floods generate negative social, economic and ecological effects. Among the ecological effects are the degradation of the riverbed, the destruction of vegetation, the modification of the soil and, sometimes, the general geographical area. All the ecosystems are damaged. In the present situation, the main modifications of the riverbed were manifest downstream where portions of the riverbed were eroded and other river branches were formed. The geography of the area was changed by the disappearance of the gardens and the occurrence of silting-up or even swamp-holes. A special impact occurred on the vegetable crops as many of the agricultural fields were covered with mud and the most of the full-grown plants were destroyed. One month after the event, an impressive image occurred: the corn straws, in their effort to vegetate had developed by stem growth while the top was caught in the mud and the inhabitants, in their efforts to save the crop, were cutting down the top of the plants to allow them to grow.

The flood destroyed the vegetation in three ways:

- by changes due to the soil as a consequence of its saturation with water ;
- by the physical transition of water over the plants; and
- by the chronic effects associated to environmental changes and reactivity.

The major impacts of the flood were:

- poor soil aeration: the flooded parts suffered oxygen constraints because the water occupied the macropores around the root of the plant thus losing contact with the gaseous component of the soil. The oxygen became available only at the top; the plants accumulated thus more carbon-dioxide, methane, hydrogen and nitrogen. The structure of the soil led to the reduction of cohesion, and the dissolving of the organic component and the metals, dispersing them in clay particles. These effects resulted in the gradual colour reversion of the plants, especially where these phenomena have extended, and eventually in their death.
- the aerobic respiration of plants: this is due to the fact that flooding causes the replacement of the aerobic micro-organisms with the anaerobic ones from the soil. The anaerobic organisms are mainly bacteria. These bacteria cause the de-nitrification and reduction of magnesium, sulphur, iron, depriving the plants of the necessary elements. Furthermore, diseases and pest growth are favoured.
- reduction of the chemical activity: the floods reduce the redox potential, and increase the *pH* of the soil. The decomposing of the organic matter in the normal soil keeps the cations and anions in relation to the essential elements while in the flooded soil, this decomposing was made through anaerobe bacteria. These are less diverse and less efficient in the decomposition process. As a result of the partial decomposing of the organic matter thin layers of mud appeared. In the normal soil, the decomposing of the organic matter will result in CO_2 and humus, with the CO_2 escaping to the atmosphere and the humus components turning into clay, aluminium and iron oxides. This process gives the soil structure and the nitrogen is transformed in ammonium ions which then turn into nitrate, and the sulphur compounds are oxidized into sulphate. In the flooded soil, the decomposition of the organic matter is no longer transformed to CO_2 and humus, but to methane and a greater number of other highly volatile elements, that mark the place with a penetrating and unpleasant smell (from hydrocarbons, ethane, and phenolic acid for example). These elements were eliminated as gas bubbles or floated through the water towards the surface, causing the unpleasant smell in the atmosphere. The plants were affected in all their growth components.
- the action on the subsurface waters: Water is an important environmental factor because most of the metabolic processes of living beings are developing in the watery environment. Water is also the transmission path of numerous infectious diseases: bacterial, viral parasitical, and therefore technologies were developed to correct the physical, chemical and microbiological parameters. The pollution of water, regardless of whether it is underground or at the surface, represents a problem even under normal circumstances, but it becomes a major problem in cases of calamities that affect its quality. As previously outlined, most of the population in Tirlisua village is supplied with water from the underground water table through fountain-type individual

installations and a low percentage used six depth sources. These sources were collected in a basin from which they were distributed to approximately 20 families and several utilities, by means of a pipe network.

In this paper we analyzed only some of the potability indicators of the water stipulated in legislation and so chose the following chemical indicators: ammonia nitrites and nitrates, substances resulting from the decomposition of organic matter and that show different stages of pollution with organic substances in terms of recent, medium-term or chronic pollution. The nitrates are directly connected to the acute methemoglobin effects on the breast-fed baby reflecting an increased risk for becoming ill under these crisis conditions. The presence of the ammonia denotes a recent pollution in terms of hours or days; the nitrites illustrate pollution older than days or weeks, as well as nitrates that denote a chronic pollution, sometimes stable and old. In the Tirlisua event, the flash-flood wave swept away all the household annexes, involving huge quantities of domestic waste, residues from animals and humans, important quantities of highly microbiologically-contaminated organic matter, with human and animal micro-organisms. Furthermore, the flash-flood caused the drowning of 70% of the livestock of the locality, with the bodies of the dead animals increasing the polluting load of the soil. The flood-wave heavily saturated and polluted the underground water including the water table and polluted the soil.

As microbiological indicators the following were chosen: Mesophile germs which develop at 37^0 C, germs that are specific to humans and warm blooded animals. The larger their number the stronger the supposition that among them there are also found pathogenic germs; coliform germs are found in the human and animal digestive tube. Their validity (as an indicator) in water is very close to the typhoid and paratyphoid group intensifying their value as a sanitary indicator. Likewise, their resistance to chlorine or other common disinfectants is greater in comparison to other germs; the enterococcus comes from the human digestive tube but in smaller number compared with coliforms. They are more water-resistant than the faecal coliforms and cannot stand the microbial variability phenomenon. Moreover, the enterococci present are types that are specific to humans and so allow their differentiation in the type of water pollution.

The quality of the underground water is an indicator of the environmental state but also an indicator of the disease exposure of the population. Before the occurrence of the catastrophe of 20th June, four water samples were collected in May 2006, one from the elementary school and three from private fountains, which exhibited old pollution with organic matter of the water sources, and even powerful microbiological pollution in one of the individual fountains.

The first post-impact water sampling and analyses were performed in the two weeks after the pollution produced by the flood, after the waters began to withdraw, and the first measures of fountain-cleaning and sanitation were applied. Water samples were analyzed both from the area affected by the flood, as well as from the unaffected area, for the identification of the source that might have been used at least for domestic purpose. Figure 2 shows that each source was polluted, and the majority indicated a powerful microbiological pollution, but aspects of the decomposition of the organic matter deposited on the soil are also present. The presence of the nitrites and nitrates shows that the pollution is a few weeks old. The detailed analysis shows the existence of all types of organic pollution. The presence of ammonia shows that pollution is a cyclic process in which new organic matter is deposited and follows the cycle of chemical decomposition.

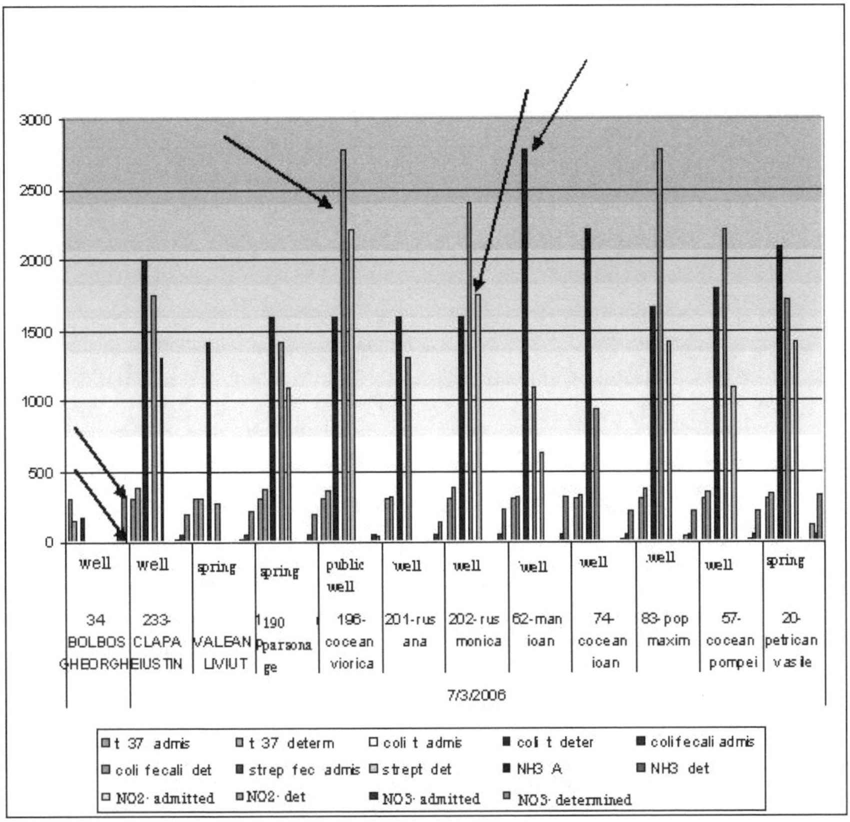

Figure 2. Results water sample 03.07.2006

The data from 3rd July 2006 show that all the parameters of the samples recorded important excess values, betraying an extreme of both microbiological and chemical pollution. The pollution is present in all samples, both in the ones harvested from the area affected by the flash-flood, and in the area where there was no flood, but where, by soil saturation with water, the pollution was extended to the water table. On 5th July 2006, after the drainage and disinfection of the fountains, the samples collected from the recovered water table show no improvements of the microbiological or chemical parameters, which express the gravity of the pollution. In July 2007 the microbiological contamination remained high. After almost one month from the event, the chemical state of the water, even with all the soil and water disinfection and drainage methods used (four tones of chloramines were consumed), the situation did not improve.

3.2. The Economic Impact

The total value of the damage caused by the floods in Tirlisua amounts to 106.675,294 thousand lei RON (1 Euro=3.7 RON April 7, 2008). The rush of water from Tirlisua, besides the losses of human lives, also affected roads, public utility networks, agricultural fields, houses and household annexes, socio-economic objectives (Table 1).

Table 1. Utilities-damages. (Source: Bistrita Nasaud county authority report, 2006)

DAMAGE		Value (RON)
FLOOD PROTECTION		
- Protection of the banks	1.205 km	3072.732
COMMUNICATIONS		
- County roads	23.06 km	17230
- Country roads	27.65 km	36779
- Streets	15.298 km	7709
NETWORK/PUBLIC UTILITIES		
- Electrical network	22 km	1800
- Communication system/optical fibre	2.150 km	86.562
- Water sources - damaged fountains	462	199.93

The cost of the damage could not be covered by the community, even with help from the county authorities, and so national involvement, by allotting governmental resources for recovery, were absolutely necessary. The value of the agricultural losses, considering that the community is based upon this type of activity in order to support themselves, was of 1480.5 thousand RON, only for the fields, without taking into account the large number of animals and birds lost during the flood. For the rebuilding of the houses and household annexes 3409.7 thousands RON were necessary; 38 houses were destroyed of which 14 were swept away by the flash-flood, leaving 14 families without homes; 61 houses were seriously damaged together with a much larger number of household annexes. As a consequence of these losses, six families changed their residence and left the locality; the others have accepted the social support of the relatives and went to live with them. The infrastructure was severely damaged and required the greatest amount of funds for recovery.

3.3. The Impact on Human Health

The impact on human health was analyzed in terms of immediate mortality and morbidity. The mortality was analyzed in terms of raw mortality rate due to the impact and is reported as the number of deaths per 1000 inhabitants. This rate was compared with values recorded in the period previous to the catastrophic event. The morbidity was analyzed in terms of the type of diseases recorded and the number of cases of sick persons recorded in the first two weeks from the impact. This period was chosen to be relevant because the impact produced a post-traumatic shock. The assessment of this period was important also because in this interval the human organism did not develop mental mechanisms of adaptation so that, the direct psychic impact can be better appreciated through the physiological modifications manifested as organ pains, (diagnosed by the General Practitioner) and less through the mental symptomatology which develops subsequently. During the impact 13 persons lost their lives and three were reported missing. The raw mortality rate (RBM) of the impact was of 4.11 for 1000 inhabitants which means that a quarter of the RBM value from the course of a year was produced during the impact. The RBM between 1990 and 2000 ranged between 12.4 and 14.1 deaths for 1000 inhabitants. The great mortality rate justified the framing of the event in the group of natural catastrophes.

Morbidity analysis showed that between 21st June 2006 and 5th July, 2006, 1,262 consultations were performed, whereas in the previous 12 months 8234 consultations were made. This means that, in just two weeks, 15.32% of the total of one year consultation was given, without including here the consultations given for prophylactic purposes. The stress on the psychic is characterized by the presence in antecedents of the traumatic events, the persistent reliving of the event by intrusive memories, symptoms of hyperactivity; the effects of the disorder produce a clinical suffering or functional organ deterioration. These illnesses are interpreted as pain without localization, cephalalgias, modifications of cardiac frequency, etc. A characteristic of the created situation is the fact that, in the absence of the complex para-clinical research methods (such as ultrasound, radiology, and laboratory tests) and under the great influx of requests, the conditions of providing medical care in the sanitary centre (which was improvised in the kindergarten class of the Community Home because the medical dispensary had been destroyed), determined that in many situations the diagnostic was recorded only as a sign or symptom of disease and not a disease *per se*. Likewise, the application of the international disease codes was not possible, and many diagnoses were only presumptive. For these reasons, we recorded the cases as they are mentioned in the primary records, without correcting them. The morbidity was structured on diseases illustrating the psychological impact on the persons directly or indirectly (Figure 3), diseases showing the physical impact interpreted by traumatisms of physical nature, diseases that reflect the impact of the environmental conditions (such as allergies, skin diseases, and infectious diseases) and other types of diseases.

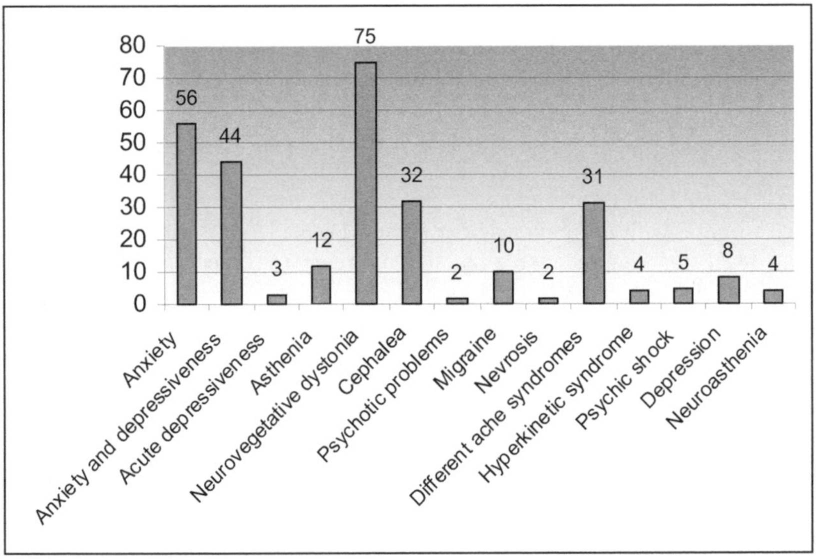

Figure 3. Health state in relation with traumatic state, period 21.06.06-05.07.06

The indirect diseases of the shock state ended in three vascular-cerebral strokes in only two weeks, whereas during the entire year of 2005 only two such cases were recorded. There were also recorded three angina pectoris, with all cases requiring hospitalization and emergency medical care. The traumatisms had various degrees of severity, from crashing thoracic traumatisms to excoriations. However, these are only a few of the aspects of the health impact recorded in Tirlisua, as a result of the catastrophe of 20th June 2006.

4. Conclusions

The community was not ready to cope with such a catastrophe, thinking in a pre-event manner: "it can happen elsewhere but not here". Such a concept surprised the inhabitants and the authorities, making them more vulnerable and the destruction produced by the impact was significant. The place is, by its very location, vulnerable from the point of view of the relief to such unpredictable and extreme phenomena, but this vulnerability was not mitigated, but accentuated by the inappropriate organization and by the massive deforestation activities. Although the authorities responded promptly and efficiently after the impact, and human solidarity was shown to reduce the negative impacts of the flash-flood, irrecoverable losses were caused and complete recovery takes years.

The impact of the phenomenon on human health affected the environment and the economic situation of the inhabitants, but important social consequences will further affect the inhabitants. The water supply of the population showed that the immediate perspectives after the floods, within the interval 1 to 2 weeks, were unfavourable for a decent lifestyle as the population were not provided with drinking and domestic water resources. In August and September 2006 the signs of the recent pollution with organic matter disappeared but the chronic pollution and a powerful microbiological pollution remained. In November 2006, five months after the impact, only two fountains were treated and rendered suitable for public use. One year after the event, the issue of water supply in the locality was not completely solved: the captured springs were released for use, but the fountains were used only for domestic purposes, as the water table still presented traces of microbiological pollution (such as cols and streptococcus of human provenance), and the pollution from nitrates was still present.

The Tirlisua lesson will remain in the collective memory of the locality. A new approach to cope with the environment should be made, and not only here. Nature manifests itself in an unpredictable manner and requires respect; therefore the Tirlisua lesson tells us that its manifestations should be anticipated and managed in the favour of man. And as a consequence of the Tirlisua events, the strategy applied by the Somes-Tisa Water Department for the quality-quantity integrated management of the water resources takes into account the request expressed by the formula "more space for the rivers" when promoting, developing and permitting projects. It also takes into consideration the need for non-structural measures of flood management. Since 2006, the level of the area managed by the Somes-Tisza Water Department began with the implementation of the national DESWAT project that intends to prevent disasters caused by water. This is achieved by an automatic system for the purchase, transmission and processing of data which will provide information in due time on the state of the water courses, and the development of prognosis with early anticipation and warning-alert systems for the protection of population and goods (Fetea et al., 2006).

In conclusion, the answer to the question 'how safe is safe enough?" should be given by considering a series of aspects specific to each situation, but at the same time making reference to the circumstances which are common and general to all risk situations (Jonkman et al., 2003). Consequently, the interdisciplinary character of the approach of the issues related to the risk becomes necessary. For instance, the psychological analysis of the way in which the population perceives the danger to which it is subjected in the potentially floodable areas, can help establish the methods for raising its awareness of the correct risk assessment. As a consequence, the holistic approach of the assessment and management of flood risks could represent a successful solution, which emphasizes the integration principle, according to which the environmental protection requirements should be presented when defining and implementing other policies within the European Community.

References

Bistrita Nasaud County Authority Report, 2006, available at: http://www.prefecturabn.ro/

Dworak, Th., Görlac, B., 2006, Flood risk management in Europe – the development of a common EU policy, *International Journal of River Basin Management,* vol.3 (no.2): 97-103, available at http://www.ecologic.de

European Parliament, Council of Ministers, 2000, Directive 2000/60/EC on the establishing of a framework for Community action in the field of water policy, Official Journal L 327, 22/12/2000, p. 0001 – 0073.

European Parliament, Council of Ministers, 2007, Directive 2007/60/EC on the assessment and management of flood risks, Official Journal L 288, 06/11/2007, p. 0027 – 0034.

Fetea, P., Sarb, M., Hasmasan, T., Ciogolia, D., 2006, Viitura exceptionala produsa in perioda 20-23 iunie in bazinul hidrografic al raului Ilisua si impactul ei asupra mediului, *Revista EcoTerra*, 10: 26-28.

Ionescu, St., 2006, Riscul nostru cel de toate zilele. Inundatii si cutremure, Editura Matrix Rom, Bucuresti, 62.

Jaap Bouma, J., François, D., Troch, P., 2005, Risk assessment and water management, *Environmental Modelling & Software, Elsevier*, 20: 141-51.

Jonkman, S.N., van Gelder, P.H.A.J.M., Vrijling, J.K., 2003, An overview of quantitative risk measures for loss of life and economic damage, *Journal of Hazardous Materials*, *Elsevier*, A99: 1–30.

Memorium of Bistrita-Nasaud county presentation, 2006, available at:
http://www.prefecturabn.ro/Sit.urgenta/Memoriu_prez_jud_BN.doc

Theme 3

Mining/Industrial Hazards/Risks

Optimisation of Disaster Forecasting and Prevention Measures in the Context of Human and Social Dynamics
I. Apostol et al. (Eds.)
IOS Press, 2009.
doi:10.3233/978-1-58603-948-6-115

Emergency Planning for Tailings Dams

Wilhelm G. COLDEWEY
University of Münster, Institute of Geology and Palaeontology, Department of Applied Geology, Germany

Abstract. Tailings dams present a tremendous risk to the environment and the human health. For this reason emergency planning for all installations of tailings storage facilities are necessary. Recommendations for measurements concerning the safety of tailings dams are given.

Keywords. Tailings dams, hazards, emergency planning

Introduction

Accidents such as Baia Mare (30.01.2000) and Aznalcóllar, Spain (25.04.1998) have shown that failures and incidents at tailings management facilities can have tremendous consequences for the environment, to human health and to the social acceptance of mining activities. Such accidents in any part of the world have the potential to rapidly impact the social acceptance for all other operations of the mining industry in general. Such failures from tailing management facilities have contributed to transboundary pollution via the mass movement of wastes in the form of suspended solids and dissolved materials (generally, tailings contain heavy metals and hazardous and/or toxic compounds).

1. Definition of Tailings Management Facilities

The following definitions largely follow ICOLD Bulletin 106 (1996).

Tailings are the fine-grained waste material remaining after the metals and minerals recoverable with the technical processes applied have been extracted. The material is rejected at the "tail end" of the process.

A *Tailings Storage Facility* can include a tailings dam (impoundment and pond), decant structures and spillways.

A *Tailings Dam* refers to a tailings embankment or a tailings disposal dam. The term 'tailings dam' encompasses embankments, dam walls or other impounding structures, designed to enable the tailings to settle and to retain tailings and process water, and which are constructed in a controlled manner.

A *Tailings Impoundment* is the storage space/volume created by the tailings dam/dams where tailings are deposited and stored. The extent of the impoundment is bounded by the tailings dams and/or natural boundaries.

2. Risk Assessment

Sources of risks at the tailing management facilities can be (UNECE):

- direct: location, tailings and site criteria,
- indirect: management.

For each tailings management facilities the hazards should be evaluated. Furthermore the assessment of great consequences and the probability of it happening again (risk assessment) must be carried out.

Based on the risk assessment three classes of risk can be distinguished:

- green – low consequence and probability,
- yellow – medium consequence and probability,
- red – high consequences and probability.

Depending on the risk classes the construction of the tailings management facility can be permitted (green class), permitted with additional conditions (measures taken to reduce the consequence or the probability to move from the yellow to green class), or it should be prohibited (red class) unless economically acceptable measures can be taken to reduce the consequences or the probability.

The hazard and risk assessment, especially taking into account the direct sources of risks, should be the precondition for planning and designing a safe tailings management facility. The hazard and risk assessment should be constantly reviewed throughout all the phases of the life-cycle of the tailing management facility.

3. Safety Aspects

While planning and designing a safe tailings management facility particular attention should be directed to:

- tailings pond: the following parameters need to be assessed accurately:
  - stability of sludge (slurry density);
  - level of the groundwater;
  - geological situation;
  - hydrogeological situation;
  - hydrological situation; and
  - geophysical situation.
- tailings dam: the following parameters need to be assessed accurately:
  - slope stability of the dam;
  - stability of the tailing material (induced liquefaction);
  - erosion to the dam (suffusion and outside erosion); and
  - slope sliding.

The dam-raising method should be chosen with regard to the local conditions (such as seismicity, tailings composition, and severe climate).

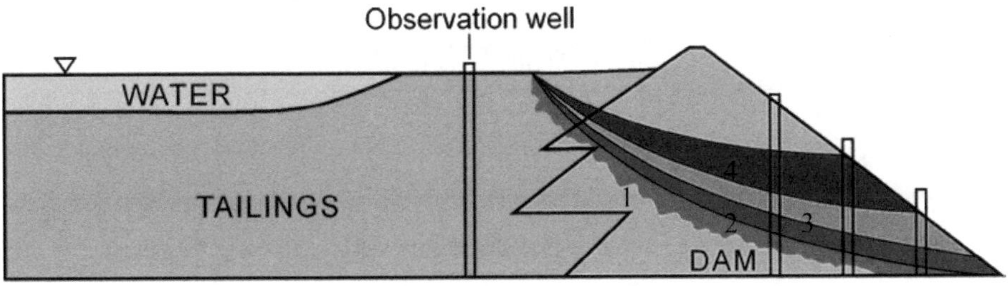

Figure 1. Observation of water in the dam with general action plan.

4. Measurements for Dam Safety

For safety reasons the following aspects must be observed:

- water level in the tailings dam;
- water level in the tailings pond;
- water outflow;
- weather conditions;
- seismic activity; and
- dam movement.

4.1. Observation of Water in the Dam

The measurement of the water levels in the dam and in the tailings is done by observation wells.

With regard to the height of the water in the dam different actions are necessary as follows (see Fig. 1):

1. no action (green – Zone 1);
2. daily water level control (blue – Zone 2);
3. hourly control and stop production (orange – Zone 3); and
4. stop production and take countermeasures (red – Zone 4).

The actions zones indicated on Fig. 1 will be dependent on the dam construction.

4.2. Observation of Water Levels in the Tailings Pond

Overflowing water can cause dam erosion and so it is necessary to observe the height of the impounded water level (see Fig. 2).

In the case of an overflow the water level must be lowered by counter-measures such as by pumping.

4.3. Observation of Water Outflow from Dam

The dam must be inspected regularly to check if there is any outflow from the dam (Fig. 3). An outflow can cause inner erosion to the dam and is very dangerous for the stability. If there is any outflow then counter-measures must be undertaken such as lowering the water level in the pond or sealing of the outflow.

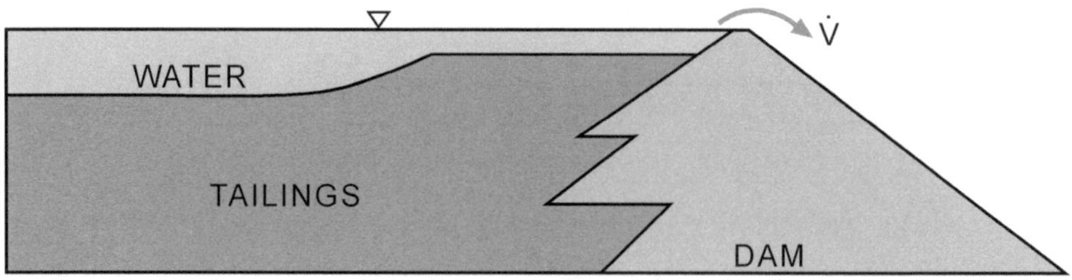

Figure 2. Overflow of a dam.

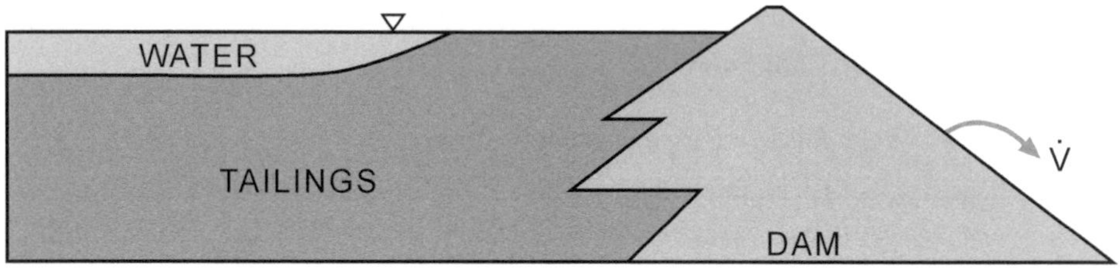

Figure 3. Outflow of a dam.

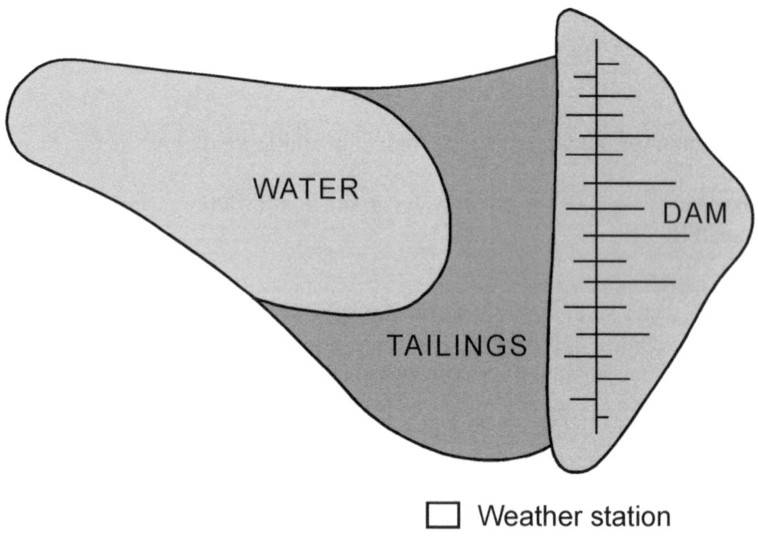

Figure 4. Observation of the weather.

4.4. Weather Observations

Heavy rainfall can be dangerous for the dam through erosion effects or by raising the impounded water level and, hence, an increase in pressure on the dam. Also, protracted periods of rain can weaken the dam stability. Thus, a weather station is essential (Fig. 4). Based on rainfall records for Central Europe, actions can be classified as follows:

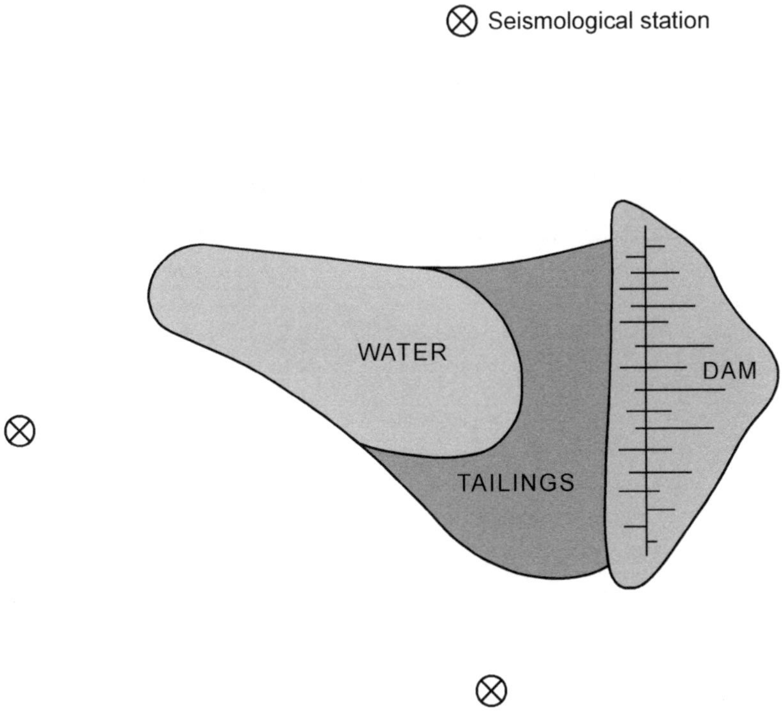

Figure 5. Observation of seismic activities.

45 $l/(m^3 \cdot d)$	= no action
45–60 $l/(m^3 \cdot d)$	= field control
60–80 $l/(m^3 \cdot d)$	= safety actions
80–125 $l/(m^3 \cdot d)$	= possible emergency actions
> 125 $l/(m^3 \cdot d)$	= emergency actions

4.5. Observation of Seismic Activity

When a tailings storage facility is located in an active seismic area a seismological station is necessary (Fig. 5). The impact of any seismic activity must be evaluated by an expert in dam stability to assess compliance with any seismic criteria used in the dam design and construction.

4.6. Observation of Dam Movement

Movement of the dam can occur for different reasons, for example through seismic activity or heavy rainfall. Internal movements are measured by inclinometers (Fig. 6) and can be classified as follows for conditions in Central Europe:

- linear movement (green – Arrow 1)
- movement acceleration: 1 mm/3 months = field control (blue – Arrow 2)
  5 mm/3 months = safety actions (orange – Arrow 3)
  10 mm/3 months = emergency actions (red – Arrow 4)

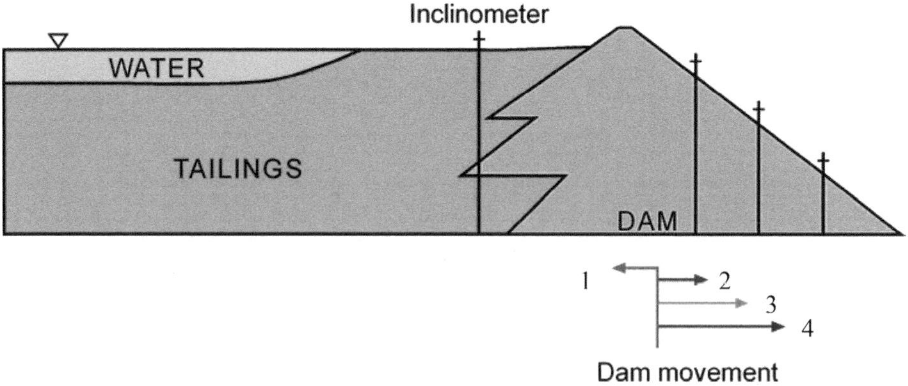

Figure 6. Observation with inclinometer.

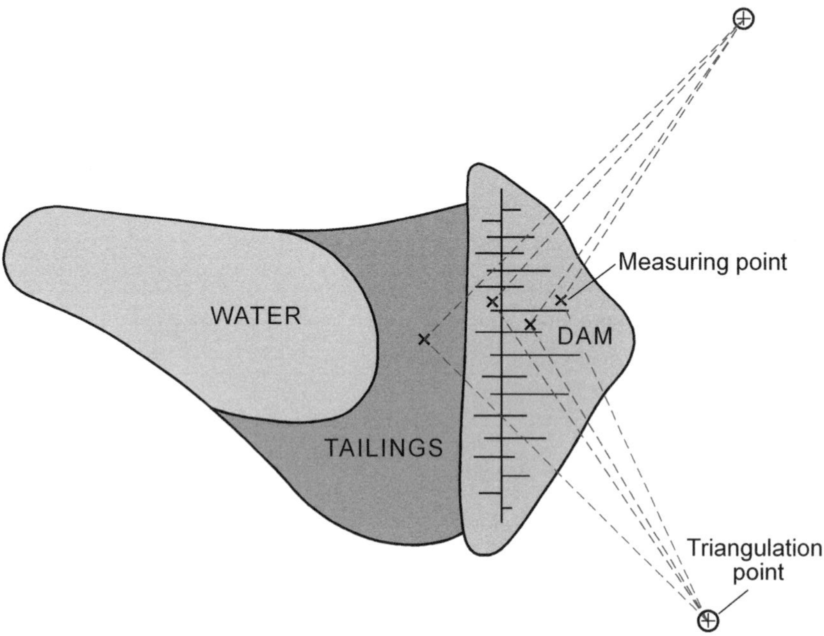

Figure 7. Observation by surveying.

External movement can be measured by conventional ground surveying equipment including automatic laser monitoring (Fig. 7).

4.7. Other Measurements

It may be appropriate to carry out other measurements depending on such aspects as type of dam, tailings material, meteorological conditions and seismology factors.

5. Concluding comments

Installations related to tailings ponds and dams are a high risk for the population and the environment. Approaches to minimize these dangers are given. Further recommendation concerning to the local condition should be worked out.

Acknowledgement

The contribution by Dr. Lutz Benner, Bochum (Germany), to this paper is acknowledged.

References

[1] ICOLD (1996): Guide to tailings dams and impoundments. – ICOLD-Bulletin, 106.
[2] UNECE (in preparation): UNECE safety guidelines and good practices for tailings management facilities.

I. Apostol et al. (Eds.)
IOS Press, 2009.
doi:10.3233/978-1-58603-948-6-122

Environmental Protection Measure Assessment in Affected Area of Ponds Collecting Waste Mine-water in Western Donbass

Galyna P. YEVGRASHKINA [1], Dmytro V. RUDAKOV[2], Mykola M. KHARYTONOV[3]
[1] *Dnepropetrovsk National University, Gagarina av. 44, Dnepropetrovsk, 49600, Ukraine, envteam@ukr.net,*
[2] *National Mining University, K. Marx av., 19, Dnepropetrovsk, 49005, Ukraine, dmi3rud@mail.ru*
[3] *Dnepropetrovsk State Agrarian University, Voroshilov st.25, Dnepropetrovsk, 49600, Ukraine, mykola_kh@yahoo.com*

Abstract. The main factors determining groundwater pollution in Western Donbass are analyzed and estimated quantitatively. The developed 2-D unsteady transport model with the un-ordered macro-dispersion scheme was proved to be effective in modelling solute transport in aquifers. The proposed complex of protective measures based on carried-out research makes it possible to reduce mineralization of groundwater and surface waters.

Keywords. waste mine-waters, ponds, solute transport, modelling, protection measures

Introduction

Western Donbass is located in the highly industrialised part of Ukraine. Most mines are sited in the well-watered Samara river valley with comparatively penetrative rocks, such that coal mining is accompanied by intensive pumping of saline mine-water. In 2006 about 42 million m^3 of water was pumped with mineralization ranging from 2.2 to 33.6g/l at the average daily rate of 114,830m^3. This water is discharged mainly into five collecting ponds constructed in the Cloughs of Kos'minna, Taranova, Svidovok, Nikolina and Stukanova in 1960-70s (Figure 1). The ponds have appeared to be one of the sources contaminating groundwater and surface waters as most of them were built without the special bases that could prevent salt transport into aquifers. Besides, mining has led to subsidence at the surface and, eventually, to the flooding of large areas. The environmental state is aggravated by leaching toxic substances from waste rocks accumulated in slag heaps, which contaminates soils and groundwater.

The aim of the presented research [1] is to assess groundwater pollution around accumulating ponds using mathematical modelling of solute transport in aquifers and to develop protective measures suitable to the Western Donbass conditions.

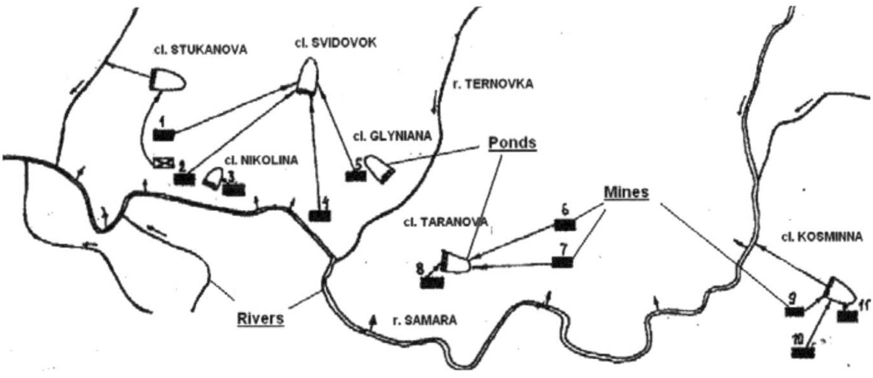

Figure 1. Scheme of mine water pond location in Western Donbass.

Key: Mines are (1) "Cosmos heroes", (2) "Blagodatna", (3) "Pavlograds'ka", (4) "Zahidnodonbas'ka", (5) "Ternivs'ka", (6) Dniprovs'ka, (7) named after Stashkov, (8) "Samars'ka", (9) "Stepova", (10) "Pershotravneva", (11) "Yuvileina".

1. Geological and Hydrogeological Conditions

The geological cross-sections around all ponds are identical. The ponds, located mainly on slopes, do not have direct connections to aquifers (Figure 2). The low permeable cover soil layer consists of quaternary clay loam up to 10m in thickness. Underlying quaternary sands, Neogene's sediments with clayey inclusions are characterized by average hydraulic conductivity of 4m/day that reaches 10m/day in some areas around the "Taranova" Clough. The general thickness of sand is about 15–25 m. Sandy deposits are bedded by clays, chalky clays and sandstones with significantly lower conductivity, so clay layers form the weakly penetrative confining layer. The active porosity of sandy deposits, depending on the clay inclusion content, ranges from 0.175 to 0.32, which amounts 0.5 to 0.8 of the total porosity. The natural water salinity in the upper aquifer away the ponds does not exceed 1g/l.

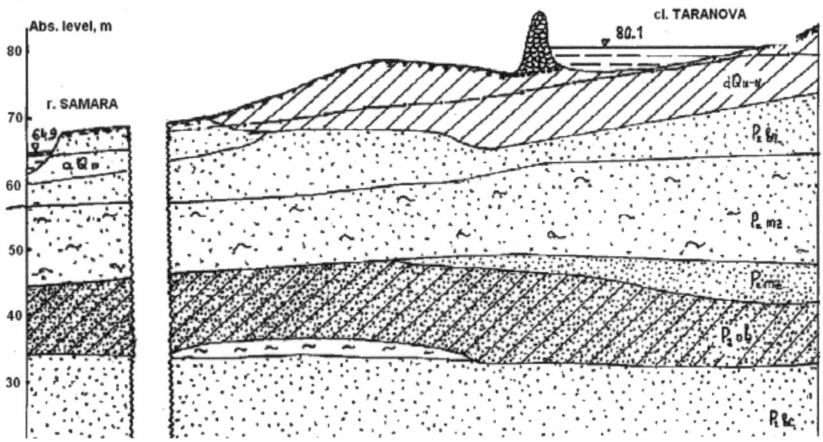

Figure 2. Cross-section along the profile "River Samara – Clough Taranova" (Horizontal scale = 500 x Vertical scale)

The area affected by the ponds is 'lowland' with a raised groundwater head. Frequently occurring floods, accompanied by significant increase in velocities of ground water flow, accelerate solute transport in aquifers. Migration of salts depends also on rock permeability, the watered area within ponds, mineralization and discharges of mine-water.

2. Field Studies

The experimental studies included the processing of data obtained from the constantly modified local monitoring network and estimating regional trends in the groundwater head deduced by analysing the regional data. The hydrogeological assessment showed the stabilization and redistribution of differently flooded area between 1980 and 2005 (Figure 3).

The chemical content of water in the ponds, aquifers and soils was investigated over more than ten years. The analysis of available data showed that the water is similar to sea water diluted in the proportion 1:3 and enriched with *Ca* ions (Figure 4). According to the monitoring data, the ratio $R=C_{Na+}/C_{Cl-}$, where C is concentration, amounts to 1.8–2.1 in the ponds and decreases in aquifers because of Na^+ adsorption and local hydro-chemical conditions as the distances from ponds grow. Generally, sorption does not affect solute transport, essentially except for *Na* ions.

This conclusion does not relate to the pond "Svidovok" which has a clay bottom that was specially treated with colloids and salts in 1984. Up to 1989 the groundwater head rose essentially, which was confirmed by observation in a well (No. 22313) near the pond, but without any change of mineralization. In fact, the clay bottom had been delaying solute transport for five years, which is in good agreement with the experimental data pointing to an increase in permeability of clays and their destruction after long-term leakage of salt water containing *Ca*, *Cl*, and *Na* [2].

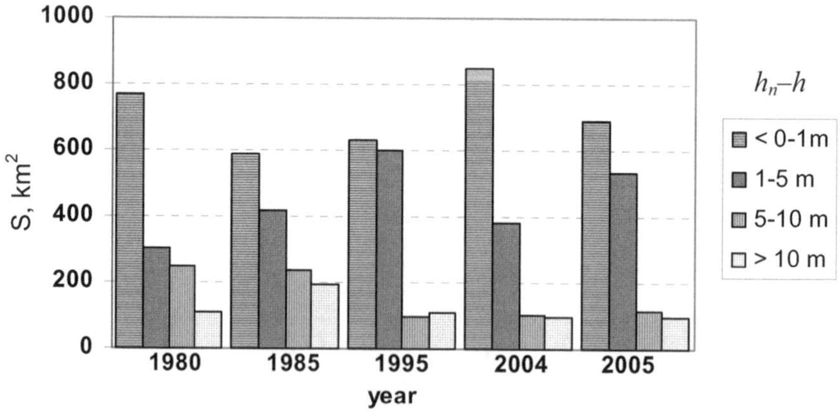

Figure 3. The time change of the area S (km^2) in the Buchak aquifer where the actual groundwater head h is lower than its natural level h_n

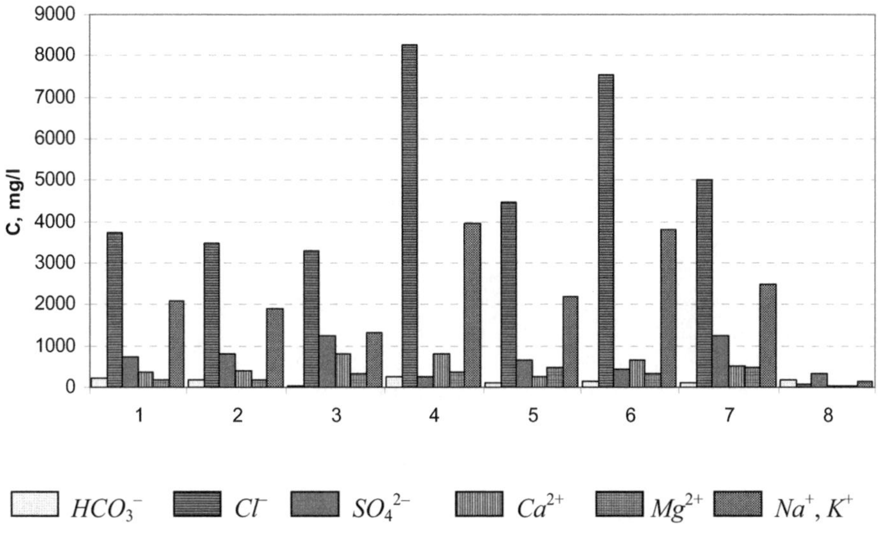

Figure 4. Concentrations of macro-components (mg/l) in ponds and groundwater:
(1) Pond "Nikolina"; (2) Monitoring well 180m from Pond "Nikolina"; (3) Monitoring well 420m from Pond "Nikolina"; (4) Pond "Taranova"; (5) Pond "Stukanova"; (6) Pond "Svidovok", (7) Monitoring well 300m from Pond "Svidovok" (1990); and (8) the same well (1988)

Monitoring near the "Taranova" Clough (Figure 5) has demonstrated increased mineralization of groundwater that can be explained by the high water salinity in the pond, up to 26g/l in some years. The average mineralization has reached 3.8–4.6g/l exceeding 10g/l in some wells; the area with mineralization of more 3g/l made 15km^2 in 1995. The pollution zone had a strip form exceeding 2-4 times the pond width near the dam.

3. Mathematical Modelling

The hydrogeological conditions are suitable for the use of a two-dimensional model for estimating the macro-component transport in groundwater. The model is based on the 2-D equation governing unsteady transport of a non-reactive soluble substance [3], taking into account mass exchange with underlying and overlying beds:

$$\frac{\partial}{\partial x}\left(D_x\frac{\partial C}{\partial x}\right)+\frac{\partial}{\partial y}\left(D_y\frac{\partial C}{\partial y}\right)-\frac{\partial}{\partial x}(V_xC)-\frac{\partial}{\partial y}(V_yC)-V_l'\frac{\partial C}{\partial z}+V_{l+1}''\frac{\partial C}{\partial z}=n\frac{\partial C}{\partial t}. \qquad (1)$$

Here D_x and D_y are the dispersion coefficients in directions x and y, C is mineralization of groundwater, V_x and V_y are the flow velocity components, and n is active porosity.

Mass fluxes through the top and bottom boundaries of the aquifer are calculated by the formula:

$$V_l' \frac{\partial C}{\partial z} = \frac{k_0}{m_0}\bigg|_{l-1} (H_{l-1} - H_l) \frac{C_{l-1} - C_l}{m_{0,l-1}}, \quad V_l'' \frac{\partial C}{\partial z} = \frac{k_0}{m_0}\bigg|_{l} (H_l - H_{l+1}) \frac{C_l - C_{l+1}}{m_{0,l}}, \qquad (2)$$

where V_l' and V_l'' are flow velocities on the aquifer's top and bottom boundaries, k_0, m_0 are conductivity and thickness of the overlying and underlying clay layers, H_l and C_l are, respectively, groundwater head and water salinity values in the l-th aquifer.

Eqns. (1) and (2) are solved by the finite difference method [4]. The developed software uses the alternative directions scheme.

The key problem in modelling was the estimation of parameters where there is a lack of field data. The main difficulties making complicated accurate comparisons between calculated and field data were the following: (1) changes in parameters of mining and pumping rates of water used for local needs; (2) hydrological changes including formation of new channels, bed deformation etc.; (3) transformation of the monitoring network; and (4) increasing leakage through the clayey bottom of the ponds. Firstly the transport model was adapted to local hydrogeological conditions using the field and experimental data. The scheme of "un-ordered macro-dispersion" (5) has provided the best agreement to the field data in comparison with the analytical solution of the transport equation assuming unlimited adsorption capacity and the simple advection scheme.

After algorithm substantiation, the finite-difference solution of Eqn. (1) was used to determine the crucial dispersion parameter D substituting the monitoring data instead of calculated values in grid nodes. According to the N.N. Verigin's formula [6] the component D_x can be calculated as

$$D_x = \frac{V_x \Delta x}{\ln(\overline{C} - 1)}, \quad \overline{C} = \frac{C_{i+1,j} - C_{i-1,j}}{C_{i,j} - C_{i-1,j}}, \qquad (3)$$

where i and j are grid indices. The component D_y is determined by analogy.

Then Eq. (1) was solved with the estimated coefficients varying the hydrogeological parameters. Inverse modelling has demonstrated that the macro-dispersion parameter D ranges from 2 to 36m^2/day when groundwater flow velocity varies within the interval 0.008-0.03m/day. The best correlation between calculated and field data was reached at D = 3m^2/day and n = 0.175 near the pond in the "Taranova" Clough, at D = 25-30m^2/day and n = 0.19 for the pond in the "Stukanova" Clough, and at D = 2.04m^2/day for the pond "Svidovok".

4. Prediction of Macro-component Transport in Aquifers

The solute transport was predicted for the ponds located in the Cloughs of "Taranova", "Stukanova" and "Svidovok". The pond in the "Kos'minna" Clough containing low salinity water does not affect essentially the groundwater quality. The pond in the "Nikolina" Clough some 600m from the river's new channel was surrounded by heaps of salty mine waste rocks. It conditioned seepage and underground run-off of water from the pond to the Samara river without any essential change in mineralization.

The forecast carried out in 1983-1986 was verified by field data in 1995. During this period the number of wells, their quantity, location and other parameters have been changed, as well as the volume and water levels in the ponds, mineralization and discharge rates. Also the bed and the water level of the Samara river have been transformed. Therefore, there was no exact concurrence in the results of the forecasts and the field data up 1995, which complicated validation of the modelling due to difficulties in comparing with field data.

Meanwhile, the area of groundwater pollution has been increased in recent years significantly. Theoretical and field studies have revealed the general trend of increasing groundwater pollution in Western Donbass. Particularly, modelling has proved the actual trend of rising groundwater levels and increased mineralization near the "Taranova" Clough (Figure 6). At the same time, due to the reduction of mineralization in discharged mine-water from near-by mines during 1985-1991, solute transport was slowed down here essentially.

The next stage of the research was the prediction of solute transport till 2015. It was established that the increased salinity zone of more than 3g/l will enlarge significantly and reach the closest rivers such as Ternovka and Samara. The expected lengths of pollution zones in aquifers near the ponds are brought together in Table 1.

Table 1 – Predicted maximal distance of polluted zone in groundwater with mineralization exceeding 3g/l

Pond name	"Taranova"	"Stukanova"	"Svidovok"
Distance	900m	2,200m	800m

5. Protective Measures

The proposed scheme for reducing salinity increases in groundwater suggests a differentiation of mine-water discharge depending on its mineralization and pond locations in the same clough. The upper pond is intended to accumulate the brackish part of mine-water. The second pond down-gradient should have a clay colloidal-salt bottom able to adsorb macro-components that functions effectively during five years. For the next five years mine-water is to be discharged to the third pond having the similar bottom, while the destructed clay layer of the second pond is regenerated.

The performance of the proposed scheme can be demonstrated in the case of the "Taranova" Clough. The first pond accumulates waters pumped from the mine named after Stashkov (see Figure 1) with mineralization of 2.66g/l. This water, with regard to its quality, is suitable for irrigation and fish farming. The second and third ponds having the clay bottom are used alternatively during a period of 5 years to accumulate water pumped from the mines "Samars'ka" and "Dniprovs'ka" with mineralization of 7.5 and 6.6g/l respectively. If mine-water is discharged from the mine "Pavlograds'ka" to the pond in the "Nikolina" Clough then usage of the first pond is not necessary. The same approach suits to the ponds in the "Stukanova" Clough and "Svidivok" taking into account the pending closure of the mine "Ternivs'ka".

Generally, regarding to the adsorption and hydrochemistry studies, the self-purification capacity of aquifers near the ponds related to macro-components should not be overestimated. This fact has to be kept in mind in planning and accomplishing the protective measures aimed at minimization of water pollution in Western Donbass.

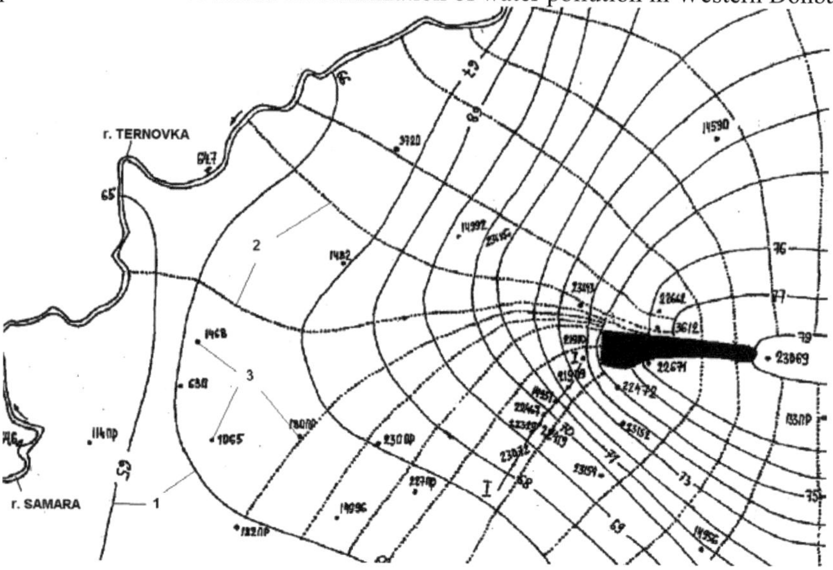

Figure 5. Groundwater monitoring network around the pond in the "Taranova" Clough
(1) equipotentials, (2) streamlines, (3) monitoring wells

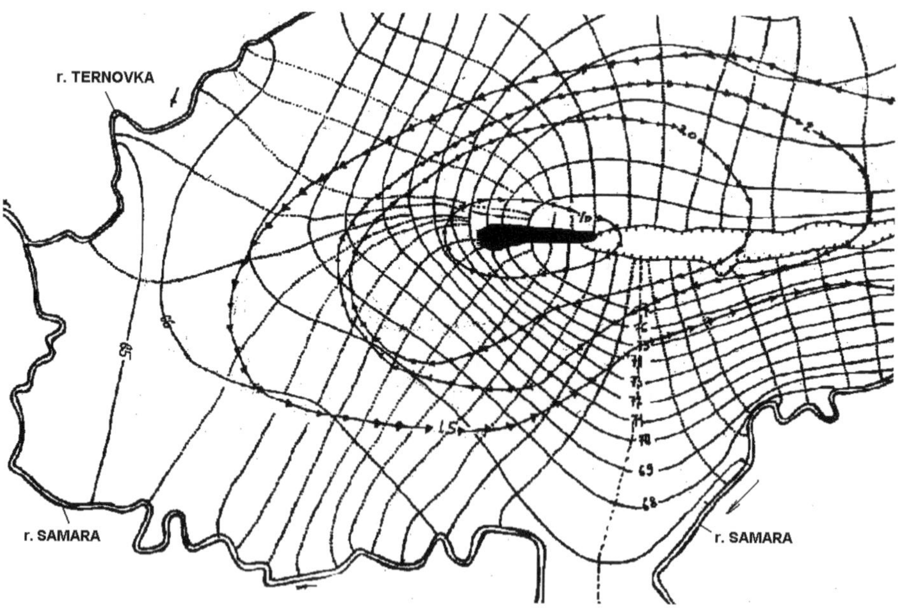

Figure 6. Mineralization of groundwater in the upper aquifer
near the "Taranova" Clough predicted to the end of 1995.
Key: —♦—♦—♦— contours of equal concentration (1.5, 2, 3, 10 g/l)

6. Conclusions

1. Long-term transport of macro-components from the ponds accumulating mine-water has led to essential increases in mineralization of groundwater and surface waters in Western Donbass.
2. Transport phenomena in sandy and loamy aquifers are described satisfactorily by the developed unordered macro-dispersion scheme, which is confirmed by comparisons with the monitoring data. The scheme accuracy increases if the dispersion coefficient is estimated according to inverse hydrogeological modelling.
3. Bottom clay layers treated with colloids and salts have proved to be of short-term effectiveness only due to high mineralization and unfavourable salt compositions in discharged mine-water. However, their delaying effect is expedient for use together with other protective measures.
4. The prediction results are in the good agreement with the monitoring data. The main differences consist of the following: changing the conditions of deposit exploitation near the "Taranova" Clough has slowed down solute transport compared with the forecast; and, on the other hand, reconstruction of the pond in the "Stukanov" Clough, followed by a growing discharge rate, has accelerated migration and increased mineralization of groundwater.
5. The discrepancies between predicted mineralization and field data are explained mostly by the reasons of planning and mining condition transformation rather than being problems of hydrogeological estimation. It makes more difficult the correct long-term forecasting of natural and man-changed systems with badly controlled and predicted parameters.
6. The protective measures mitigating after-effects of mining in Western Donbass should also include: (1) regulated mine-water discharge with regard to its mineralization; (2) deepening river beds to prevent flooding; and (3) neutralization of phyto-toxic substances in the upper layer of mine waste heaps that inhibit leaching and solute transport in soils.

References

[1] G.P. Yevgrashkina. *The Influence of Mining on Hydro-Geological and Soil-Ameliorative Conditions of Territories.* Dnipropetrovsk, Monolit, 2003. (in Russian)

[2] *Ameliorative hydrogeology* / I.Ye. Zhernov, A.G. Soldak, P.Yu. Kusch, O.O. Gryza. Kyiv, Vyscha Shkola. 1971. (in Ukrainian)

[3] J. Bear, D. Zaslavsky, S. Irmay. *Physical Principles of Percolation and Seepage,* UNESCO. 1968.

[4] A.A. Samarsky. *A Theory of Difference Schemes.* Moscow: Nauka. 1977. (in Russian)

[5] V.A. Mironenko, V.G. Rumynin, V.K. Uchaiev. *Ground water protection in mining regions.* Leningrad, Nedra, 1980. (in Russian)

[6] *Hydrodynamic, physical and chemical properties of mine rocks.* S.V. Vasilyev, N.N. Verigin, V.S. Sarkisyan, B.S. Sherzhukov. Moscow, Nedra, 1977. (in Russian)

I. Apostol et al. (Eds.)
IOS Press, 2009.
doi:10.3233/978-1-58603-948-6-130

Management of Risks Associated with Mining Wastes (Tailings Dams and Waste Heaps)

Oana–Cristina MODOI[a], Lucrina ŞTEFĂNESCU[a], Sanda MĂRGINEAN[a,b], Corina ARGHIUŞ[c] and Alexandru OZUNU[a]

[a] *Babeş-Bolyai University, Faculty of Environmental Sciences, 4 Stefan cel Mare Street, Cluj-Napoca, Romania, Postal code 400192, Tel: +40 264-405300 Extension 5444, Fax: +40 264-599444, E-mail: aozunu@enviro.ubbcluj.ro*

[b] *Regional Centre for Major Industrial Accidents Prevention, 67 Donath Street, Cluj-Napoca, Romania, Postal code 400293, Tel: +40 264 420590, extension 19, Fax: +40 264 420667*

[c] *Babeş-Bolyai University, Faculty of Geography, Clinicilor 5-7, 40006 Cluj-Napoca, Romania, Tel: +40-264-40.53.00, Fax: +40 264 59.19.06*

Abstract. The paper presents significant aspects regarding risk management related to wastes in the Romanian mining industry. The national and international legislative framework is approached thematically, focusing on the new European Directive regarding mining wastes (Directive 2006/21/EC on the management of waste from the extractive industries – the mining waste directive). The two studied subjects of interest are: waste heaps and tailing dams, and plans for the preliminary analysis of the risks induced upon the environment. Best practice environmental management requires mitigation of risks of environmental damage from unexpected incidents occurring in the mining industry. The management of mining wastes represents a very significant component of the mine closure and conservation process, in view of the ecological reconstruction of the area and reuse of the affected land. Consequently, an adequate waste management system in the mining industry determines medium and long-term benefits for the environment, for economic efficiency and for the mining communities.

Keywords. Mining waste management, risk assessment, tailings disposal facilities

Introduction

As Romania has an old tradition in mining, it is necessary to take the appropriate measures for the rehabilitation of the tailings disposal facilities and for the pollution prevention on the sites with a high degree of toxicity. In 2004 the Romanian Government approved the Strategy of the Mining Sector for 2004–2010 which highlights the need to support the efficient application by the mine administrators of the measures regarding the mitigation of risks associated with tailings dams and waste heaps.

The Romanian Government has completed a Mining Sector Environmental Assessment (MSEA) which sets the basis for the evaluation of state- or privately-owned mines all over the country. MSEA identifies the main environmental problems caused by the mining operations. The amplitude and effects of these activities, and especially of the technical accidents produced in this field, have had a significant impact upon the environment (i.e. water, air, soil, subsoil, vegetation, and fauna) and the population (though injuries, panic, and health risks).

Romania is confronted also with the risk of pollution due to mining accidents. The accidents at the tailings dams in the Maramures region in 2000 (namely, the Aurul mine on January 30th and the Baia Borşa mine on March 20) have demonstrated that it is necessary to direct the concerns for safety and environment towards mining activities.

1. The Legislative Framework on Environmental Safety and Wastes Management

Within the European Union, the extractive industry generating among the largest of industrial wastes has drawn the attention of the legislative bodies to the need for adequate legislation (Hámor, 2004). Thus, a study started on the applicability of the international environmental directives and efforts were made for a pan-European surveillance of this industry. The focus is on the sustainable operation of the extractive industry, according to the sustainable development concept. The main directives in this field are: the SEVESO Directives (Original Seveso Directive 82/501/EEC ("Seveso I") Council Directive of 24 June 1982 on the major-accident hazards of certain industrial activities – 82/501/EEC; Seveso II Directive 96/82/EC; Directive 2003/105/EC of the European Parliament and of the Council amending Council Directive 96/82/EC on the control of major-accident hazards involving dangerous substances; and Seveso III 2003/105/CE). We need to mention that Romania has a total of 333 Seveso facilities.

More specific directives are the Directive 2006/12/EC of the European Parliament and of the Council of April 5, 2006 on Waste and Directive 2006/21/EC on The management of waste from the extractive industries.

1.1. Romanian Legislation on Waste Management

Romania, as a member of the European Union, has engaged to adopt and implement the environmental directives. The Seveso II Directive was transposed by Governmental Decision 95/2003, which became effective in August 2003. The Framework Directive 75/442/EEC of the European Council regarding the wastes was transposed into Romanian legislation by Governmental Ordinance O.U.G. 78/2000 approved and completed in 2001 by Law 426/2001. As a result of the implementation of this law, "The National Strategy for Wastes Management" was developed.

Other legislative provisions regulating the management of wastes in general and in the extractive industry in particular are:

- Governmental Decision no. 1470/2004, regarding the approval of the National Waste Management Strategy and of the National Wastes Management Plan;
- Law no. 263/2005 regarding the regime of chemical dangerous substances and products;
- Governmental Decision no. 349/2005 regarding the storage of wastes; and
- Governmental Decision no. 856/2002 regarding the records of waste management and the approval of the wastes list, including dangerous wastes.

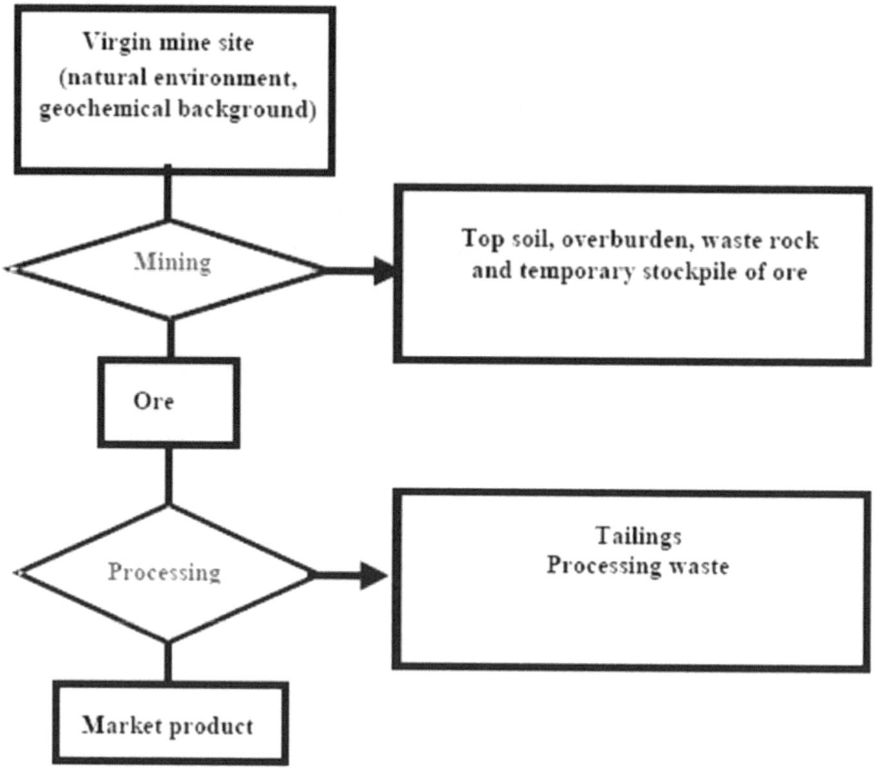

Figure 1. Mining waste types (Source P. Charbonnier, 2001, pp. 7).

2. Situation with Industrial Wastes

Mining waste can be defined as a part of the materials that result from the exploration, mining and processing of substances governed by legislation on mines and quarries. It may consist of natural materials without any modification other than crushing (ordinary mining waste, unusable mineralized materials) or of natural materials, processed to varying degrees during the ore-processing and enrichment phases, and possibly containing chemical, inorganic and organic additives (see Fig. 1). Overburden and topsoil are classified as waste.

This waste can affect the environment through one or more of the following intrinsic criteria:

- its chemical and mineralogical composition;
- its physical properties;
- its volume and the surface occupied; and
- the waste disposal method.
- Besides these parameters, one must also take into account extrinsic parameters such as:
- climatic conditions liable to modify the disposal conditions;

Table 1. Main categories of industrial wastes generated in 2004 in Romania

Type of industrial wastes	Amount of wastes (million tons)
Mine tailings	36
Ashes and slag from power-stations	6.4
Metallurgical wastes	2.6
Residual sludge	2.5
Chemical wastes	2.2
Ferrous wastes	1.9
Construction wastes	3

- geographic and geological location; and
- existing targets liable to be affected (man and his environment).

Thus, identification of the environmental risks associated with the exploitation of mines and quarries, and with ore processing, at the scale of the European Union not only requires the characterization and quantification of the different types of waste, as well as a knowledge of the processes used, but also an assessment of the vulnerability of the specific environments contingent upon the geological and hydrogeological conditions and peripheral targets.

2.1. Scale of Industrial Wastes in Romania

Of the 77 million tons of solid wastes generated in 2004, about 69 million tons have been industrial wastes (including mine tailings). The amount of mine tailings was 36 million tons (approx. 52%), and the quantity of other industrial wastes was 33 million tons.

The amount of mine tailings had an uncertain evolution over the years, according to the nature of the extractive activities; as a general trend it may be stated that the tailings quantity has recorded a continuous decrease.

The main categories of industrial wastes generated in 2004 in Romania are presented in Table 1. A special category of industrial wastes is represented by the dangerous substances.

2.2. Storage of Industrial Wastes in Romania

The management of industrial wastes consists in the valorification (such as recycling), storage, final disposal, and incineration. The ratio of these options is, on average, approximately the same every year:

- Disposal: 81.0%;
- Valorification: 15.0%;
- Temporary storage: 3.3%; and
- Incineration 0.7%.

At the end of the year 2004 there were more than 18 million tons of wastes in stock within the economic units generating wastes.

3. Industrial Wastes Disposal Facilities in Romania

Disposal of wastes on barren land represents the most important route for the elimination of industrial wastes in Romania,and more than 80% of the generated wastes are disposed every year.

Most of the industrial wastes disposal facilities (354) are simple (usually concrete platforms); also, there are numerous tailings dumps (251) and tailings dams (209). Most of the industrial disposal facilities (approximately 76%) cover small surface areas (up to 5 ha).

Table 2 presents the situation of the industrial wastes deposits, on categories, as well as the surfaces they occupy.

Of the total industrial wastes deposits, at least 50 do not have any kind of environmental protection facilities, and most of them are only fenced. Some of the deposits have one or more special facilities (such as waterproofing, drains, protective channel, and monitoring drilling), but very few have all the arrangements needed to achieve the necessary conditions for environmental quality protection.

Table 2. Situation of the industrial wastes disposal facilities

Industrial deposits	Tailings dams	Tailings dumps	Slag and ash dumps	Simple disposal facilities	Underground disposal facilities	Total
Number	209	251	108	354	29	951
Occupied surface (ha)	2,466	5,932	2,823	748	17	11,986

(Data provided by the Romanian Ministry of Environment and Sustainable Development)

Figure 2. D 3 Tailings dam – Maramures county, NW of Romania.

Figure 3. Tailings dump located above the Colbu 2 tailings dam, Maramures. (Courtesy of the Romanian Ministry of Environment and Sustainable Development)

4. Risks Associated with Disposal of Wastes from the Extractive Industry

Romania has a large number of abandoned tailings dams and tailings dumps, which also represent contamination sources. Accidents like those in the North-West of Romania in 2000, from the Aurul S.A. gold mine and Baia Borsa mine, the collapse of the tailings dams may have serious and devastating consequences.

The main problems regarding the safe operation of tailings dams occur even in the engineering phase and the failures are due to the following main factors:

1. **Engineering errors:** operation of closed regime water circuits; absence of some failure compartments intended for water discharge.
2. **Misevaluation of external stresses:** rainfall; run-offs from slopes; flows of watercourses crossing the dam; stability of embankments; non-observance of the particle sizes of deposited tailings; non-observance of shore lengths; absence or malfunction of drainage systems.
3. **Operation problems:**
   - non-observance of provisions regarding tailings disposal manner in tailings dam
   - failures of electrical power supply system
   - failures of hydro-transport pipes caused by their wear
   - failures of clear water discharge system
   - inadequate materials for emergency response in case of failure (material and reagents to neutralize dangerous substances)
   - insufficiency of exploitation personnel
   - deficiencies in the operation of information system
   - delayed execution of works identified by surveys.

The risks identified at mine tailings disposal facilities deal mainly with continuous contamination due to acid waters generated by tailings dumps; acid water discharges from the embankments of tailings dams; and stability of tailings dumps. Figure 4 presents the map of the main tailings disposal facility hotspots in Romania.

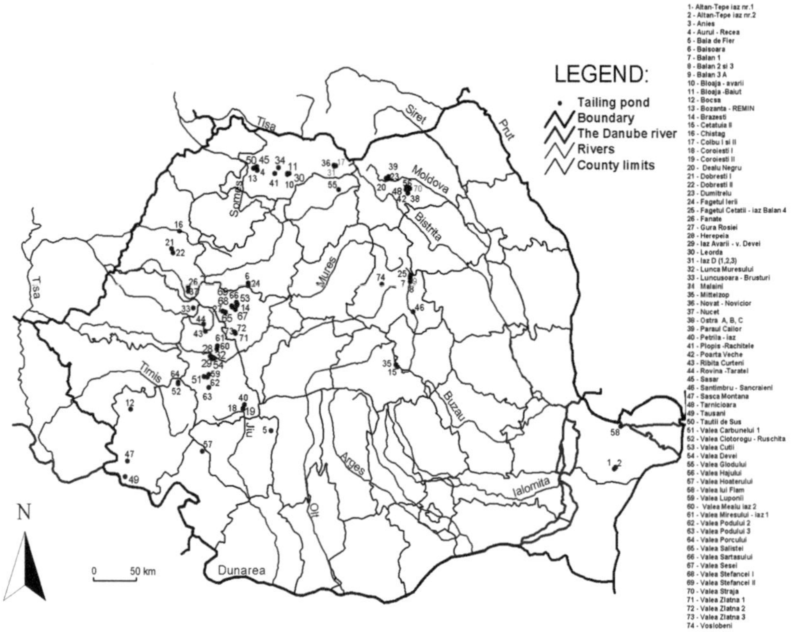

Figure. 4. The main tailings disposal facilities hotspots in Romania.
(Data provided by the Romanian Ministry of Environment and Sustainable Development)

The factors contributing to a high risk of accidental water pollution as a result of mining activities in Romania in general, and in the Tisza River basin in particular, are the following:

- *Bad management of tailings disposal facilities (tailings dams and waste dumps).* There are a few risks associated with the tailings ponds, among which we mention: failure of the dam structure and discharge of wastes; instability of waste dump slopes; bad water management and seepage of contaminated water into underground and surface waters; dust emissions on the shores of the tailings dams, both airborne and stored in watercourses; soil erosion associated with wind and rainfall; dangerous chemical substances and heavy metals contamination. The tailings dams present the greatest danger in terms of unexpected collapses, as happened at the mines in Baia Mare and Borsa in 2000 due to overloading. The unstable waste dumps located upstream of the tailing dams contribute also to the failure risk.
- *Insufficient knowledge of pollution degree and risk induced by mines and lack of reliable information.* There is an acute lack of reliable data regarding the main hypotheses taken into account when originally designing the tailings disposal facilities.
- *The limited institutional capacity to implement the regulatory tools and standards.* The continuous enhancement of mining accident consequences represents neither the result of the lack of tools, regulatory standards nor of the lack of institutional organization. They are due mainly to the implementation and application deficiencies of the current regime. Both the human and the financial resources are insufficient, the awareness and understanding of the com-

plex environmental factors in the mining sector are limited; and the entities which support and regulate the tailings dams do not have the specific experience.

– *The low efficiency in reacting to the emergency situations.* Central and local administrations do not have the necessary preparedness capacity to approach the emergency situations or to monitor the emergency preparedness activities, as well as the due timing of remedial actions.

The governmental programme for the period 2004–2010 for the mining sector includes the allocation of $150.7 million (US) for the ecological rehabilitation of the mines which will be closed or sold to private companies, including the improvement of the tailings disposal facilities safety and management.

5. Assessment of Risks Linked to Mining Wastes

The notion of environmental impact of mining activities is only fully meaningful if it includes a change in the initial environmental parameters due to such activities. These parameters, which govern the "quality of the environment", may have several components, such as chemical composition of the waters, soils, etc., biological diversity, and visual aesthetic qualities.

The major risks linked to mining waste for the environment are twofold:

- Risks associated with not only potential pollutant sources (*e.g.* acidity and heavy metals in non-ferrous metallic ores) but also the specific environmental context and the presence of targets in the event of release. The possible risks from the potential pollutant source (such as acidity and heavy metals) in waste is dependent not only on the mineral characterization of the solid but also on the quality of the potential leachates, the direct environment (i.e. soil, groundwater, surface water and air) and the potential targets (i.e. humans, fauna and flora). The realization of a Geographic Information System (GIS) specific to mining waste quantities and their pollution potential in different environmental contexts would thus constitute a tool in the assessment of risks linked of such materials.
- Risks associated with the stability of the tailings dam, as indicated by the recent spectacular accidents in Spain (Aznalcollar) and Romania (Baia Mare). As regards the potential risk from tailings dams, it will be necessary to evaluate on each site the stability of tailings dams. Particular parameters, such as exceptional climatologic conditions, should be carefully taken into consideration during the evaluation. In addition, common minimum safety standards for the design, construction, operation, and monitoring should be developed and applied. These minimum safety standards could be built on the know-how of the profession.

Figure 5 outlines the principal elements in a typical risk assessment model.

This risk assessment model is best understood by working through the steps listed below:

1. **Identify the hazard** – How an accident might happen? Consider what or how things could go wrong when the activity is carried out.

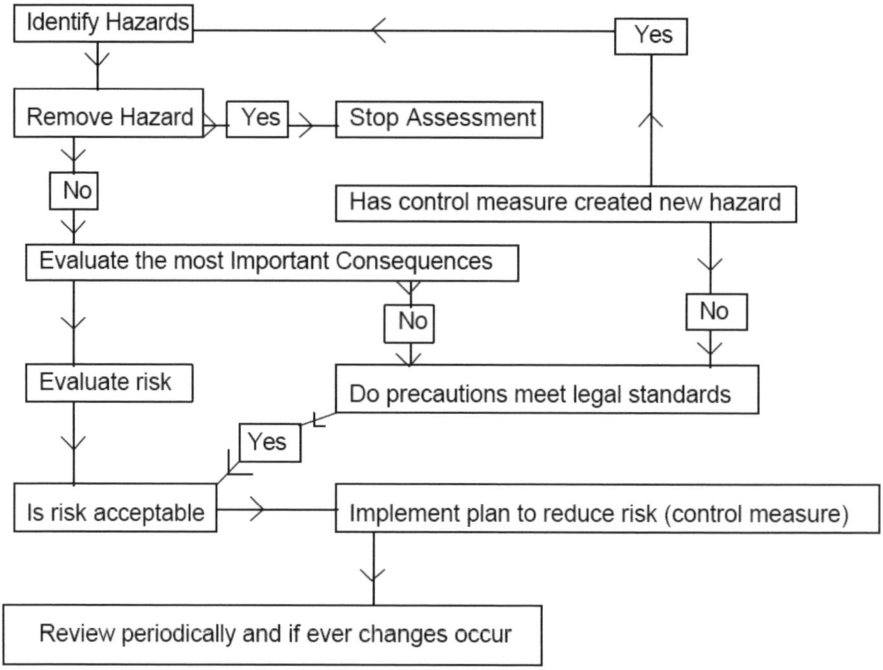

Figure 5. Model of risk assessment.
(Source Safety and Health Commission for the Mining and other Extractive Industries, 1998, pp. 4)

2. **Identify who is at risk** – Who is involved in the activity? Who else could be at risk?
3. **Remove the hazard** – Can the activity be carried out in another way so as to eliminate the hazard.
4. **Evaluate the risk** – How likely is an accident to happen? How serious would the injury be if there is an accident while carrying out the activity?
5. **Decide on control measures** – Look at what measures have been taken already to ensure that persons do not have an accident. For example, have suitable and sufficient guards been fitted? Decide whether anything else needs to be done. For example, it may be necessary to provide extra training in the safe use of machinery and only allow trained workers to use it.
6. **Record the assessment** – The risk assessment should be recorded.
7. **Review** – The assessment will need to be reviewed every time there are changes in the workplace, for example new members of staff, new equipment, new systems of work and new location.

6. Risk Management

6.1. Assessment of Quantities of Mining Waste Generated

The first step in risk management is to assess the quantities of mining waste generated since the start of the mining activities: substances mined, typology of the ore deposits, the operating systems and processes used. These also include information and data on the quantities of ores extracted and processed, the quantities of marketable products generated, and the quantities of residual waste.

- *Inventory of mining sites*: identification and location of the sites, the associated mining steps, the operating companies, any known accidents, and pollution, and any known on-site studies related to overall product and waste management.
- *Description and analysis of mining systems*: the operations performed on a mine site to exploit and use an ore deposit can be divided in three main steps:
  - access to the ore deposit (clearance, and galleries producing barren waste)
  - in situ ore extraction and selection, and
  - ore-processing.

In risk assessment is necessary to identify typical systems accounting for the main techniques of mining and processing mineral substances which are, or were, used in mining industry:

- The topographic, geological and hydrogeological situations, as well as the geometry and
- Morphology of the ore deposit, determine the mining method used for its exploitation;
- The chemical composition and mineralogy of the ore deposit determine to a large extent
- The processing; and
- The reserves and economic conditions determine the production rate.

- *Typology and quantities of waste generated by the different mining steps*: each mining step is liable to generate mining waste, normally with different *physical* and chemical properties. Their respective volumes, especially for access to the ore deposit, depend on the type of mining method and the type of the raw material. Similarly, their chemical composition depends on the type of ore, its geological envelope, and its processing.

6.2. Identification and Analysis of Potential Environmental Impacts Associated with Mining Waste Management

At this stage these steps can be followed:

- the identification of potential pollution resulting from mining waste and a quick identification of the impact of this type of activity on human health and the environment;
- compilation of the requested information on specific sites for a simplified assessment of the risks associated with mining waste and environmental repercussions (knowledge of geological and hydro-geological aspects, characteristics of the pollution source, etc.);
- identification of potential pollution generated by mining waste; and
- necessary information for a simplified assessment of mining-related risks.

6.3. Mining-Waste Management Practices and Identification of the Need of Improvement

- Design of tailings and waste rock facilities – inventory with actual practices in waste management; and
- Examples of accidents connected with the disposal of mining waste and redevelopment (yard, heaps or dumps).

6.4. Inventory and Analysis of Legislation

Mines in all European Union countries are governed by a set of laws, generally combined in a Mining Code. The numerous regulatory texts, laws and standards, reveal that mines are a matter of concern to the national administration. Mining waste is governed by general waste laws and the extent to which environmental concerns are addressed in these national laws, varies between member states.

7. Improvement of Waste Management

As one of the biggest issues facing the minerals industry, tailings management will benefit from adopting a more holistic view. Managers need to understand the potential impact which begins with reagent input and ends with the establishment of viable alternative land uses on decommissioned tailings storage facilities. In the risk management of mining it is important to take into consideration the adoption of a stewardship approach to the safe management of tailings.

Best practice

Commit to the implementation of best practice in tailings management. The term "best practice" describes a management approach involving a commitment to achieve outcomes beyond those expected for regulatory compliance. To achieve best practice, an operator would be expected to have developed management systems that ensure the identification of opportunities for improvement and to see that change is implemented, monitored, and evaluated. In achieving best practice it is expected that operators would have a commitment to minimizing wastes and a well-developed strategy to ensure continual improvement. Best practice should also incorporate the systematic evaluation of new technologies and opportunities for change. Finally, it is inherent in the notion of best practice that community expectations will be met and so the strategy should also involve communication with key stakeholders.

Waste minimization

In order to minimize the potential impacts on the environment, it is recommended that mineral treatment and tailings management practices are selected. The principles of waste minimization include the reduction of waste stream volumes and/or toxicity. Beneficial outcomes can be achieved by process changes that reduce the generation of wastes, re-direct wastes to useful purposes, or modify wastes so they are more benign to the environment. It is now widely accepted that industry should implement the principles of waste minimization in the management of production activities. This means that managers use the waste management hierarchy (Table 3) as a guide to assist them in making decisions. Waste characterization is a key prerequisite to adopting one or

Table 3. Waste Management Hierarchy

Hierarchy	Keyword	Description
1	Avoidance	Where possible, processes or materials should be changed to eliminate the generation of the waste
2	Reuse	Some wastes may be useful as feedstock for other processes
3	Recycling	The raw materials contained in the waste may be reusable for further production
4	Recovery of energy	Wastes may be useful as fuel for energy production or substitution
5	Treatment	It may be possible to make wastes innocuous by further treatment or processing
6	Containment	Secure storage of wastes in facilities that are isolated from the environment is often preferable to discharge
7	Disposal	Discharge of waste to the environment under controlled conditions, and in a manner which does not harm the beneficial uses, is the final alternative

(Source Ministerial Council on Mineral and Petroleum Resources Minerals Council of Australia, 2003, pp. 3)

more of these strategies. Avoidance or elimination of wastes is not practical in most cases for mine tailings, although it may be possible to reduce the volume of tailings at some mines. In addition, some technologies which offer promise for elimination of tailings wastes, such as in-situ solution mining, introduce other environmental risks.

In general, however, operators should demonstrate commitment to the improvement of tailings management practice and the better use of tailings wherever feasible. This approach should commence at the project design stage with a rigorous examination of available technologies. Regular re-evaluation of tailings disposal strategies should then be a part of the ongoing management system for the operation.

In general, managers should use the following hierarchy to guide decisions about appropriate processes, technology, and management of wastes.

Ongoing improvement

Tailings are the largest process waste stream resulting from mineral processing. To minimize the environmental risks associated with the management and storage of tailings it is important that operators are committed to continual improvement of their practices and processes. This implies a commitment to the identification, development, and implementation of new and innovative technologies. Effective management for continual improvement has a number of key elements including:

- setting appropriate objectives and goals;
- monitoring operational outcomes;
- evaluating the results;
- identifying changes in technology or widely accepted standards;
- implementing appropriate changes to practices or equipment; and
- setting new objectives and goals.

This is an iterative process and requires a long-term commitment to improvement by all levels of management. The implementation of significant improvement to an existing operation requires careful planning, and will often involve incremental change over a long period. In addition to good operational management systems, these constraints underscore the need for careful planning prior to development of a facility.

Innovation

Strategic research can be a powerful tool for identification and development of new technologies for the management of tailings. Research projects can be conducted at a small scale appropriate for individual operations, or may be much larger and only practical as collaborative projects sponsored by many stakeholders.

Operators should evaluate their needs and decide on an appropriate research strategy for their circumstances. Where mine geology or local conditions create unique challenges, it is appropriate that the operators develop site specific research projects to address those issues. It may also be desirable for individual companies to fund research directed to improvements in technologies or processes specific to their operations. However, all operators should also consider participation in broader research programmes aimed at more general improvement in the management of tailings. Operators should consider forming strategic alliances to identify and promote useful research. Operators should also maintain processes for regular review of relevant technical literature to ensure that they are aware of emerging technologies and can direct their own research or development resources to the best effect. There is a need to encourage and improve information exchanges on a variety of tailings related policy and technical issues. This will improve the ability of both industry and the regulator to locate the practical guidance they need to make effective day-to-day decisions. The outcomes from sponsored research and work by others should be evaluated regularly.

Current management and operational practices should be reviewed in light of any new technologies becoming available and strategies for change developed when appropriate.

Benchmarking

To ensure best practice is achieved it is essential that effective communication is maintained with both the wider industry and the community. Operators should maintain close contact with other companies and industry organizations. Current tailings management systems should be regularly benchmarked against those in operation at other sites and those regarded in the wider industry as “state of the art” in technology and operation.

Knowledge and expertise relating to tailings management should be shared, and other operators should be encouraged to also implement best practice technologies and systems. While technically sound operational systems may provide good protection for the environment, they may not be fully adequate if key concerns held in the local community are not also addressed. Mechanisms for community consultation should be used to ensure management has a good understanding of community expectations and concerns in relation to tailings management. Those concerns should also be considered in any programme for ongoing improvement.

8. Conclusions

All these considerations lead to the conclusion that waste management needs to adopt specific measures that are adequate for each phase of waste elimination from the environment. The observance of these measures should be the purpose of the monitoring of environmental factors affected by the presence of wastes.

The problems raised by wastes management in Romania may be summarized as follows:

- disposal of wastes on barren land is the most important path towards their final disposal;
- the existing deposits are sometimes located in sensitive places (such as near residential areas, near surface or underground watercourses or near recreation areas);
- the wastes disposal facilities are not appropriately arranged for environmental protection, leading to water and soil pollution in the respective areas;
- the current wastes disposal facilities, especially for domestic wastes, are not appropriately operated: they are not consolidated and covered periodically with inert materials for fire prevention purposes; there is no strict control of the quality and quantity of wastes; the main and secondary roads used by the waste transportation vehicles are not appropriately maintained; the transportation vehicles are not washed when leaving the wastes disposal facilities; and many disposal facilities are not fenced or provided with appropriate warnings; and
- the land occupied by the wastes disposal facilities is considered degraded land which is no longer used for agricultural purposes. Nowadays, in Romania, more than 12,000 ha of land are affected by domestic and industrial wastes storage.

References

Charbonnier, P., 2001, Management of mining, quarrying and ore-processing waste in the european union – Study made for DG Environment, European Commission, BRGM/RP-50319-FR, pp. 88.

Eftimie Adriana, 2003, Prevenirea şi remedierea poluării în diferite sectoare industriale, Studiul pilot al NATO/CCMS.

Fodor, D., Baican, G., 2001, Environmental impact of the mining activities, Deva, Infomin, pp. 47-76.

Hámor, T., 2004, Sustainable Mining in the European Union: The Legislative Aspect, Environmental Management, Vol. 33, no. 2, pp. 252-261.

Vibeke Burchard, 2004, Construirea Capacităţii pentru Implementarea Acquis-ului de Mediu la Nivel Local şi Regional, IPPC şi BAT – Privire generală asupra surselor BAT, EuropeAid/116215/CSV/PHA.

Commission of the European Communities, 2000, Communication from The Commission, Safe operation of mining activities: a follow-up to recent mining accidents, COM(2000) 664, Brussels.

Directive 2006/21/EC on the management of waste from the extractive industries (the mining waste directive).

Directive 2006/12/EC, Waste Framework Directive.

Ministerial Council on Mineral and Petroleum Resources Minerals Council of Australia, 2003, Strategic Framework for Tailings Management, ISBN 0 642 72243 9, pp. 39.

Official Gazette no. 411/07.05.2004, The strategy of the mining industry for the period 2004–2010, approved by Governmental Decision no. 615 of April 21, 2004.

Safety and Health Commission for the Mining and other Extractive Industries, 1998, Guidance for carrying out risk assessment at surface mining operations, doc. no 5995/2/98-en, pp. 31.

United Nations University, 2006, Measuring Vulnerability to Natural Hazards: Towards Disaster Resilient Societies, Editor Jörn Birkmann.

I. Apostol et al. (Eds.)
IOS Press, 2009.
doi:10.3233/978-1-58603-948-6-144

Some Results from Dynamic Monitoring Linked to Mining: Case Studies in Bulgaria (Provadia) and Belarus (Starobin)

I. PASKALEVA [a], A. ARONOV [b], G. VALEV [c], R. SEROGLAZOV [b], M. KOUTEVA [a] and T. ARONOVA [b]
[a] *Central Laboratory of Seismic Mechanics and Earthquake Engineering, Bulgarian Academy of Sciences, "Acad. G. Bonchev", str., block 3, Sofia 1113, Bulgaria, E-mail: paskalev@geophys.bas.bg, paskalev_2002@yahoo.com*
[b] *Centre of Geophysical Monitoring (CGM) of National Academy of Sciences of Belarus, Kuprevich str. 7, 220141, Minsk, Republic of Belarus*
E-mail: centr@cgm.org.by
[c] *Geoprecise bul. H. Smirnenski, N1, Sofia, 1000, Bulgaria*
E-mail: geoprecis@abv.bg

Abstract. This work focuses on the assessment of seismic risk issues associated with the potassium salt deposit of Provadia and Soligorsk. The long-term studies, 1983–2007, of the only terrestrial Bulgarian salt deposit (Provadia, $\varphi = 43.06^0$N, $\lambda = 27.45^0$E) and Belarusian (Starobin $\varphi = 52.84^0$N, $\lambda = 27.47^0$E) in connection with the observed higher seismic activity and probable manifestations of technogenic seismicity in the region is presented. The characteristic features of the seismic processes are identified by the curves of recurrence of seismic events with an energy range of 4÷8. A quasi-periodic character of the seismicity activation over time against the general trend of increasing seismicity activation is established. It is shown that zones of epicentres of seismic events are larger that mining areas. Some differences in the pattern of seismic processes, such as: seismic activity in the range of small energies (K = 4–8) is higher in the Soligorsk region; events of the higher energy class K > 9 characteristic of the Provadia region are carried out.

Keywords. Geophysics, geodesy, induced seismcity, surface subsidence

Introduction

Worldwide, the problem of induced seismicity is very difficult for management and control, i.e. earthquakes caused by mining, exploration and extraction of mineral resources. There are observations showing that in conditions of compressional tectonics the extraction of fluids (oil and gas) from the Earth's crust it is possible to trigger big tectonic earthquakes. Therefore it is very important to have accurate and timely information about the seismicity and the tectonic processes in regions prone to tectonic and induced seismicity from independent sources of information for both the public and industrial society, and for the regional and Civil Defence authorities as well.

As used here, "*induced*" describes seismicity resulting from an activity that causes a stress change that is comparable in magnitude with the ambient shear stress acting on a fault to cause a slip, whereas "*triggered*" is used if the stress change is only a small fraction of the ambient level (McGarr et al., 2002). By "*stimulated*" we refer generally to seismicity either triggered or induced by human activities. Most of the case histories reviewed by McGarr et al., 2002 related to triggered rather than induced seismicity.

In many regions the large urbanized areas are overlapped by the regions of manifestations of different natural disasters. This is one of the reasons that has provoked the bilateral international project exploring the problems of natural and technogenic hazards and their possible mitigation (Aronov et al., 2005). There is a trend for quantitative assessment of the natural phenomena and their consequences, analyzing different scenarios. A great problem is the formulation of the optimal measures for the possible loss reduction and particularly their effective implementation. The size of an activity that perturbs the crustal state of stress appears to be a good predictor of the maximum credible earthquake for that operation. This is evidently the case for all types of induced and triggered earthquake sequences reviewed by McGarr et al., 2002, including those due to mining, quarries, liquid injection, large reservoir impoundment, and oil and gas exploitation. It is important to note, however, that many large-scale activities result in little or no recorded seismicity. The case histories reviewed (McGarr et al., 2002, Seroglazov R., 2003) help to support the following conclusion: the seismogenic continental crust is in a state of stress close to failure, whether the tectonic setting is active or inactive. Because of this, perturbations of the state of stress toward failure, even as small as 0.01 MPa, may trigger seismicity. One implication of this is that, for a given activity, triggered earthquakes are as likely in stable as in active tectonic settings, but within stable regions (Starobin and in some extend Provadia) such events are more obvious because the background seismicity is comparatively low. Extensive mine excavations, especially at depths of several hundred metres, amplify the ambient stresses substantially so as to bring the strengthened rock mass back to the point of failure in localized regions.

The upgrade of the existing monitoring systems has been done following the prescription of the European norm EUROCODE 7 1993 (first edition). EUROCODE 7 is the first comprehensive norm on geotechnical works adopted, and it already has been copied by the Japanese standardization committee. EUROCODE 7 includes a section on monitoring as a means of obtaining additional control on the quality of the works and to check, during the works, the hypothesis adopted for the design. The goals of this norm are to ensure good maintenance conditions for the construction after the realization of the project, all in agreement with the specified requirements.

The first provision is to acknowledge the interest of the *Observational Method* (Section 2.7), in cases "*when prediction of geotechnical behaviour is difficult*". The basis of the *Observational Method* is to design the project from hypotheses which are not over-conservative, and to control the process by an extensive monitoring scheme in order to ensure that the actual behaviour of the construction lies within acceptable limits. Intervals between observations must be sufficiently short to allow for modifications of the design or the construction sequence. In the *Observational Method*, monitoring plays an active role in the construction process, and the extra costs generated by instrumentation and data processing are normally largely compensated by the economy generated on the project.

The second, and not least, provision of EUROCODE 7 is to require the creation for every project of a Geotechnical Design Report (Section 2.8), with a specific mention to monitoring:

1. The assumptions, data, methods of calculation and results of the verification of safety and serviceability shall be recorded in a Geotechnical Design Report.
2. The Geotechnical Design Report shall include a plan of supervision and monitoring, as appropriate. Items which require checking during construction or which require maintenance after construction shall be clearly identified in the report. When the required checks have been carried out during construction, they shall be recorded in an addendum to the report.

This monitoring is the object of detailed recommendations (Section 4.5). The objectives assigned to monitoring are to check the validity of predictions of performance made during the design, and to ensure that the structure will continue to perform as required after completion (Section 4.5.1). Monitoring may be extended after the construction period (Section 4.5.6): The length of any post-construction monitoring period may be altered as a result of observations made during construction. For structures which may impact unfavorably on appreciable parts of the surrounding physical environment, or for which **failure may involve abnormal risks** to property or life, monitoring should be required for more than ten years after construction is complete, or throughout the life of the structure. The EUROCODE 7 also insists on the necessity to process monitoring results (Section 4.5.7): The results obtained from monitoring shall be always evaluated and interpreted and this shall normally be done in a quantitative manner.

1. General Information for the Target Deposits

1.1. The Mirovo Salt Deposit

The Mirovo salt deposit near the town of Provadia is the only terrestrial salt deposit in Bulgaria. The exploitation of the salt dome started in 1956 using a leaching method that extracts saline brine using a telescopic borehole system to the surface. There are 43 underground chambers with diameters varying between 100 m and 140 m and heights in the range 50–200 m. The indirect leaching method is designed and used such that water is injected in the annulus between the two production strings close to the roof of the cavern. In its horizontal and vertical flow the water dissolves NaCl from the roof and the walls of the cavern. A leaching technology in stages and layers is designed and introduced in production. The height of the developed stage varies from 5 m to10 m and the diameter is 100 m. In 1991 a cavern technology of leaching was introduced in a group of caverns with water being injected at a specified depth of the cavern (Valev et al., 2000). The salt is extracted by leaching (solution) from three depths, 700 m, 1000 m and 1200 m (Fig. 1A and Fig. 1B), using a telescopic borehole system circulating water at a well-head pressure of 50 bars. The roofs of the chambers are controlled by a floating oil layer.

The Provadia region is located in seismically quiet part of the country. According to the Code for Design and Construction in Seismic Regions (1987) the Provadia region falls to the zone with seismic coefficient Ks = 0.10 g. Since the beginning of the last century no strong or moderate seismicity have been reported till 1980; there is also no historical evidence of any strong earthquakes. During the last 15–20 years the population in the region has become increasingly alarmed by two problems, namely increased seismic activity and large surface subsidence, which influence the salt body,

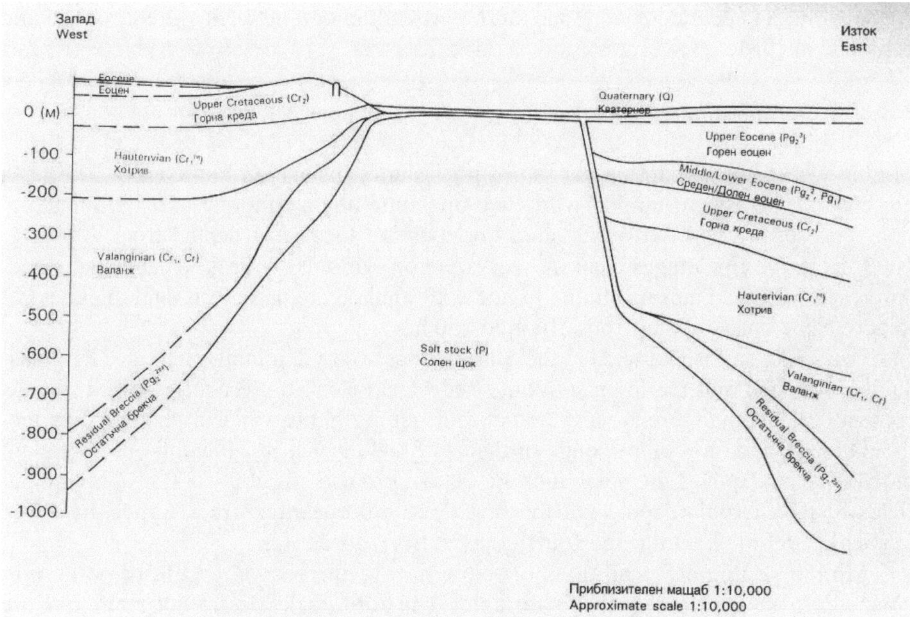

Figure 1A. Cross-section W-E of the Provadia salt deposit.

Figure 1B. Sketch of underground leaching technology.

the whole underground chamber pillar system and the salt extraction installation, and all the equipment in the target area, as well as in the neighbourhood villages. Since 1980 several moderate earthquakes with magnitude 4.0 to 4.5 were generated in this region and caused damage to the neighbouring villages. For the purpose of monitoring the seismicity, subsidence and strong ground motion in the region, a local seismological network, a geodetic network and strong ground motion network were designed and deployed in 1981.

1.2. The Starobin Potassium Deposit

Mining works at the Starobin deposit of potassium salts started in the early 1960s. At present four potassium mining works are operating and a fifth one is under construction. The second and third potassium horizons are located at depths from 400 m to 1000 m. In the first stages clearing works were performed with the chamber-method. At present various kinds of mining by the wall-method are used. The total thickness of impermeable layers ranges from 210 m to 250 m.

Potassium salt is extracted at the Starobin deposit by a mining method. The potassium Horizon II and the lower sylvinite bed of Horizon III are being worked out. At present the potassium ore is recovered at four mining fields which are bounded by tectonic blocks, and only in the southern flanks of the deposit, by pinching-out potassium horizons. The depth of the potassium ore extraction works is 400–1000 m. The mining fields of the Starobin deposits at the first three mining sites are stripped by three vertical boreholes, and at the fourth one by five boreholes.

Mining geological conditions of extraction of the second and third potassium horizons in general are quite favourable. The roof rocks are rather rigid and are prone to fracturing, and the soil is swelling. Water-impermeable strata over Horizon II consist of saliferous deposits with a thickness ranging from 0 m to 130 m and rocks of clayey-marlaceous strata (CMS) up to 100 m thick. The total thickness of these strata in the mining fields of the Industrial Amalgamation (IA) "Belaruskaly" ranges from 210 m to 250 m.

Various systems of mining development work are used at the deposit fields. Development drifts follow, as a rule, the potassium beds and the amount of mining from those drifts constitutes 15–20% of the bulk mining. The main type of mining equipment is a PK-8 with a rotor power unit of continuous operation performing the simultaneous destruction of the whole face area and arranging the extraction with an arched roof shape and a rectangular bed shape. The width and height of the single working are 3 m × 3 m. Mines are aerated using local pressure-type ventilators.

Extraction of minerals at the first stage of deposit-mining began with room-mining systems on rigid pillars. A few methods of room-mining were used was due to insufficient knowledge of mines' geological and hydrogeological conditions where there were a risk of possible water penetration into the mines. The main methods of this mining system are shown in Fig. 5. In general, more than 30 methods were tried out.

As a result of the experience gained and research conducted (All-Union Scientific Research Institute of Geology, JSV "BelGORKHIMPROM"), new versions of the room-method were developed, tried out and adopted. These were rooms with a capacity for roof lowering onto flexible pillars up to 1.5 m in width (Fig. 2A/B). When the lower part of clayey-marlaceous strata was determined to be impervious to water, a method of pillar-mining of sylvinite beds with complete roof-caving was adopted. This method suggests selective mining of sylvinite beds by two (upper and lower) extraction long walls in one extraction pillar. This mining method decreased losses connected with extraction and increased the production.

Under conditions of the potassium Horizon II, two-bed selective mining by four-miner long walls was used together with bulk mining. Two sylvinite beds were worked

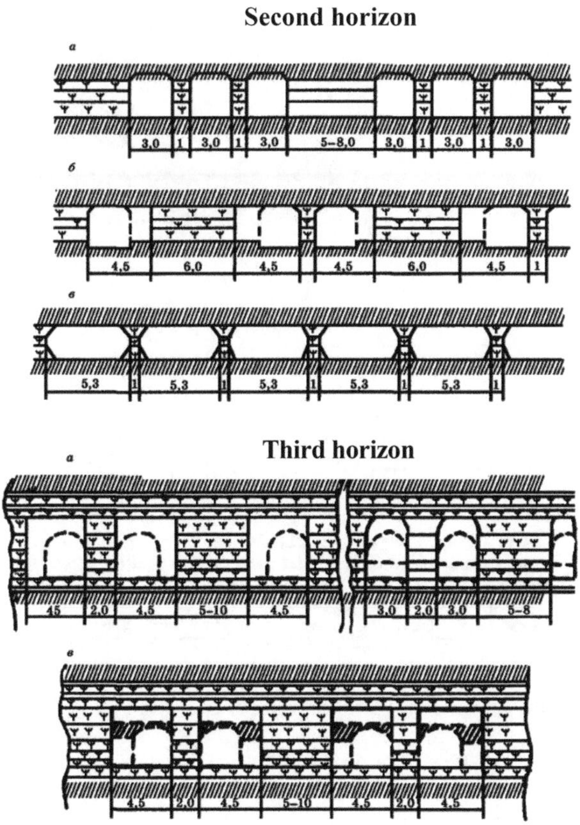

Figure 2A. Main methods of room-mining at the Starobin deposit (*Yermolenko, Bordon,* 1993).

out by two duplex 180–200 m long walls. The upper sylvinite bed is extracted 4.9–6.5 m in advance of the lower one.

The pillar system was used when the fourth sylvinite bed of Horizon III was extracted by the wall-method. The underlying second and third beds were mined by the room-system with miners "Ural-10KS". Such techniques (i.e. a combination of the pillar and room systems) were applied for the first time in the world and were named 'integrated system' (Fig. 2B).

The Horizon III mining practice was improved by two walls. The main distinction of this practice from that applied at Horizon II is that it consists of the separate development and mining of the upper (i.e. fourth) sylvinite bed from 1.0 m to 1.3 m in thickness, and the lower second and third beds their total thickness ranging from 1.9 m to 2.1 m. Development working of the lower wall is carried out under the fourth sylvinite bed extracted before. An intermediate rock salt layer (between beds 3 and 4) 1.1–1.3 m in thickness that is left in the worked-out space, serves as a protection from roof rocks caved as a result of an advance extraction of the fourth bed.

Prospect development of mining techniques at the Starobin deposit consists of a purposeful integration of new selective extraction processes, an increase of productivity of working, and drifting complexes.

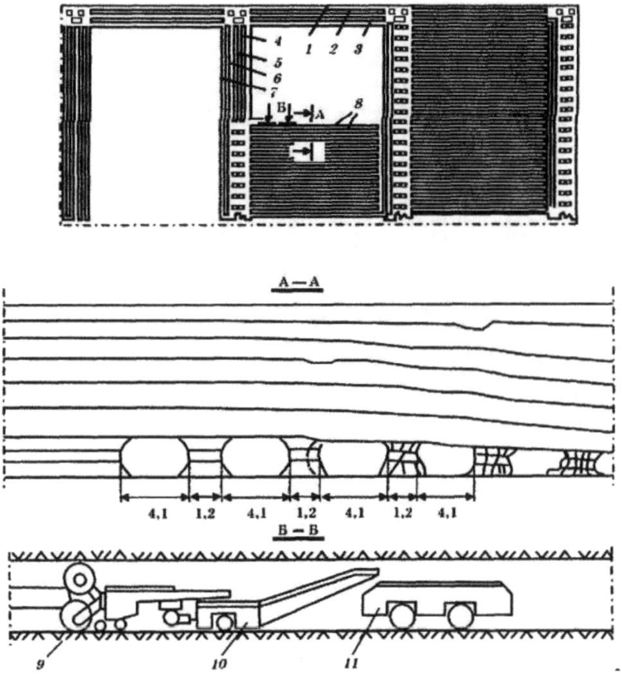

Figure 2B. Room-mining with lowering pillars and mineral extraction by a miner "Ural-10KS" *(Tomchin, Smychnik,* 1998):
Key: *1, 2, 3* – panel-spalling drifts (transport, belt and airway); *4* – heading; *5, 6, 7* – block drifts (belt, transport and airway); *8* – second working rooms; *9* – drifts miner; *10* – bunker-loader; *11* – self-propelled car.

An underground technological complex, which envisages commercial working of the rock salt bed under conditions of an operating potassium ore mine, was developed for the first time in Belarus at the Starobin deposit. The rock-salt bed underlying the potassium Horizon II was opened by inclined drifts in the mining field of the first mining site. A depth of mining work at the industrial testing site ranges from 420 m to 480 m (Fig. 2C).

Like in the case of the Provadia in the Starobin mining region, negative ecological consequences are due to considerable deformations of the ground over worked-out underground mines, vast areas occupied by wastes of potassium production, as well as phenomena of induced seismicity. Salt dust formed as a result of salt pouring off, and packaging, is a serious hazard to the environment.

2. Geological Settings and Existing Monitoring Systems in the Vicinity of the Provadia and Starobin Deposits

2.1. Provadia

Generally, the lithosphere of the Balkans shows a rather complex tectonic structure indicating a complicated stressed pattern and differential movement of the micro plates. The thickness of the crust (i.e. the Mohorovicic discontinuity depth) varies between

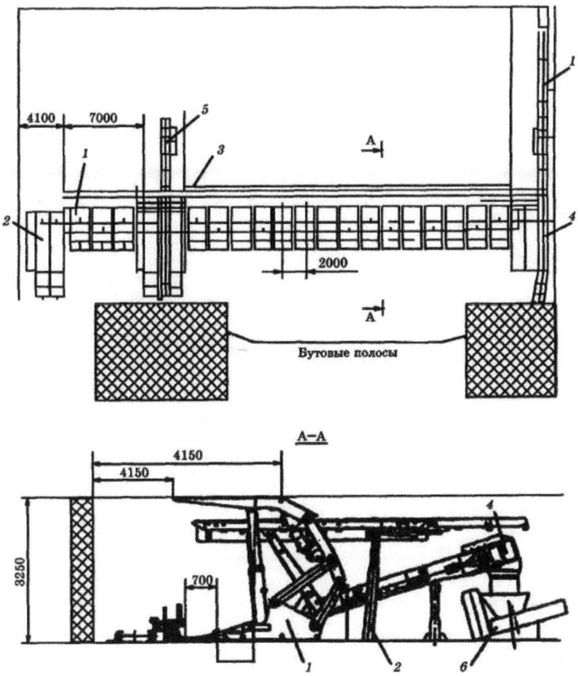

Figure 2C. Equipment for selective mining of the potassium bed -305 m:
Key: *1* – face support section; *2* – support section; *3* – long wall conveyor; *4* – gate conveyor; *5* – loading elevator; *6* – thrower; *7* – power train.

30 km and 36 km, with a maximum in the western and southern parts of the Moesian mega block (to which the salt deposit belongs), and a minimum in the depression and shelf section of the Black Sea. The salt deposit is the only terrestrial salt deposit in Bulgaria and is several kilometres across at its surface exposure, extending and broadening to depths of, perhaps, 3,500 m. It passes trough the whole sediment complex including Perm to upper Eocene and is covered by Quaternary deposits. Starting from 12–20 m under the surfaces, the salt, with a shape of a frustum of a cone, reaches depths of 3,500–4,000 m, where a salt layer is formed. The deposit is imbedded in Cretaceous limestones and dolomites, and Paleocene marlstones and it is covered by quaternary sediments. The salt body is in contact with the surrounding rocks by the so-called residual breccias. The rock salt massif, built up predominantly by halite, is rather non-homogeneous. For example, the compressive strength varies from 8.5 MPa to 30 MPa with average values Rc = 14 – 16 MPa (lab tests) and Rc mass = 5.5 – 5.8 MPa (massif).

The neotectonic and recent features of the Provadia region's formation are conditioned by its location. The recent vertical movements (from state geodetic network data) of this region are influenced more by Balkanidi development features than the Moesian platform ones. There is a positive correlation between the neotectonic and recent vertical movements, showing the continuation of the general trends of the tectonic regime; this occurred during the neotectonic stage of the development of the region in the recent epoch (Paskaleva et al., 1992 and 1994). Since 1981 two local observation systems have been built in the region of the salt deposit; one a geodetic and

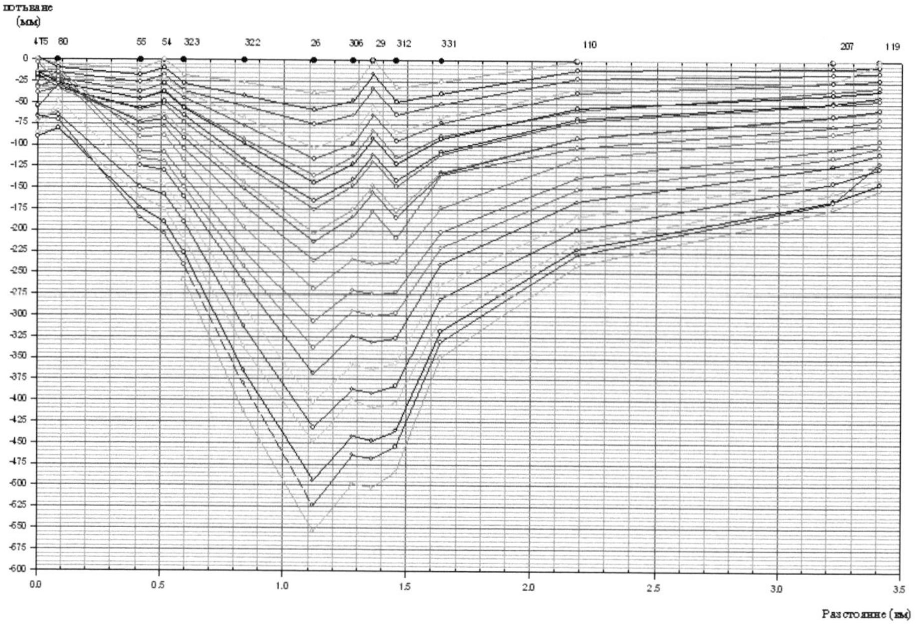

Figure 3A. Accumulated surface subsidence in [*mm*] along the profile crossing the salt deposit with a length of 3.5 km for the period 1989–2007 – Provadia salt deposit.

strong ground motion network, and later in 1994 a local seismological network for Provadia encompassing a larger tectonic region started operation.

2.1.1. Geodetic Monitoring

During the period 1981–2007 extensive geodetic measurements have been carried out to monitor the vertical and horizontal deformations assigned to the company Geoprecise Engineering Ltd. The main purpose of the network is to establish 3-D control, and to monitor shifts, deformations and the most significant site displacements. The aim is to examine to what extent the exploitation of rock-salt is connected to the local seismic activity because, recently, earthquakes have been occurring there more frequently and with increasing intensity. The network covers the area of the wells and outside of them.

Measuring the vertical deformations and subsidence involved a geodetic network of 26 stable pillars and 200 levelling benchmarks situated on two almost perpendicular lines crossing each other in the centre of the salt mirror profiles. The observations show that there are movements along some faults in the salt deposit region. The central part of the deposit goes down and the velocity decreases with distance from the centre. This means that the movements along the faults limit the deposit continuity. The geodetic observations have been processed on both large and close grid networks. Accumulated surface subsidence in [*mm*] along the profile crossing the salt deposit, with a length of 3.5 km for the period 1989–2007, is shown in Fig. 3A. For the same time period the 3-D view of the accumulated surface subsidence in [*mm*] is given in Fig. 3B. A constant subsidence with average velocity 3–4.5 cm per year and horizontal block movements have been observed in this region for the recent few years. The largest subsidence is observed in the central part where the top of salt body is located.

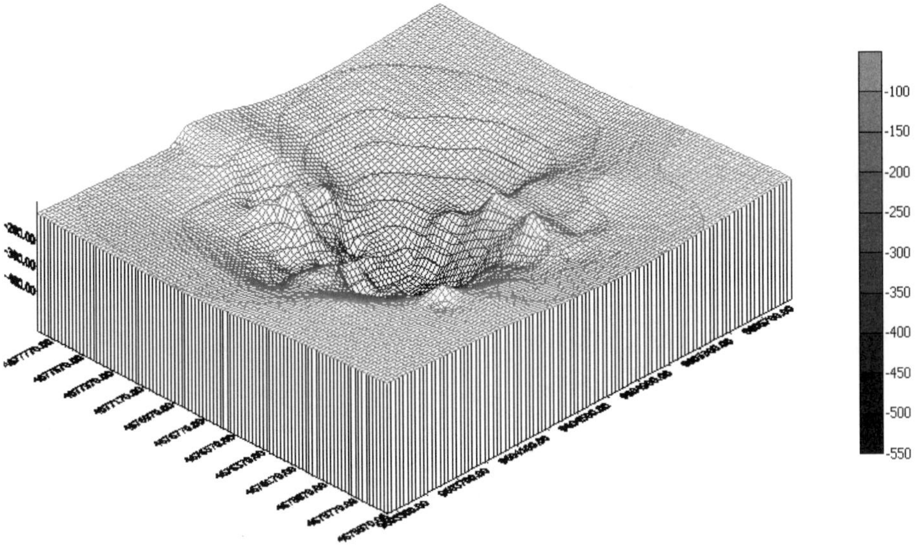

Figure 3B. 3-D view of the accumulated surface subsidence in [*mm*] for the period 1989–2007 – Provadia salt deposit.

2.1.2. Seismic Monitoring

In the Provadia deposit area, situated in the north-eastern part of Bulgaria, seismic observations, performed by a local network of five stations (LSN), have been carried out since 1994. The Provadia seismological station was established in 1990 as a part of the National Seismological Network (Fig. 4A) and now it is equipped with broad-band sensor KS2000 (Nikolova et al., 2007) with a frequency range 120 sec – 50 Hz, gain 2000 and 3 channel Reftek 130-01. However, a single very well equipped station is not sufficient to enhance significantly the accuracy of earthquake locations representing the low magnitude local seismicity. Therefore, it has been necessary to upgrade the LSN in the region, and recently four new stations were installed, located at distances of 5 km to 35 km from Provadia.

The seismicity record of the period 1900 and 1970 (Rangelov, 1994) has shown that the Provadia region can be considered to belong to a zone with moderate seismicity that is controlled mainly by the Devnya fault to the north of Provadia. The region is characterized by compression stresses on a NE-SW axis which is in agreement with the general situation in northern Bulgaria (Knoll et al., 1995; Karaguleva, 1974). According to the potential seismic source map, the maximum expected magnitude at this site is M = 5.6–6.0 corresponding to focal depths 5–10 km. Within the time, since 1900, there were only a few events felt near the Mirovo salt deposit (in 1901, with M = 3.8 and epicentral intensity I_0 = V MSK – 64; 1901, M = 3.6, I_0 = V; 1902, M = 3.6, I_0 = V; 1903, M = 2.6, I_0 = III). Up to 1964 there were no other data for the seismicity of the Mirovo district. Generally the seismic regime of the whole Bulgarian territory is characterized by a recurrence rate with relatively low slope, which is 0.36 for the period 1900–1930 and 0.34 for the period 1931–1970.

An epicentral map of the seismic events in the Provadia region is given in Fig. 4B, where the locations of the operating local seismic stations are also shown. The distribution of the epicentres suggests that they are tending to a linear arrangement, SW-NE,

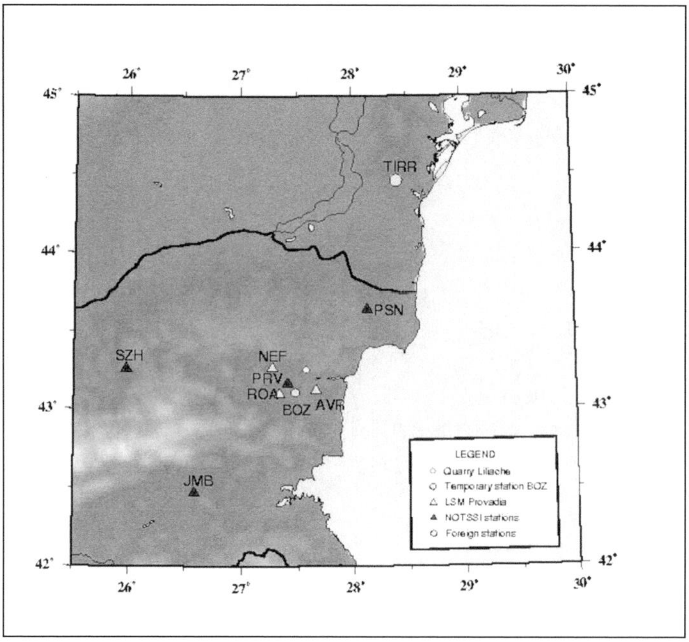

Figure 4A. Seismic network in the region of Provadia – *Nikolova, S. (2007).*

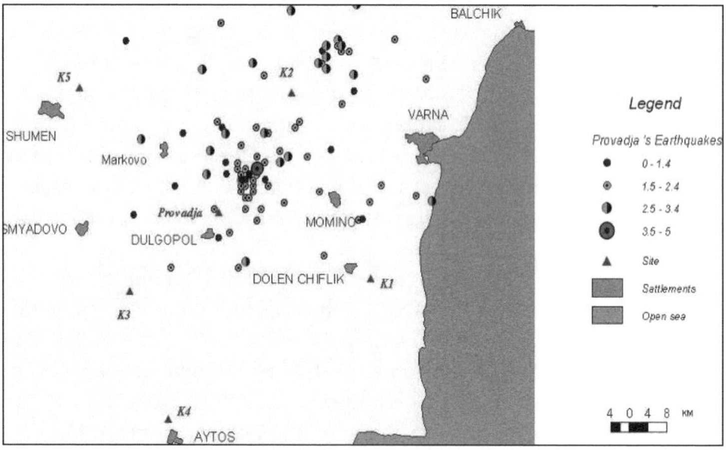

Figure 4B. Map of epicentres of seismic events in the Provadia region.

Key: Circle with central dot shows the strongest earthquake of this region with M = 4.4 in December 18, 2003.

oriented along the profile joining the settlements of Dulgopol and Dobrich (outside the map fragment). Most of the epicentres are confined to a 20 km zone across the salt deposit.

Since 1981 CLSMEE runs a strong ground motion network with five SMA-1 instruments (i.e. analogue registration stations equipped by accelerometers SMA-1) with maximum ranges 0.5 g and 1 g, a sensitivity ± 0,0050 g and damping of about 60%.

Later, in 1993, a digital RefTek instrument was installed and calibrated by the GTU Ingenieur Buro Knoll specialists. About 81 events within an epicentral distance up to 27 km occurred in the last 20 years [Paskaleva et al., 1992a, b, 2007; Nenov et al., 1990]. More than 200-strong ground motion components registered by the SMA-1 instruments were processed. All the records are with relatively short duration (0.12–2.97 sec.) and have been "saturated" with high frequency vibrations. According to the response spectra SA (T) for the accelerations and 5% damping, the maximum periods (T) have been in the range 0.085–0.2 sec. for the vertical and 0.1–0.57 sec. for the horizontal components. The dynamic effect SA (T)/Amax of the single events, obtained from the response spectra with 5% damping, varies from 1.2 to 6.0 for the horizontal components and reaches 7.0 for the vertical components. The ratio between the peak accelerations, vertical and horizontal, is 0.17 to 2.26. This ratio shows the predominant influence of the vertical component and confirms the local origin of the earthquakes. The duration of the intense part of the accelerograms is about 3 seconds. Such a short duration means that these events act as single short-time impulse load on the chamber-pillar system. The fact, that the peak vertical accelerations are often larger, compared to the horizontal ones (i.e. 50% of the registrations) has to be considered, when dynamic analysis of the stress-strain state of the system is performed, since there is a possibility for pillar failure due to vertical crack occurrence. The available records can be efficiently used for the vulnerability analysis of the structures in the region of Provadia, and for pillar capacity re-estimations.

2.2. Starobin

The Starobin deposit of potassium salts is situated in the north-western centroclinal part of the Pripyat Trough, which is a sub-latitudinally striking palaeorift. The Saliferous strata have interbedded members of potassium salt and carbonate-clayey rocks, as well as of sandstone and siltstone layers. The deposit territory is of clearly defined block structure. It is bounded by the sub-regional Liakhovichi and Glusk faults with amplitudes of 150–350 m on the north, and by a set of faults forming the Southern tectonic zone on the south. The Central, North-Western, Northern (Guliayevo) faults are running immediately within the deposit and divide the territory into the eastern, central, western and north-eastern block. Four potassium horizons are found within the saliferous strata; Horizons II and III are mined. The second potassium horizons occur in the depth range 250–700 m and the third potassium horizons range from 350 m to 1000 m.

In the region of the Starobin deposit the crystalline basement occurs at depth of 1700–2100 m. Upper-Proterozoic formations overlie it conformably and are represented by Riphean and Vendian complexes. The Paleozoic erathem involves the middle and upper divisions of the Devonian. Saliferous strata with abundant potassium horizons are of Lebedian-Streshin age (D^2_3 lb-str). A thickness of sub salt clayey-marlaceous deposits varies between 230 m and 560 m. The Mesozoic erathem comprises deposits of the Jurassic and Upper Cretaceous and the Palaeogene is represented mainly by sandstones. The Neogene is restricted in area. Quaternary deposits are 20 m to 150 m thick.

Some active faults were identified from a series of aerospace surveys and geological and geophysical data obtained in the region of the Starobin deposit. Systems of lineaments distinguished show correlation with the pre-platform and platform faults, as well as with dislocations of a disjunctive nature. The Stockhodsk-Mogilev super regional fault belongs to the pre-platform age structures, and the Liakhovichi, Chervonaya Sloboda, Rechitsa, Glusk and Mikashevichi faults are structures originated at the platform stage of evolution.

When potassium ores were mined in the Starobin deposit region a number of geological features was revealed, such as:

1. rock fissuring;
2. zones of rupture dislocations;
3. zones where sylvinite was replaced by potassium salt in productive beds;
4. subsidence troughs and their associated gas-dynamics phenomena; and
5. squeezing-out brine inflows into mine works (Vysotsky et al., 2003).

As regards seismic processes, their effects on mining works were not appreciable until the present.

2.2.1. Seismic Monitoring

According to a division of the East European Platform-west into seismotectonic regions, the territory of the Starobin deposit of potassium salts is related to the Pripyat potentially seismic super zone with a magnitude M = 4.0 and a focus depth H = 5km (Garetsky et al., 1997).

As for induced earthquakes, the first of these was recorded in 1978 ($K = 9$ $I_0 = V$) by the seismic station "Minsk" located a distance of 170 km away. Continuous instrumental seismic observations in the deposit region were carried out in 1983 by equipment with short period seismographs. Operating frequency bands were 1–10 Hz with analog recording. Besides, within 1983–2000 observations were carried out by self-contained seismic instruments with operating frequency bands of 0.5–60 Hz and duration of independent work of 20 hours, and the information was recorded on magnetic tape. About 1000 seismic events were recorded in the region within this period (Aronov et al., 2003).

A seismologic telemetric complex was installed in 2004 and the information was transmitted into the computer. At the first stage the complex was composed of four observation stations but a total of six observation stations were envisaged. Each station was instrumented with three-component short-period seismic detectors with capacitance-type transducer and magnetic-electrical feedback. The operating frequency band was 1–70 Hz. The information was continuously transmitted to a computer in the real-time network and then to a computer when it was accumulated, processed and stored. Observation stations were located both in mines, and on the ground surface. A dynamic range was at least 120 dB. A reception range was at least 30 km.

A three-level database was the result of long-term seismic monitoring based on an automated telemetric complex:

- level 1 contains general geological and geophysical information on the territory under monitoring, specific data on the seismograph network, tectonic blocks, velocity models, seismic wave travel time curves, etc.;
- level 2 contains digital seismograms (arrival times and amplitudes of seismic phases, major wave groups with maximum amplitudes, etc.) of recorded seismic events; and
- level 3 contains the main results of interpretation, i.e. space and time, energy; parameters of foci of seismic events as well as some other parameters.

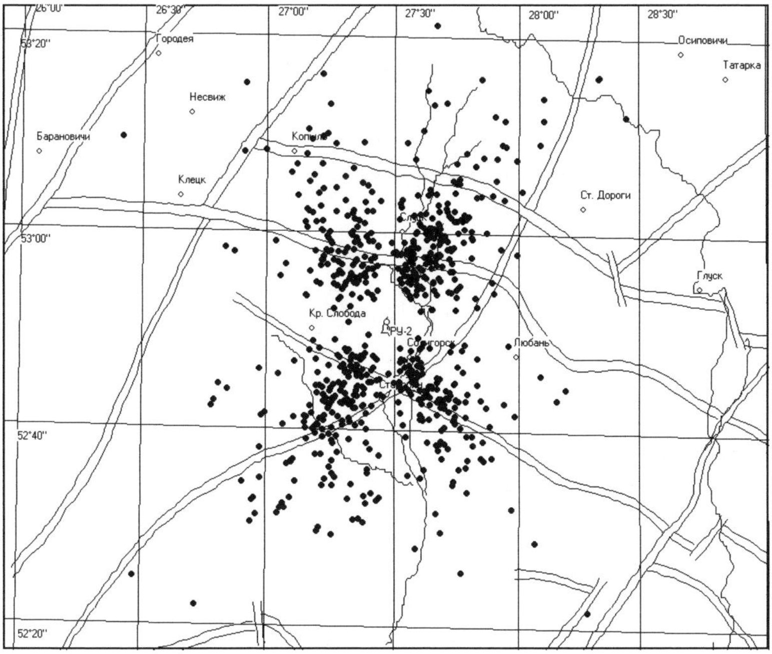

Figure 5. Map of epicentres of seismic events in the Soligorsk region.

The data are considered to be most important for studying geodynamics of the Soligorsk industrial region.

The map of epicentres of the Soligorsk seismic events recorded in the period 1983 till 2004 is presented in Fig. 4. Over this period more than 1000 seismic events have been registered in the area monitored, four of which produced a tangible effect, namely: 10 May 1978; 2 December 1983; 17 October 1985; and 15 March 1998. The energetic class of those seismic events which is connected to the magnitude correlating $K = 1.8M + 4$ is located within the diapason of 8.0–9.5 class (level). The intensity of soil shaking rose by up to 4–5 scores. All the earthquakes were accompanied by macro-feelings: rumble, window glass rattling, swaying of hanging objects, furniture and floor creaking on the ground floors of wooden structures. Scattered plaster cracks were also observed and during the earthquakes in 1978 and 1998 roof collapses took place.

The map (Fig. 5) also shows rupture dislocations active at the present-day. The strongest seismic events recorded recently in the Soligorsk region are confined to zones of tectonic disturbances which are still active at the present. These are a set of the North-Pripyat super regional faults, the Chervonaya Sloboda and Stokhodsk-Mogilev fault systems. This is also evidenced by the prevailing fracturing orientation measured immediately in mines and by the regional stress system of the East European Platform-west.

At present, continuous seismic observations at the Soligorsk geodynamic testing ground are carried out with an automated telemetric seismic complex intended for prompt monitoring of space and time distribution of seismicity and assessment of the geodynamic environment by outlining seismically active areas (tectonic blocks).

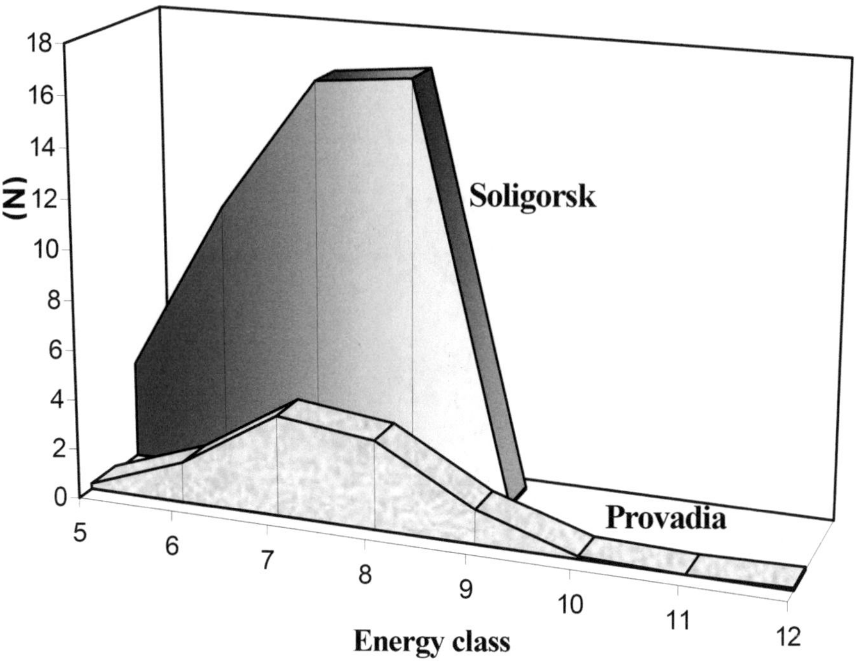

Figure 6A. The curve of recurrence for the Soligorsk and Provadia regions.

3. Comparison of the Results from Target Deposits

Figure 6A represents a recurrence curve, computed using the method of intervals for both regions. The average annual value of the number of seismic events is plotted as ordinates. This curve shows that in the region of Soligorsk the seismic activity is higher for the Energy Class range 5–9 and the curve's shape corresponding to this range is almost uniform. It should be noted that this range of energy corresponds to rock – tectonic shocks according to the Malovichko's classification (Malovichko et al., 2000). A quasi-similarity of the curves suggests also the similar origin of seismic events.

The range of technogenic earthquakes (Fig. 5) is more abundant compared with background seismicity. One earthquake, which took place between the settlements of Markovo and Momino in December 18, 2003, had a magnitude of 4.4 ($K \approx 12$). The time distribution of the seismicity in Provadia and Soligorsk regions is given in Fig. 6B, while Fig. 6C shows the total annual values of the seismic energy (E) for these regions.

A quasi-periodic character of a number of seismic events with the general increasing trend is characteristic of the Provadia and Soligorsk regions. The aggregated annual energy value is in general agreement with the number of events. This regularity has been slightly disturbed after 1996 in the Soligorsk region, when the number of events increased and the summed annual energy has been tending to decrease. In the Soligorsk region the events of the less energy class have been dominant, suggesting a technogenic effect. Besides, it might be supposed that the seismicity in the Soligorsk has been influenced (though to a lesser extent) by the other factors, e.g. lunar – solar tides

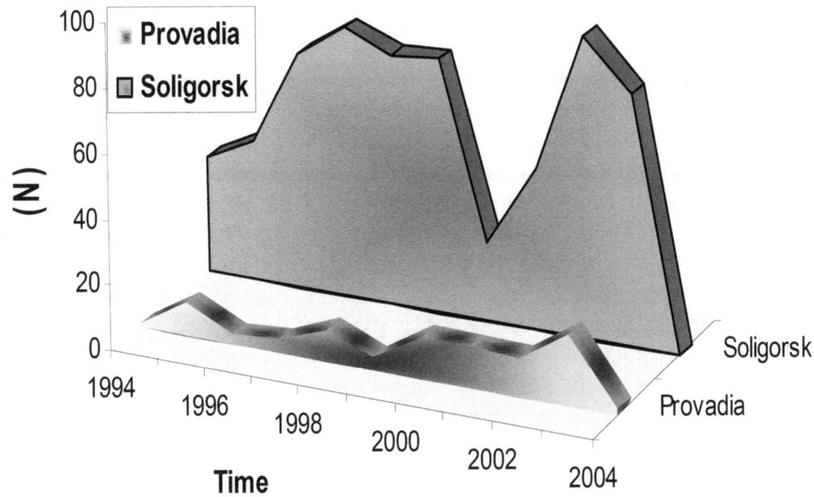

Figure 6B. Annual number of seismic events (N) for the Provadia and Soligorsk.

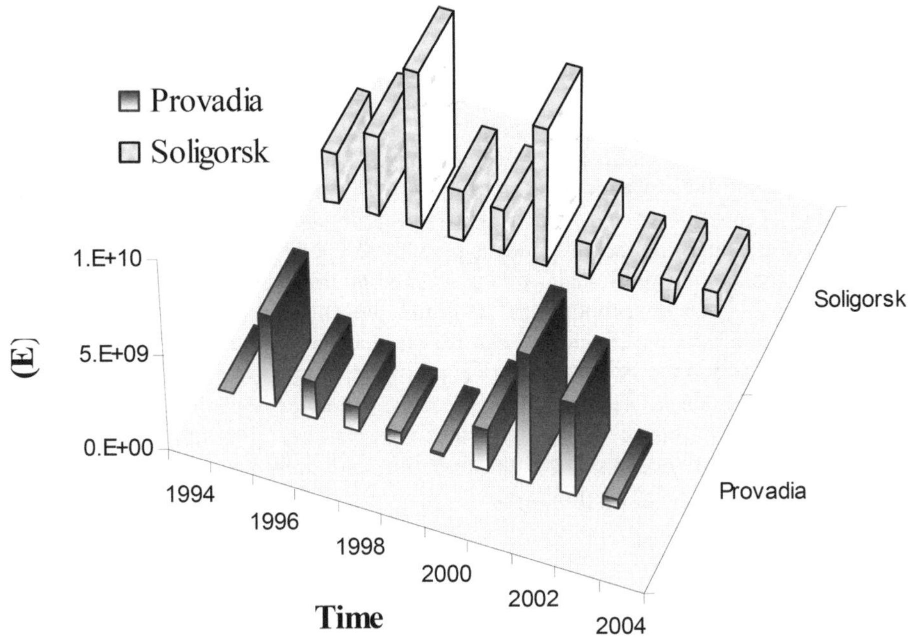

Figure 6C. Annual values of seismic energy (E) in [Joules] for the Soligorsk and Provadia regions.

(Seroglazov, 2003). In the Provadia region the tectonic processes seem to play the more important role, which is confirmed by the stronger seismic events that occur there. This is due to the fact that the Provadia region is situated in the seismically active zone.

However, in the Provadia region the seismic events with Energy Class 9 and higher, i.e. in the energy range related to technogenic earthquakes is more abundant.

4. Conclusions

The exact place and time of a disaster related to subsidence cannot usually be predicted with any degree of certainty. This is true for both slow subsidence related to fluid withdrawal and sudden subsidence related to sinkhole formation or mine collapse. Mitigation is the best approach to these hazards, and areas susceptible to such problems should be well known but it is not yet the case. In the areas located above known mining operations or former mining operations, a broad data base of information is needed to construct the risk maps. They can be used as a guide for land use and planning, or for controlling the mining activities. This is more important in areas of dense populated regions, and therefore new and cost-effective methods are necessary to cover the needs of permanent monitoring and for the regular update of the risk maps.

Seismic processes in the regions of the potassium salt deposits of Provadia and Soligorsk have shown the following characteristic features: a) the identity of the recurrence curves of the seismic events of the Energy Class range 4–8; b) a quasi-periodic character of the seismicity activation in time against the general trend of seismicity activation increasing; c) the zones of epicentres of seismic events are larger that mining areas.

There are some differences in the seismic process patterns: a) seismic activity in the range of small energies ($K = 4 \div 8$) is higher in the Soligorsk region; and b) events of the higher Energy Class $K > 9$ are characteristic of the Provadia region.

In both deposits the epicentral areas of seismic events overstep the limits of mine workings. This is typical of the stimulated seismicity zones.

The monitoring of the potassium salt deposits of Provadia and Soligorsk is of high scientific and economic interest. The accumulated seismic and technological data will allow to the scientific and engineering community to study more effectively the problems of the seismic hazard and risk assessment, earthquake preparedness and accident prevention in the target regions. By means of such observations it would be possible to correlate the generation mechanism of the seismic events with the local and regional tectonics as well as with technological activities within the mine. Seismic investigations, as well as geodetic measurements at the surface, can be used to calibrate the numerical models and to verify the obtained results. Relying on the numerical modelling, it is possible to obtain information about both the global deformation with respect to the geological conditions, tectonic stress state, seismic loading and mining processes, and the radius of the mining-induced deformation and stress redistribution regarding the development of induced seismic events.

5. Acknowledgements

This work is partly sponsored by the bilateral project "An investigation into the character of the induced seismicity manifestations and the elaboration of methods for the seismic hazard assessment (case studies Starobin deposit in Belarus and Provadja in Bulgaria)" 2004, PROMIRA Project Bulgarian National Science Fund Ministry of Education and Science – "Environmental Monitoring Implement for Risk Assessment of natural and man-made hazard)" and NATO SfP-980468 "Harmonization of the Seismic Hazard and Risk Reduction in Countries (Moldova, Romania, NE Bulgaria) Influenced by Vrancea Earthquakes".

References

[1] Aronov A., Seroglazov R., Aronova T., Paskaleva I., (2005). Typical features of induced seismic processes in mineral mining regions (illustrated with the potassium salt deposit in Belarus and Bulgaria). Problems of water resources, geothermics and geoecology. Proceedings of the International Sci. Conf. dedicated to the 100th anniversary of the birth of academician, Professor G. V. Bogomolov. Minsk, 1st–3rd June, 2005. In two volumes. Volume II. Geoecology, Geology and Hydrogeology of potassium salt deposits, oil – gas occurrence and oil – gas Hydrogeology. General geological problems. – Minsk: IG & G NAS of Belarus, 2005, 141 – 143 (in Russian).

[2] Aronov A.G., Seroglazov R.R., Aronova T.I., (2003). Belarus. The earthquakes of North Eurasia in 1997, Obninsk, 172-180 (in Russian).

[3] *Bulgarian Code for Design and Construction in Seismic Regions (1987).*

[4] Garetsky R.G., Aizberg R.Ye., Aronov A.G., Karabanov A.K., Safronov O.N., (1997). General seismic division of the Belarusian-Baltic region. Dokladi Akad. Nauk Belarusi. Vol. 41, N4, pp. 98-102 (in Russian).

[5] *CEN Committee TC250/SC7 – 'Geotechnical Design' EUROCODE 7 provisions.*

[6] Karaguleva J., (1974). Tectonic map. Proc. of the Sem. on the Seismotectonic map of the Balkan Region, 69-72, Dubrovnik, Skopje – UNESCO.

[7] Knoll P., Schreiber B., Kowalle G., Rother Kl., Paskaleva I., Kouteva M., (1995). Analysis of the Dynamic Stability of a System of Caverns in the Salt Diapir of Providian/Bulgaria, Proc. of the 8th Int.Congress on Rock Mechanics, September 25-30, 1995 Tokyo, Japan.

[8] Malovichko A.A., Zavialov A.D., Kozyrev A.A., (2000). Rock shocks. In: Natural hazards of Russia. Vol. 2. Seismic hazards. Moscow. Kruk Publ., 243-293 (in Russian).

[9] McGarr, A., Simpson, D., Seeber L., (2002). Case Histories of Induced and Triggered Seismicity in International hand book of engineering seismology, Vol. 81A, 647-661.

[10] McGarr, A., Simpson, D., (1997) in Rockbursts and Seismicity in Mines, 385-36, Balkema.

[11] Nenov, D., Georgiev, G., Paskaleva, I., Lee, V., Trifunac, M., (1990). Strong ground motion data in EQINFOS: Accelerograms recorded in Bulgaria between 1981–1987, Rep. No 90-02, Univ. of South. California 1990.

[12] Nikolayev A.V., (1994). Problems of induced seismicity. In: Induced seismicity. Moscow, Nauka Publ., 5-15 (in Russian).

[13] Nikolova, S., (2007). Progress Report on "*Environmental Monitoring Implement for Risk Assessment of natural and man-made hazard* (PROMIRA Project)", Fond materials of GFI-BAS and MON-Sofia (In Bulgarian).

[14] Paskaleva, I., Manev, G., Kouteva, M., (1992a). Induced Seismicity at Mirovo salt Deposit, Bulgaria, Acta Montana 1992, Ser.A, No 2(888), 193-200, Praha 1992.

[15] Paskaleva I., Manev G., Kouteva M., (1992b). Assessment of the situation at the Mirovo Salt Deposit and some measures for longterm safe exploitation, Journal of Mining and Geology, vol. 1-2, 1992 (in Bulgarian).

[16] Paskaleva, I., (2007). Progress Report on "*Environmental Monitoring Implement for Risk Assessment of natural and man-made hazard* (PROMIRA Project)", Fond materials of CLSMEE-BAS and MON-Sofia (In Bulgarian).

[17] Paskaleva I., Ranguelov B., Knoll P., Kowalle G., Kouteva M., (1994). Investigations related to tectonic and possible induced seismicity of a salt mine district in Bulgaria. Proceedings and Activity reports 1992–1994, Vol. 3. XXIV General Assembly ESC, Athens 1994.

[18] Ponomarev V.S., Romashev A.N., Turuntayev S.V., (1994). Regular features in disturbance of high energy media in induced seismicity manifestations. In: Induced seismicity. Moscow, Nauka Publ., 1994, 73-91 (in Russian).

[19] Ranguelov B., Paskaleva I., Kouteva M., (1994). First results of Tectonic-induced Seismicity Interaction on the Example of Mirovo Salt Deposit, Bulgaria, Comptes rendus de l'Academie bulgare des Sciences, Tome 47, No 7, 43-46.

[20] *Reports on Complex Investigations of the Geodinamical Processes, Deformations on the Surface and Local Earthquakes around Mirovo Salt Deposit, led by Central Lab. of Seismic Mechanics and Earthq. Eng. Sofia, GEOSOL, Provadia.*

[21] *Reports of the GEOPRECISE ENGINEERING Ltd. On Geodynamical Investigations of Mirovo Rock Salt Deposit by Geodetic Methods. Annual Reports, 1990–2007. GEOSOL Ltd. Provadia.*

[22] *Report 1, 2 of the INGENIEURBURO KNOLL. (1994–1995) Rock mechanical and seismological investigations in the area of the MIROVO Salt deposit, Bulgaria.*

[23] *Report the potassic salt deposits of Belarus: geology and mineral resources rational use // the Belarusian scientific research and designing Institute of mining and chemical industry, the Belarusian state university, Minsk, 2003, 263 p.*

[24] Seroglazov R.R., (2003). Earth tides as an initiating factor of seismicity in the Soligorsk seismoactive zone. Lithosphere, 2003, N2 (19). Minsk, 85-94.
[25] *Skip, B. (1995–1996), Soil Mechamics Limitited, Reports on the Geomechanical stability of the mining system Mirovo salt deposit, Provadia, Bulgaria Interm reports, GEOSOL LIMITED, Provadia.*
[26] Tomchin L., Smychnik A., (1998). Experience of development of the Starobin deposit of potassium salt // Mining magazine. – 1998. -№ 11-12, 79-84.
[27] Valev, G., Janev, G., Rajnov, G., (2000). Geodynamic research of the Mirovo salt deposit near Provadia, NE Bulgaria, Special symposium on sinkholes and unusual subsidence over solution-mined caverns and salt and potash mines, 15-18 Octob., 2000, San Antonio, Texas, USA.
[28] Vysotsky E.A., Gubin V.N., Smychnik A.D., Shemet S.D., Yashin I.A., (2003). Potassium salt deposits of Belarus: geology and efficient management of mineral resources. Minsk, 263p (in Russian).
[29] Yermolenko V.A., Bordon V.Ye., (1993). Belarusian agricultural ore: geology, economy, ecology. – Minsk: The Institute of geology, geochemistry and geophysics, 183 p.

Optimisation of Disaster Forecasting and Prevention Measures in the Context of Human and Social Dynamics
I. Apostol et al. (Eds.)
IOS Press, 2009.
doi:10.3233/978-1-58603-948-6-163

Mining Dump Rehabilitation: the Potential Role of Bigeminate-legged Millipeds (Diplopoda) and Artificial Mixed-Soil Habitats

SHORT TITLE: A STUDY OF THE ECOLOGICAL INTER-RELATIONSHIPS OF BIGEMINATE-LEGGED MILLIPEDS (*DIPLOPODA*) AND ARTIFICIAL MIXED SOILS

O. Pakhomov, Y. Kulbachko, O. Didur, I. Loza

Dnepropetrovsk National University, Gagarina av., 72, 49010, Dnepropetrovsk, Ukraine

Abstract. The paper is devoted to research on the interaction of artificial soil blends which may be used for soil rehabilitation of mining dumps and soil saprophages (*Diplopoda*). The mutual influence of mining spoil/rock, chernozem soil and litter of *Robinia pseudoacacia* L. on *Diplopoda* biomass and carbon dioxide emission of artificial soil blends has been studied. Mathematical models that describe the dependence of vegetable-eating milliped's biomass and carbon dioxide emission changing upon experimental factors (blends composition) are presented. A significant decrease of body mass of the millipeds inhabiting mining gobs, and an increase in response to the addition of *Robinia pseudoacacia* L. leaf litter to the soil blend, were found. Experiments testified that biological activity of soil blends increases in the presence of invertebrate saprophages. The experimental results show that this will help enhance the quality of the overall soil media and so reduce the potential polluting effects from surface run-off from mining spoil and reduce erosion risks.

Keywords: coal mining spoil; artificial soils; remediation; vegetable-eating millipeds; change in biomass; carbon dioxide emission

Introduction

One of the types of re-cultivation in landscapes with completely destroyed biota is the creation of artificial wood biogeocenosis. These damaged situations are observed in such regions of Ukraine as Western Donbass, and the coal basins of Alexandria, Krivbass, for example. The full re-cultivation in such conditions of the technogenic landscape, in particular, in the mining areas of Western Donbass, consists of making artificial edaphotopes with their optimal water-physical and agrochemical properties. At the biological stage of re-cultivation there are created preconditions for the regeneration of the autotrophic (green) and the heterotrophic part of biogeocenose (animals and edaphic microbiocenosis)[1, 2, 5–9, 11, 13, 16–20].

In natural conditions it is not always possible in the short term to estimate the advantages or disadvantages of those or others of artificial mixed soils used for the re-cultivation of soils impaired by the mining industry, and being the environment for representatives of phyto- and zoo- , microbiocenosis. The detection of the positive influence on the formation of steady culture-biogeocenosis is possible when carrying out model experiments.

1. Material and Research Methods

The subjects of the research were vegetable-eating millipeds (*Diplopoda*) in the context of the interrelationship of the activities of edaphic saprobes with their habitat formed by artificial mixed soils. The purpose of the research was to define the influence of primary incinerators of vegetation (tree waste) on the biological potency of the system, as well as to assess the production of CO_2 by a poly-component mixture with the participation of animal saprobes.

The following factors have been used in the experiment:

- mine spoils (x_1) with $pH_{aquas} = 3,5$;
- the edaphic mass of ordinary chernozem (x_2) as the substrate ($pH_{aqua} = 7,0$) containing humus; and
- a litter of Robinia leaves (x_3) with $pH_{aqua\,(1:10)} = 6,95$.

An investigation was made of such characteristics as the changes in biomass of vegetable-eating millipeds (reflecting the influence of factors on animals) in the mixed soil with a different composition, and the stream intensity of CO_2 evolved by the triple mixture of substrates (reflecting the influence of animals on mixed soil) (Fig. 1). The number of animals in all the tests was the same (i.e. nine specimens) and measurements were made of the full weight of bodies for the nine animals before and after the experimental treatment.

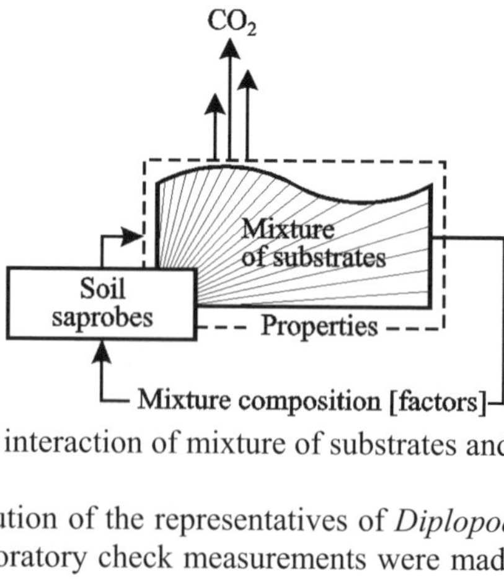

Figure 1. A model of interaction of mixture of substrates and soil saprobes

To define the contribution of the representatives of *Diplopoda* in the evolution by artificial soils of CO_2, laboratory check measurements were made with two variations:

trials without the participation of saprophages (i.e. the control), and trials with the participation of saprobes (test).

The concentrations of CO_2 (g/hour·m^2) during the aeration of mixed soils[15] were defined in a titrimetric way according to the law of multiple proportions[3, 4, 14]. In order to create in the tests the optimal hydrothermal conditions for the animal habitation, the temperature in the laboratory was maintained within the limits of 20-22°C, and a precipitation event was simulated at 35 mm per 30 days using distilled water. In each point of the plan there were two parallel tests carried out. The duration of the experiments was 30 days (from 10.10.2007 to 10.11.2007) and the experiment's plan is given in Table 1.

TABLE 1. Matrix of simplex-planning and designation of responses

Number of test	Mixture composition (unit fraction)			Dependent variable
	Mine spoil (x_1)	Soil mass of ordinary chernozem (x_2)	Litter of *Robinia* leaves (x_3)	
1	1	0	0	y_1
2	0	1	0	y_2
3	0	0	1	y_3
4	0.5	0.5	0	y_{12}
5	0.5	0	0.5	y_{13}
6	0	0.5	0.5	y_{23}
7	0.333	0.333	0.333	y_{123} (Checking point)
8	0.15	0.595	0.255	Checking point
9	0.3	0.49	0.21	Checking point

Variables x_i (i = 1, 2, ... q) given in the planning matrix are proportions (i.e. abundance ratio contents) of mixture i-components and meet the condition:

$$\sum_{1 \le i \le q} x_i = 1 \quad (x_i \ge 0).$$

A certain composition mixture corresponds to each point of the plan. To check the adequacy of the resultant model, tests were performed at three checking points (tests number 7, 8, and 9).

2. Results and their Discussion

2.1. A Change in Biomass of Diplopoda Representatives

The results of measurements of the biomass of animals are given in Table 2. In the right-hand column are the indicated average values of a mass increase of the animal which reflect the difference between the animal's mass before and after the experiment.

TABLE 2. Results of planning (a change in biomass of a saprobe-animal)

Number of test	Mixture composition (unit fraction)			Response
	Mine spoil (x_1)	Soil mass of ordinary chernozem (x_2)	Litter of *Robinia* leaves (x_3)	Average rise of mass per month, g ($n = 9$ specimens) (y)
1	1	0	0	–0.48
2	0	1	0	0.04
3	0	0	1	1.04
4	0.5	0.5	0	–0.04
5	0.5	0	0.,5	–0.12
6	0	0.5	0.5	0.94
7	0.333	0.333	0.333	0.14
8	0.15	0.595	0.255	0.36
9	0.3	0.49	0.21	–0.04

After the realization of the planning matrix the following regression equation resulted:

$$y = -0.425x_1 + 0.085x_2 + 1.04x_3 - 1.66x_1x_3$$

(R^2 considering degrees of freedom = 93.2 %),

where:

y is a mass increase of vegetable-eating millipeds;
x_1 is the relative contents of mine spoil in a mixture;
x_2 is the relative contents of chernozem in a component mixture; and
x_3 is the relative contents of leaf tree waste in the ternary, or triple, system.

Coefficients for effects of factors, and their interactions, are given in the above regression equation with $\alpha \leq 0.05$. A check of the model has shown that it is adequate (i.e. $\alpha = 0.004$).

It has been shown that while forcibly inhabiting the mine spoil, the mass of animals decreases ($-0.425x_1$) but if they inhabit a substrate such as chernozem, then a small positive increment of mass ($+0.085x_2$) is observed. The largest increase in the mass of a body is characteristic for the animals inhabiting on the litter of *Robinia pseudoacacia* L. leaves ($+1.04x_3$).

During the interaction in the mixture of mine spoil and the litter, there appears a negative effect ($-1.66x_1x_3$) which testifies that the animal's mass decreases. In terms of this effect it is concluded that the "minus" sign is exactly related to the presence of mine spoil in the mixture. The triple effect of the interaction of mine spoil, chernozem and the litter turned out to be statistically insignificant.

It has been found out that as soon as the *litter's share* in the mixture *increases* (for example, from 50% to 85%) *and the mine spoil share decreases* (from 25% to 5%) a positive increase of the body's weight of animals is observed.

When *the contents of mine spoil increase* (from 40% to 85%) *and decrease in the mixture's share of the litter* (from 20% to 5%), there occurs a regular loss in weight of the body of animals. Such a reaction of organisms is the result of the deterioration of ecological conditions of the habitation. The observable *effect from the influence of the edaphic mass of chernozem* on the weight of saprobes is an order lower than that from the litter's influence. Its value is small and varies in a very narrow range.

According to the resultant regression equation the response surface has been constructed (Fig. 2) which allows the visual estimation of the influence of factors on a change in the animal's biomass. It can be seen that the response has a clear area of maximal increase in the body weight of vegetable-eating millipeds and an area of their maximal loss in weight. A variation in the experiment for the composition of the mixture of mine spoil from 0% to 20%, chernozem from 0% to 75%, and the litter from 20% to about 100%, resulted with the area of maximal increase of the body weight of the animals. On the contrary, the animals lose weight if the content of mine spoil in the mixture increases, and thus, the litter content and the edaphic mass of chernozem decrease. The loss of weight fixed in the experiment is explained by the fact that mine spoil are acidic ($pH = 3.5$). Therefore, with a change in the mixture composition to the larger content of mine spoil, there are created negative conditions for the animals' existence.

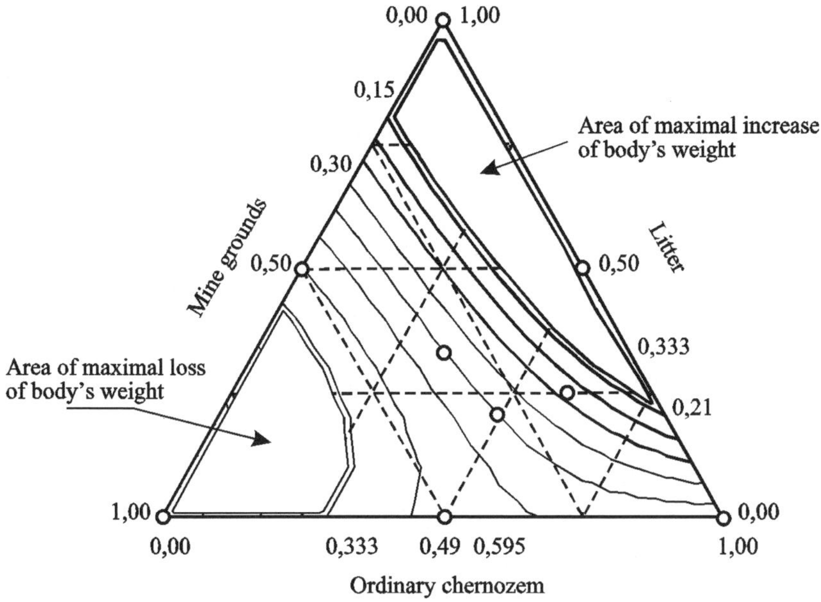

Figure 2. Areas of extreme changes of biomass on the projection of response surface for the investigated ternary system

2.2. A Change of Carbon Dioxide Emission by Mixed Soil

The results of tests in the identification of the contribution of saprobes (*Diplopoda*) in the biological potency, in particular, in the "breathing" of model soils are indicated in Table 3. Statistical data processing has shown that, in the experiment, the contribution made in the evolution of CO_2 by the animals is authentic ($\alpha = 0,05$): the production of CO_2 with the participation of animals is higher (on 20th and 27th day) than without them.

TABLE 3. Influence of animals-saprobes on the evolution of CO_2

Number of test	Mixture composition, unit fraction			Response offset* and ratio C/T (n = 9 species per test)			
	Mine spoil (x_1)	Edaphic mass of ordinary chernozem (x_2)	Litter of *Robinia* leaves (x_3)	20th day		27th day	
1	1	0	0	+	1.35	+	1.32
2	0	1	0	+	1.84	+	1.53
3	0	0	1	+	1.30	+	1.32
4	0.5	0.5	0	+	1.84	+	1.19
5	0.5	0	0.5	+	1.42	+	1.12
6	0	0.5	0.5	+	1.52	–	0.75
7	0.333	0.333	0.333	+	1.66	+	1.37
8	0.15	0.595	0.255	+	2.68	+	1.32
9	0.3	0.49	0.21	+	1.52	+	1.35

**A response offset is the ratio difference in the test without animals (C is control) and with them (T is Test).*

After the development of the planning matrix for CO_2 streams on 20th day from the commencement of the experiment, the following regression equation was derived:

$$y = 0.519x_1 + 0.599x_2 + 0.585x_3 + 0.869x_2x_3 + 3.945x_1x_2x_3$$

(R^2 considering of freedom degrees = 91.40 %),

where

y is the intensity of CO_2 stream produced with the participation of representatives of saprobes by a mixture of substrates, g/hour·m^2;

x_1 is the relative content of mine spoil in the admixture;

x_2 is the relative content of chernozem in a component mixture;

x_3 is the relative content of the litter of Robinia leaves in a three-component system; and

x_2x_3 and $x_1x_2x_3$ are interactions of factors.

The coefficients of influent factors and their interactions are given in the regression equation with $\alpha \leq 0{,}05$. The value of the quotient of determination (R^2) of the resultant equation testifies to a suitable work capacity of the model itself. A statistical check has shown that it is adequate ($\alpha = 0.04$).

It has been established that for the intensity of CO_2 stream fixed on the 20th day the significant influence of all used components of the mixture is authentic: mine spoil (x_1), ordinary chernozem (x_2) and the forest floor of *Robinia pseudoacacia* L. leaves (x_3). Their combined effect is also authentic; and the same is true for the interaction of chernozem and the litter (x_1x_3), as well as for a triple interaction of mine spoil/rocks, chernozem and the litter ($x_1x_2x_3$). Double effects x_1x_2 and x_1x_3 are not significant statistically and in the regression equation they are not quoted.

A comparison of the value of regression coefficients shows that the contribution to the emissions of carbon dioxide by such components as edaphic mass of chernozem (+0.599) and the litter (+0.585), exceeds the contribution of the effect brought by mine spoil (+0.519).

The dependence of the CO_2 stream on the factors chosen in our experiment has a non-linear character. It is defined by the influence of a double interaction of the edaphic mass of chernozem and the leafy tree waste (x_2x_3), as well as a triple effect of all components of the mixture ($x_1x_2x_3$).

If the composition of components of the mixture in the experiment changes as follows: the contents of mine grounds from 0% to 40%, chernozem, from 25% to 70%, and the litter from 30% to about 75%, then the area of maximal stream of CO_2 responds with it (Fig. 2). In other words, for the mixed soils of little use for the existence of representatives *Diplopoda* (in this case such a substrate is mine spoil), it is possible to reach a larger emission value for carbon dioxide. This emission (biological potency) may be rise due to an addition in the mixture of chernozem mass and the litter of *Robinia pseudoacacia* L. leaves.

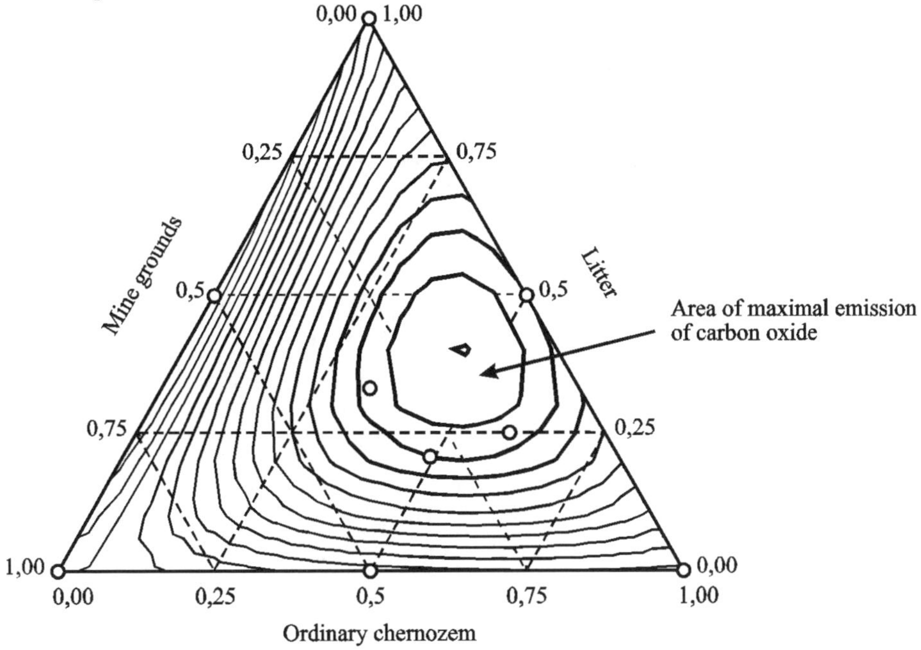

Figure 3. Contour curves for carbon oxide emission by mixed soils

3. Conclusions

A mathematical model was developed that described the dependence of changes in the biomass of vegetable-eating millipeds on the experimental factors. An authentic decrease was observed in the biomass of millipeds inhabiting mine spoil, as well as an increase during an addition to the components of the mixture (to mine spoil and chernozem) of the tree waste of Robinia leaves. It has been found out that the presence of animal-saprobes in artificial mixed soils promotes a more intensive evolution of carbon dioxide as one of characteristics of the biological potency of soils.

Calculations for the mathematical model described the production rate of CO_2 from the multi-component composition of the mixed soil. In the performed tests with the participation of representatives of edaphic saprobes, there has been defined a statistically authentic influence of mine spoil, ordinary chernozem and the litter on the nature of aerobic edaphic.

It has been shown that mine spoil makes the least contribution to the biological potency of mixed soils but when the litter is added, the evolution rate of CO_2 increases, and that can promote the formation of a favourable habitat for animals and so improve their structural-functional characteristics. This, in turn, will help enhance the quality of the overall soil media and so reduce the potential polluting effects from surface run-off from mining spoil areas and reduce erosion risks.

References

1. N. A. Belova, Ecology, Micromorphology and Anthropogenesis of the Forest Soils in the Steppe Zone of Ukraine (Dnepropetrovsk, the DGU (Dnepropetrovsk State University), 1997).
2. A. P. Travleev, V. A. Ovchinnikov, V. N. Zverkovskiy et al., The Biogeocenotic Cover of Western Donbass, its Technogenic Dynamics and Optimization. An Educational Book. (Dnepropetrovsk, the DGU, 1988).
3. A. P. Vasiliev, Analytical Chemistry in 2 parts. Part 1. Gravimetric and Titrimetric Methods of Analysis (Moscow, High School, 1989).
4. N. L. Glinka, Tasks and Exercises in General Chemistry (Leningrad, Chemistry, 1987).
5. Yu. I. Gritsan, The Environmental Principles of Transformed Influence of Sylva on the Steppe Environment. A Monograph (Dnepropetrovsk, the DGU, 2000).
6. V. N. Zverkovskiy, N. A. Polyshchenko, The Bioenvironmental and Agrotechnical Features of the Forest Recultivation of Mining Dumps in Western Donbass, The Items of Steppe Silvics and Forest Recultivation of Lands (Dnepropetrovsk. The Editorial and Advisory Department (EAD) of DGU), 195-201 (2006).
7. V. N. Zverkovskiy, N. A. Polyashchenko, The Forest Recultivation as an Effective Method of Exploration of the Impaired Lands, The Problems of Forest Recultivation of the Impaired Lands of Ukraine. Theses of Reports of the International Conference (Dnepropetrovsk, the DGU), 8-12 (2006)
8. V. N. Zverkovskiy, N. A. Belova, N. P. Tupika, Some Items of Creation of Wood Biogeocenose Cultures on Reclamated Soils of Western Donbass. Biogeocenology, Anthropogenic Changes of Vegetation and Their Prediction. Theses of Reports of the 2d Republican Meeting (Kiev, Naukova Dumka), 1978. p. 165.
9. V. N. Zverkovskiy, The Peculiarities of the Development of Growing Stock in a Long-Term Experiment in Reclamation of Spoil Heap in the Mine "Pavlogradskaya", The Items of Steppe Silvics and the Forest Recultivation of Lands (Dnepropetrovsk. The Editorial and Advisory Department of (EAD) DGU), 21-30 (2002).
10. I. G. Zedginidze, A Planning of the Experiment for a Research of Multicomponent Systems (Moscow, Science, 1976).
11. A. F. Kulik, E. P. Suslova, The Influence of Allelopathic Factors on Soil-Building Processes of the Reclaimed Pieces of Ground in Western Donbass. The Biogeocoenological Researches of the Forests in Technogenic Landscapes of Ukraine's Steppe (Dnepropetrovsk, the DGU), 4-9 (1989).
12. E. N. Lvovskiy, The Statistical Methods of Elaboration for Empirical Formulas. Studies. A Study Guide for Technical Higher Educational Institutions (Moscow, High School, 1988).
13. A. N. Masyuk, The Structural and Functional Organization of Sea Buckthorn Plantations, Anthropogenic Influences on the Forest Ecosystems of the Steppe Zone (Dnepropetrovsk, the DGU), 101-112 (1990).

14. S. K. Piskareva, K. M. Barashkova, K. M. Olshanova, Analytical Chemistry (Moscow, High School, 1994).
15. A Workshop in Agrochemistry. A Study Guide. Edited by Academician of the Russian Academy of Agricultural Sciences Mineev (Moscow, The Publishing House of the Moscow State University (MGU), 2001).
16. A. P. Travleev, The Scientific Bases of Technogenic Biogeocenology, The Biogeocenotic Researches of Forests in Technogenic Landscapes of Ukraine's Steppe (Dnepropetrovsk, the DGU), 4-9 (1989).
17. A. P. Travleev, The Theory of Forest Recultivation of Impaired Lands in Western Donbass in the Region of Dnepropetrovsk, Soil Science (Ukraine), 16 (1-2), 19-29 (2005).
18. I. Kh. Uzbek, T. I. Galagan, The Physical and Chemical Properties of Edaphotopes in the Technogenic Landscapes and Their Ecological-Economical Meaning, Soil Science (Ukraine), 5 (1-2) 102-106 (2004).
19. N. N. Tsvetkova, V. N. Zverkovskiy, N. P. Tupika, N. V. Voloshina, The Dynamics of Microelement Structure of Fill Grounds in Western Donbass, The Anthropogenic Influences on the Forest Ecosystem of the Steppe Zone (Dnepropetrovsk, the DGU), 4-10 (1990).
20. V. I. Shemavnyov, I. P. Chaban, V. O. Zabaluyev, The Technogenic Territories: A Rational Use in Agriculture, The Problems of Forest Recultivation of the Impaired Lands of Ukraine. Theses of Reports of the International Conference (Dnepropetrovsk, the DNU), 41-44 (2006).

I. Apostol et al. (Eds.)
IOS Press, 2009.
doi:10.3233/978-1-58603-948-6-172

The Potential of Liquid Rocket Fuel for Regional Catastrophes and Prevention Solutions

Prof. Dr.-Ing. Wolfgang SPYRA
Brandenburg University of Technology in Cottbus, Germany

Abstract. The risks of storage of the hazardous chemicals Melanj, Samin and Izonit used as liquid rocket fuel are described. The historical background of conversion, general problems are explained and focused on the storage situation of these missile propellants in Azerbaijan. The toxicology and environmental effects of the stored chemicals are combined with a risk assessment and risk management. The result is a remediation potential developed under the responsibility of NATO and NAMSA.

Keywords. Melanj-, Samin- and Izonit-storage, conversion, Azerbaijan, risk assessment, risk management

Introduction

Forces are equipped with weapons, instruments and material to protect their nation and to fight successfully. The tasks of the forces vary from injuring people up to killing them and destroying military potential and civil infrastructure. So they dispose of various kinds of speciality materials like explosives or nuclear material which contain the risky part of the equipment. The propellant which drives the weapons is also risky. In some contexts the material used is extremely risky, being toxic and explosive, for example. In some cases, the storage is a risk itself. Normally, the military forces, as user of these chemicals, have the knowledge to use this equipment in a safe way and handle the risky material with care. However, the situation changed in many countries in recent decades.

After the end of the Second World War the victorious Allied Forces disbanded and formed two powerful blocks. On one side was the Eastern Block, ruled over by the Soviet Union, and on the other side were France, Great Britain and the United States, who together formed the elements of the Western Block.

As East and West powers built political and economic alliances, the impact of this new polarized political climate spread to lands outside of Germany. The military sectors of the West formed the North Atlantic Treaty Organization (NATO). To counteract NATO, the Soviet Union established the Warsaw Pact. The line of separation dividing the east and west powers formed the so called “Iron Curtain”, the symbol of the decades-long Cold War between the eastern and western blocks.

In the time of the Cold War the construction of military objects was done under military aspects. These are a part of the defence or offence strategy of the country and have to fit within the military requirements of camouflage and optimally thrive on the military effect.

The protection of the population is carried out by effectively guarding the military objects, mostly. Environmental aspects played no decisive role neither during the construction nor the use of the objects.

A trend in the late 1980's and early 1990's towards political harmonization between the two blocks eventually lead to the collapse of the powerful bilateral structure. What was left of the Soviet Union was a large, centralized Russia, which held varying degrees of diplomatic relations with the other former Soviet states. Diplomatic relationships ranged from cases of mutual, political, and economic cooperation, to situations where former block politics were completely rejected. Political changes after the fall of the Soviet Union led domestic policies to focus on new pressing issues, most obviously acknowledging major economic instabilities.

As a consequence of economic hardships, political and economic structures of the former Soviet Republics were forced to undergo major changes. A priority was placed on budget analysis within the various governmental sectors, such as the military. Military sectors in both blocks found themselves confronted with major budget cuts. Once the Cold War ended and the level of threat and danger diminished, maintaining the massive defence departments of the past few decades was no longer justifiable. The need for large-scale military base conversion was thus born.

In those nations where the military played an extremely powerful role, relative to other governmental agencies, the process of conversion brought considerable problems. This can be understood when it is seen that the conversion process in many countries was made possible only through the help of foreign aid, and this foreign aid often came from countries which the civilian society had historically understood only as the adversary.

The conversion process means the dismantling of military potential, which includes:

- Significantly decreasing the magnitude of combat forces;
- Reducing management and operational resources; and
- The transfer of occupied lands back to the public for civilian use.

In terms of sustainable steps toward world peace stability, conversion is a positive sign. In terms of domestic peace and social tolerance of political restructuring, however, conversion measures present a considerable burden which contains a potentially "explosive" social element.

For the dismantling and rebuilding of a country to be successful specialists are needed. Soldiers are specialists. During the Cold War around 80% of all scientists in Russia were employed by the military. When such specialists are released from military duty, they find themselves in a labour market that demands neither their qualifications nor the sheer mass of labourers they present. With few exceptions, this is a global phenomenon. Since 1992, over 1.2 million military personnel have become unemployed through restructuring. As unemployment rates rise and people are unable to provide for the basics of life, this leads, inevitably, to social stress, if not to all-out domestic conflict. Therefore, military conversion applies not only to land parcel and facilities conversion, but also to a social conversion. Indeed, conversion also involves a change in conceptual thinking. New conceptual thinking leads to societal restructuring.

Practical Examples of Dangers on Conversion and Transferring Resources

It has been established that with the evacuation of military lands, the management and operational resources have been transferred to other military locations. The locations to which these resources have been transferred stand out as being especially safe against an outside attack. Indeed, it is important to assure that these management and operational resources do not fall into unauthorized hands and misused. Experience shows that in many cases the transfer of equipment from one location to another occurs without incurring additional dangers.

In some cases, munitions and combat materials are transported to other military installations for storage, at which locations munitions or hazardous facilities are already present. In other cases, military materials are transported and stored in areas outside of the military realm. In these cases the level of security may be called into question. Hence, depending on the conditions of transportation and storage, risks can either be compounded or negated. In the cases we are aware of, the only disadvantage encountered is in the case of an emergency where only experts with a specialization in dealing with this specific type of danger would be able to combat a dangerous situation. This means that a dangerous situation or an accident would not be able to be handled adequately.

Further practical examples show that often munitions which were planned to be moved to a new location were prepared for transportation, but never moved. There are further cases where, although the munitions actually were transported to a new location, the trains with over 100 freight cars full of munitions were never unloaded. Security and surveillance is also often lacking. In some cases, it may be possible for persons lacking technical knowledge or permission to gain access to the hazardous substances. An instance of misuse is, therefore, not unimaginable and builds the potential for terrorist activities.

In this report on experience the propellants of military rockets are described in terms of the hazards and risks which remain when there is no further use for them. These rockets can be driven by different kind of fuel: gas, liquids or solids as technical products, or in mixtures. Our example deals with the problem of liquid rocket propellants. Possible LRF (liquid rocket fuels) are, for example, isopropyl nitrate; kerosene and liquid oxygen; liquid hydrogen and liquid oxygen; nitrogen tetroxide and hydrazine.

The report further deals with the problems of a binary system in which the energy (fuel) and the oxidizer are separated into two different components. The fuel of this system is called Samin and the oxidizer called Melanj. This propellant was used for short- and middle-range rockets in tactical and anti-aircraft types like SCUD-and SS-types.

In the 1990s the development of new military rockets went to solid systems. Now, the liquid rocket fuel systems are not state-of-the-art anymore and many countries lost their interest in these chemicals but still store them. So the liquids are stored without any options for use. Under the conditions that the military budget is cut and many specialists left the army the general result can be described as follows: these risky materials are not under control, anymore!

The following facts were collected during a risk assessment of the liquid rocket fuel storage sites in Azerbaijan. The results, and further information, are published in the NATO Science Series under the topic "The Conversion of Liquid Rocket Fuels", too. This time the publication was indented more for military personnel. Now, the topic of conversion and remaining risks gets more and more important for civil aspects, especially the catastrophes and terrorism prevention.

1. Storage Situation in Azerbaijan

According to information made available by the Azerbaijan Ministry of Defence (MOD) liquid ballistic missile propellants (namely, fuels and oxidizers) that are no longer used by the Azerbaijani Armed Forces are stored at two major sites, military supply installations at the cities Elet and Mingecevir. In both depots in total the following substances are stored:

Oxidizers

- 1,400 metric tons of AK-20k, AK-27p, AK27i Oxidizers (73–80% nitric acid and 20–27% nitric oxides with different inhibitors/additives).

Fuels

- 450 metric tons of TQ-02 Samin (50% triethylamine + 50% xylidines);
- 25 metric tons of OT-155 Izonit (isopropyl nitrate); and
- An unknown quantity of TM-185 (20% gasoline + 80% kerosene).

Since the disintegration of the Soviet Union with the withdrawal of Soviet troops from Azerbaijan and the formation of national armed forces, these chemicals have been virtually "abandoned". Even though they are still stored on military bases and guarded by military staff, there are no interests in a military use and, obviously, there are no means of safe storage and handling for these chemicals.

For all stored chemicals, leakages are evident. In particular oxidizers are continuously leaking into the atmosphere and to the soil due to their aggressiveness in combination with water (by formation of nitric acid), thus corroding stainless steel as well as aluminium storage tanks. While the leakage of oxidizers is obvious due to macroscopic leaks in tanks, traces of leakages on the soil (oxidation of soil iron is evidenced by the brownish Fe(III)), yellowish-brown plumes of nitrous gases escaping from the tanks and a characteristic smell on the storage sites, the leakage of TQ-02 Samin, OT-155 Izonit and TM-185 is mainly detected by organoleptic means due to a strong characteristic smell at all inspected storage sites.

The detailed description of the two inspected storage sites is as follows.

Storage Site Elet

Elet (sometimes also referred to as Alyat) is a town of approximately 20,000 inhabitants (estimated), situated approximately 60km southeast of Baku on the coast of the Caspian Sea. The main road from Baku to the south of the country and to Iran as well as the two major rail tracks to Georgia and the Naxivan exclave, lead through Elet.

The groundwater table depth at the Elet storage site is not precisely known. However, according the Azerbaijani MOD, no groundwater use is known in the vicinity and despite the vicinity of the Caspian Sea the groundwater is believed to be at a depth of approximately 40m below surface underneath a confining layer.

Figure 1. An oxidizer tank at the Elet depot.
The tank is lying on the bare ground. Due to corrosion and rocks penetrating the tank hull, a large spill with the loss of all oxidizer from the tank occurred recently.

The military depot for all kinds of fuels (diesel, kerosene, gasoline, liquid missile propellants, etc.) is situated at the outskirts of Elet. The distance to the nearest residential area is approximately 350 m. Sensitive installations such as schools and hospitals are situated within a perimeter of 500 m from the depot.

Besides fuels, lubricants and other chemicals that are supplied to armed forces units from the depot, a large quantity of liquid ballistic missile propellants is stored in a separated part of the depot. According to an inventory of the stored chemicals on behalf of the Azerbaijani MOD the following types and amounts are stored at Elet in January 2003:

- 1,028 metric tons of oxidizers (AK-20k, AK-27p, AK-27i);
- 155 metric tons of Samin (TQ-02); and
- 100 kg of Izonit (OT-155).

The oxidizers are stored in 22 tanks with a volume of 40–50 m^3 each. Several tanks show severe leaks such as corroded safety valves, broken welding seams, holes due to corrosion of the tank metal plates and holes at the bottom of the tank due to stones pressing themselves through the metal plates of the tanks lying on the ground. However, due to the documented leakages, many of the tanks appear to be empty. Traces on the soil surfaces and concrete barriers surrounding the tanks indicate that several leakages (see Figs 1 and 2) with major releases of nitric acid/nitric oxides occurred recently. Concrete barriers show severe weathering (probably due to contact with nitric acid) and the bare soil around the tanks often appears wet or rusty-brown, obviously due to oxidation of Fe(II) to Fe(III) by nitric acid. Even where tanks appear to be almost empty, yellowish-brown fumes escaping from valves or leaks indicate an ongoing atmospheric

Figure 2. Detail of another oxidizer storage tank at the Elet depot that has been leaking recently.

release of nitrous gases. According to the MOD of Azerbaijan, the underground storage facilities (earth-covered shelters) do not contain any oxidizers. According to information supplied by the Azerbaijani MOD, the oxidizers where removed from the underground storage facilities because no personal protection equipment (respiratory masks, filters, protection suits, gloves, boots) is available – thus making the handling of oxidizers in underground storage facilities impossible.

In addition to the 22 storage tanks, four rail tank cars are standing on the site. According to the Azerbaijani MOD, the tank cars will be still suitable for transport purposes after an overhaul. Rail tracks that might also need overhaul before use for transportation connect the depot with the main railway route Baku – Georgia and Baku – Naxivan exclave. Approximately 70 small tanks (6 m^3 each) are also available for transportation and emergency response purposes.

Until recently, ammonia that is used as a neutralizing agent for the oxidizers, was available at the Elet depot. However, with the latest spills, the ammonia supplies have been exhausted. Further supplies of ammonia are not available in Azerbaijan. So, they should be obtained from neighbouring states, such as Turkey, separated by distances of over thousand kilometres. Recent prevention and emergency response at the depot with respect to leakages of oxidizers is thus restricted to the cutting of vegetation and ploughing of the soil in the vicinity of the oxidizer storage tanks (Fig. 3) in order to prevent any contact of the oxidizer with flammable material, and – in case of a major leakage – pumping of the oxidizer into the small storage containers.

Samin, a mixture of 50% triethylamine and 50% xylidines, is stored in conditions similar to that of the oxidizers. Approximately 15 tanks with a volume between 10m^3 and 50 m^3 each are kept separately (distance circa 100 m) from the oxidizers. Like the oxidizer tanks, the tanks are lying on the bare ground (Fig. 4) without any stand, rack

Figure 3. Oxidizer storage tanks at Elet.

Ploughing and removal of vegetation as a preventive measure were witnessed during the visit. Nitrous gases are escaping from the safety valves of the 50 m³ tank on the left.

Figure 4. Storage of Samin rocket fuel in tanks similar to those used for the storage of the oxidizer at the Elet depot.

The tanks are lying on the bare ground. A strong characteristic amine odour is surrounding the storage site.

Figure 5. Pump used for the pumping of Samin at the Elet depot.
The pump is out of order and could probably not be repaired. Replacement pumps would be necessary.

or catchment basin for leakages. A strong characteristic smell of xylidine and triethylamine indicates leakages. However, no macroscopic leaks are visible on the tanks. In contrast to the oxidizer tanks which are made from stainless steel or aluminium, the Samin storage tanks are made from steel and thus are – despite the arid climate – subject to corrosion.

Isopropyl nitrate, of which only 100 kg are reportedly stored at the Elet depot, is kept in a separate shed. Infrared footage of the ten barrels indicates that these are leaking as gas plumes seem to be emitted from the barrels from time to time. A branch of the Elet depot (approximately 15 km from the main depot) that was not visited also holds stockpiles of both oxidizers and Samin, according to information provided by the Azerbaijani MOD.

Storage Site Mingecevir

Mingecevir is a town of approximately 100,000 inhabitants (estimated). It is situated approximately 240 km east of Baku at the bayou of the Kura river from the Mingecevir reservoir, a huge drinking water reservoir. The main road from Baku to Ganca and Georgia the south, as well as the major rail track to Georgia (used for oil export), pass Mingecevir. A military depot for all kinds of fuels (diesel, kerosene, gasoline, liquid missile propellants, etc.) is situated approximately 15 km southeast of Mingecevir, about 500 m from the main railroad to Georgia. The distance to the nearest residential area is approximately 500 m. Sensitive installations such as schools and hospitals are situated within a perimeter of 1,000 m from the depot.

Besides fuels, lubricants and other chemicals that are supplied to armed forces units from the depot, a large quantity of liquid ballistic missile propellants is stored in a

Figure 6. A smaller oxidizer storage tanks at the Mingecevir depot.
A spill occurred recently, and the tank appears to be empty now. Oxidation of iron in the soil is clearly visible on the right side of the tank.

separated part of the depot. According to an inventory of the stored chemicals on behalf of the Azerbaijani MOD, the following types and amounts are stored at Mingecevir in January 2003:

- 380 metric tons of oxidizers (AK-20k, AK-27p);
- 290 metric tons of Samin (TQ-02); and
- 24 metric tons of Izonit (OT-155).

The oxidizers are stored in about 15 tanks, mostly lying on the bare ground as seen in Elet. Leaks and traces of spills similar to those in Elet were also clearly visible. Because no large backup tanks for emergency response are available, during the last leakage oxidizer from a static tank was pumped into a single rail tank car standing on the site. All tanks including the rail car show severe signs of corrosion and leakage (Fig. 6). Safety valves have typically corroded away (Fig. 7) from the tank inspection holes (Fig. 8). In contrast to the Elet depot, the storage at Mingecevir seems to have always relied on storage in surface-tanks and not in earth shelters. Nevertheless, two tanks were covered with earth. Preventive measures are restricted – as for the Elet depot – to the cutting of vegetation and ploughing of the soil in order to prevent organic (flammable) material in the vicinity of the oxidizer. For emergency response purposes (i.e. oxidizer neutralization), 17 m³ of ammonia is available in a tank situated within the range of the oxidizer storage tanks. The ammonia is frequently used to neutralise spilled oxidizer, yielding ammonia nitrate. Pumps and approximately 170 backup containers (similar to the 6 m³ containers seen in Elet) for the oxidizer are also available.

Figure 7. Close-up of caps of oxidizer storage tanks at the Mingecevir depot.
As in Elet, safety valves have corroded away and are lying on the ground. Nitrous gases are escaping continuously.

Figure 8. Close-up of a large leak at the top/cap of an oxidizer storage tank at the Mingecevir depot.
The tank is empty, now.

Figure 9. Storage of Samin and TM-185 in 50 m³ tanks in a pit at Mingecevir.
Due to the high groundwater table, the tanks have been in contact with groundwater for more than ten years. A strong characteristic Samin and kerosene/gasoline smell was observed. The groundwater table seems to be varying, speeding up corrosion of the steel tanks due to contact of wet surfaces with oxygen.

The storage facility for the approximately 240 metric tons of Samin is situated in a pit (Fig. 9) that is approximately 2.5 m deep, 100 m long and 20 m wide. Because the groundwater level is situated between 1.0 and 2.0 m below ground surface, the tanks are lying in the groundwater. Water level traces on the tank walls show that the water level is variable over time (Fig. 10). Together with the Samin, an unknown quantity of TM-185 fuel (80% kerosene, 20% gasoline) is stored in tanks in the same pit. A strong characteristic smell and floating substances on the water are evidence of leakages from the tanks. Due to the construction material of the tanks (namely, steel), severe corrosion is evident where the tanks have been exposed to water. Dark-red to violet colouring of large areas (Fig. 11) of the soil surrounding the storage pit point to larger spills of Samin in the past. It is known that Samin spills on soils yield a violet to dark-red colour of the soil, probably due to a chemical reaction of triethylamine cations with soil constituents. According to the deputy commander of the depot, no backup tanks, pumps, neutralising agents or personal protection equipment for the handling and/or emergency response with respect to Samin is available.

Isopropyl nitrate (24 metric tons) is stored in sheds that are situated about 50 m and 150 m from the Samin/TM-185 storage pit and the oxidizer storage tanks respectively. A strong characteristic smell inside the Izonit storage shed, similar to that of isopropyl alcohol, provides evidence of leakages from the Izonit barrels. On inquiry, the deputy commander of the depot declared that no means for emergency response are available with respect to isopropyl nitrate.

Figure 10. Close-up of a Samin tank in contact with groundwater in a pit at the Mingecevir depot.

Figure 11. Dark-red to violet colour of the soil surrounding the pit with Samin storage tanks at the Mingecevir depot.

The colour is probably due to a reaction of triethylamine cations with soil compounds. The contamination extends over an area of at least one hectare.

2. Risk Assessment

The risk assessment is subdivided into two parts, with the first part considering the properties of the chemicals of interest (namely, nitric oxides, triethylamine, xylidine and isopropyl nitrate) and the second part considering the chemicals with their respective properties in the situation of the two storage sites at Mingecevir and Elet.

2.1. Risk Assessment of Chemicals

Within the risk assessment there are different kinds of facts to consider. But the important facts are the properties of the chemicals in the environment which can be described by the information of chemical databases. For most chemicals, information about human toxicology is available too. All these data present an overview about the risk of the chemicals.

Most times one fact is not mentioned, namely that the chemicals, which are stored for a long time, can undergo certain reactions resulting in changes in their properties. Originally the chemicals were needed in special technical qualities. That means technical impurities by unwanted chemicals were mentioned and reduced for the intended application and maybe for short-term storage, but not for long-term storage over several decades.

Even if high purity chemicals are used, they can be affected by chemical altering like degradation and polymerisation, for example. In some cases, stabilisers or inhibitors were added to the primary chemicals to keep the necessary quality intact as long as possible. By the degradation of these substances the chemicals are getting unstable. Also, the uptake of water from the air, as occurs with hygroscopic substances, can change the properties of the chemicals.

Oxidizers

The oxidizers that are referred to as Melanj or with their military codes AK-20k, AK-27p, and AK-27i, are all compositions of nitric acid and nitric oxides with different additives and inhibitors such as iodium or fluorine. The content of nitric acid is 80% for AK-20k and 73% for AK-27i and AK27p.

Nitric Acid (HNO_3):

- Colour/Form: Colourless, yellow or red, fuming liquid.
- Odour: Characteristic acrid to sweet odour.
- CAS-No.: 7697-37-2
- Melting point: – 41.6 °C
- Boiling point: 83 °C
- Density: 1.55 g/cm³ @ 15.5 °C
- Vapour pressure : 63.1 mm Hg @ 25 °C = 8,412.5 Pa
- Solubility: Miscible with water, soluble in ether
- Corrosivity: In presence of traces of oxides nitric acid attacks all base metals except aluminium and special chromium steels.
- Environmental fate: If released into the environment, nitric acid will be diluted by soil water, surface water and/or groundwater. Hydronium ions will be neutralized by carbonate minerals and thus be eliminated quickly. However, larger spills may result in the (local) solution of soil material and acidification

of water resources. Nitrate ions will remains in the water for longer periods and may contribute significantly to eutrophication. Transport with (ground) water flow has to be considered significant because no retardation is to be expected.

- Combustibility: Nitric acid is not combustible, but will react with water or steam to produce heat. Contact of concentrated nitric with combustible materials may increase the hazard from fire and may lead to an explosion.
- Human toxicity: Contact with nitric acid or inhalation of nitric acid fumes will result in severe cauterization of skin, mucous membranes, the respiratory system (pulmonary oedema) and eyes.

Nitric Tetroxide (N_2O_4):

- Colour/Form: Colourless gas, below 22 °C yellow or colourless liquid.
- Odour: Odourless gas.
- CAS-No.: 10544-72-6
- Melting point: – 9.3 °C
- Boiling point: 21.15 °C
- Density: 1.45 g/cm³ @ 20 °C
- Vapour pressure : 646 mm Hg @ 25 °C = 86,124.7 Pa
- Solubility: Not applicable (gas)
- Corrosivity: Corrosive as liquid.
- Environmental fate: Due to high vapour pressure, only atmospheric pathway is relevant. Quick dissolution in ambient air. However gases are heavier than air and will spread along ground.
- Combustibility: Does not burn itself but will support combustion as a strong oxidising agent. May ignite combustibles (e.g. wood, paper, oil, clothing, etc.), containers may explode when heated, ruptured cylinders may rocket.
- Human toxicity: Inhalation of nitric tetroxide fumes/gas will result in slowly evolving but progressive inflammation of lungs. Will impair gas exchange and lead to cyanosis with in severe cauterization of skin, mucous membranes, the respiratory system (pulmonary oedema) and eyes.

Samin

Samin is a mixture of 50% triethylamine and 50% xylidines. It is used as a fuel for ballistic and air defence missiles.

Triethylamine ($C_6H_{15}N$):

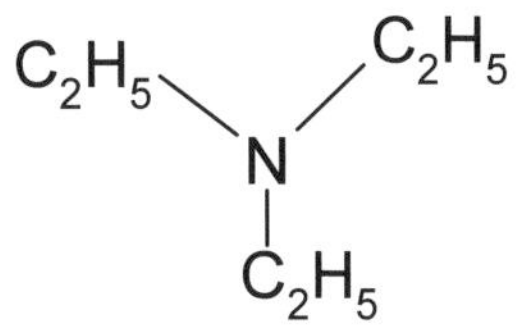

- Colour/Form: Colourless liquid.
- Odour: Strong, ammoniacal to fishy odour.
- CAS-No.: 121-44-8
- Melting point: – 114.7 °C
- Boiling point: 89.3 °C
- Density: 0,7255 g/cm³ @ 25 °C
- Vapour pressure : 57.1 mm Hg @ 25 °C = 7,612.5 Pa
- log K_{OW}: 1.45
- Solubility: Soluble in ethanol, oils and ethyl ether, very soluble in acetone, solubility in water 15,000 mg/l – 20,000 mg/l (@20 °C/65 °C)

- Corrosivity: Liquid triethylamine will attack some forms of plastics, rubber, and coatings.
- Environmental fate: Due to a high vapour pressure, triethylamine will solely exist as vapour in the ambient atmosphere and will be degraded by hydroxyl radicals (photochemically produced) with a half-life of four hours. Due to a pKa of 10.78, triethylamine will mainly exist as a cation if released to soil or water and therefore strongly adsorb to organic carbon and clay (in soil). Based on screening tests, biodegradation of triethylamine will not play a relevant role in the degradation of triethylamine. Because of a low log K_{OW}, biodegradation will not occur.
- Combustibility: Triethylamine is a flammable/combustible material that may be ignited by heat, sparks of flames. Vapours that are heavier than air, form explosive mixtures with air. Vigorous reaction with strong acids.
- Human toxicity: Mainly local effects. Eye irritant – eye contact causes severe burns. Clothing wet with triethylamine will cause skin burns. Carcinogenic (skin), temporary blue hazy vision due to exposure to ethylamines, vapours irritate nose, throat, lung, causing coughing, choking and difficult breathing.

Xylidines ($C_8H_{11}N$ – six isomers):

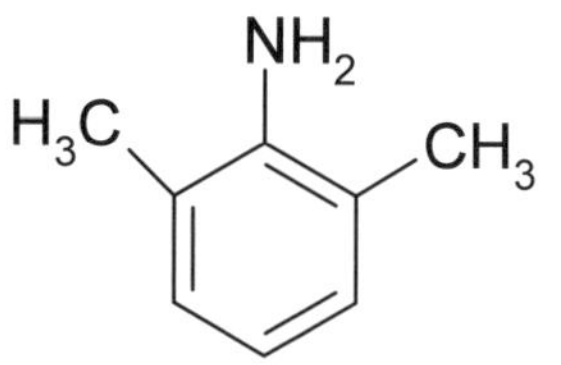

- Colour/Form: Pale yellow to brown liquid above 20 °C.
- Odour: Weak aromatic amine odour.
- CAS-No.: 1300-73-8
- Melting point: – 36 °C
- Boiling point: 213–226 °C
- Density: 0.97–0.99 g/cm³
- Vapour pressure : 0.028–0.015 mm Hg = 3.7–20 Pa
- log K_{OW}: Not available
- Solubility: Slightly soluble in water, soluble in alcohol and ether.
- Corrosivity: Liquid xylidine will attack some forms of plastics, rubber, and coatings.
- Environmental fate: Xylidines are expected to have a high to medium solubility in soil even though anilines are expected to adsorb strongly to humus or organic matter in soil due to the high reactivity of the aromatic amino groups. Due to low vapour pressures, xylidines are not expected to occur in the vapour phase or to volatilize from soil or water. In water, xylidines might strongly adsorb to suspended organic matter. Experimental data suggest that xylidines might biodegrade under aerobic conditions in soil and water.
- Combustibility: Xylidines are flammable/combustible substances that may burn but do not ignite readily. When heated, vapours may form explosive mixtures with air. Contact with metals may evolve hydrogen gas.
- Human toxicity: Possibly carcinogenic, intoxication may result in headache, dizziness, cyanosis.

Isopropyl Nitrate (2-propyl nitrate, $C_3H_7NO_3$)

- Colour/Form: Colourless liquid
- Odour: Isopropyl alcohol like odour
- CAS-No.: 1712-64-7

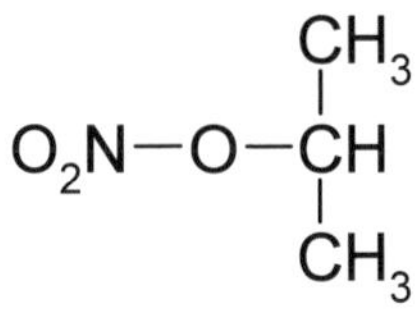

- Melting point: – 82 °C
- Boiling point: 101–102 °C
- Density: 1.047 g/cm³ @ 15.5 °C
- Vapour pressure : 40 mm Hg = 5,332.8 Pa
- Flash Point: 13 °C
- Solubility: Probably highly soluble in water
- Environmental fate: no data available
- Combustibility: Highly flammable, self-ignition possible when in contact with organic material, explosive.
- Human toxicity: No data available

2.2. Risk Assessment of Storage Sites

The recent condition of the chemicals is not known after the storage of over 30 years. The facts that no high purity chemicals were used and that Melanj is hygroscopic, may lead to a change in quality and properties of the stored chemicals. This is the reason why analyses have to be done to estimate the actual potential of the risk. During the introduction of the depots the most important safety rule was followed, namely that the oxidizers were stored separate from the fuel. So any unintended contact between both substances is not possible. This would result in setting free enormous amounts of energy, creating different conditions where the substances were set free to escape into the environment. One important reason is described subsequently and observed at the aluminium tanks.

Aluminium Tanks

Storage of the oxidizer Melanj is in 10 to 50 m³ tanks. But these tanks are not totally filled with Melanj because due to climatic conditions a safety value of remaining storage capacity has to remain. The tanks are exposed to temperatures from minus 10 °C to plus 60 °C. In this wide range of temperature an important change in the volume occurs. So the remaining space is for this volume expansion without destroying the tank hull. Even if the expansion is greater than the capacity of the tank can store, there is a safety valve to release the amount of substance which is producing an overpressure within the tank. By reaching the overpressure the substances are released in the gaseous phase into the environment. The staff of the depots reported rare observations of this process.

The tanks for the oxidizer Melanj were manufactured with a material thickness of 30 millimetres. As already mentioned, inhibitors like fluoride, iodide and phosphate were added to the oxidizer to prevent corrosion. This inhibiting property can be observed up till today. But the inhibition is restricted to the liquid part of the stored chemicals. The gaseous phase above the liquid phase does not contain inhibitors and that is why there is no protection against corrosion. In Fig. 12 the principle of the situation is shown while Fig. 13 shows real corrosive effects at the storage tank. The corrosion has progressed so far that the chemicals can escape from the storage tanks and their odour is hanging in the air.

Stainless Steel Tanks

Stainless steel is well situated for the storage of aggressive chemicals like Melanj because it is less corrosive. Practical experience shows that the property does not corrode

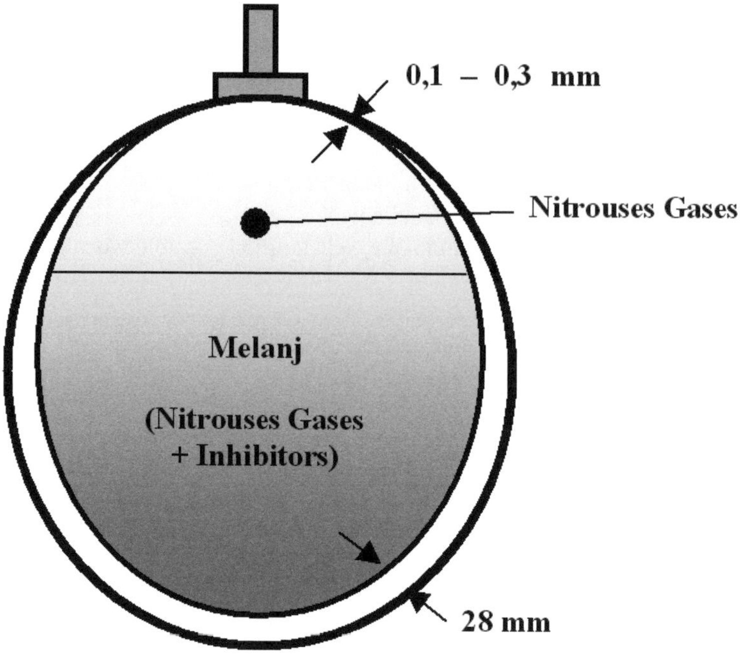

Figure 12. Wall thickness of a long-term used aluminium tank.

Figure 13. Detail of a leaking tank.
The safety valves have corroded away (due to contact with concentrated nitric acid). They were found lying on the ground.

in contact with Melanj. The problem which occur are caused at the weld seam which is the location where the Melanj causes corrosion to the storage tank earlier than to the larger surface of the steel plates. The corrosion of the weld seams leads to leakages. Through the perforated hull of the storage tanks nitrous gases can then escape. The summary of the situation is that all Melanj tanks are in desolate conditions which can lead to a disaster in an unknown time.

Risk Assessment for Elet (Alyat) Depot

The situation at the Elet depot has to be considered less dangerous than at the Mingecevir depot. The main hazard at Elet is posed by the oxidizers that are stored in leaky tanks with continuously ongoing leakage and occasional larger spills. As no neutralizing agents, personal protection equipment, pumps etc. are available, immediate action to provide the necessary emergency response equipment is necessary to prevent further leakages of nitric acid and nitric oxides.

Only the calcium carbonate-rich soil prevents some major impacts on the environment, i.e. the presence of carbonate results in neutralisation of the nitric acid as follows:

$$2\ HNO_3 + CaCO_3 \rightarrow Ca(NO_3)_2 + H_2O + CO_2$$

The storage conditions for Samin are inadequate. However, because of the dry climate and non-corrosivity of Samin against steel, the storage can be considered relatively safe on a scale of several months to a maximum of two years; this is despite leakage that is obvious with a strong characteristic amine odour in the vicinity of the Samin tanks. The quantity of isopropyl nitrate stored at the Elet facility is low (approximately 100 kg). However, uncontrolled detonation of the isopropyl nitrate after self-ignition of the leaking barrels could result in greater damage if the Samin and oxidizer storage would be affected during such an accident. Major leaks or spills of nitric acid and nitric oxides as well as Samin would pose severe risks for the local civil population living in residential areas only 350 m from the storage site. Schools and hospitals could also be affected.

Risk Assessment for Mingecevir Depot

The situation at the Mingecevir depot has to be considered to be more critical. The main threat here is posed by heavily corroded tanks filled with Samin and TM-185 that are situated in a pit in direct contact with the groundwater. Here, organoleptic observations (strong characteristic Samin odour, dark-red to violet soil) indicate recent and on-going severe leakage of triethylamine and xylidines into the environment. Due to the high solubility of triethylamine in water (15–20 g/l) and the storage of tanks in contact with groundwater, the on-going contamination of huge groundwater quantities close to a major drinking water reservoir of the Republic of Azerbaijan is almost certain.

Residents living close to the depot regularly notice the odour of Samin and oxidizers but do not – until now – protest against the situation. However, due to the short distance between the depot and surrounding civil population, effects on the population during a major accident (such as a spill or leakage) are likely. It is not known whether people in the surrounding residential areas use groundwater from wells as drinking water or process water. The fact that no emergency response equipment (personal protection equipment, pumps, backup tanks, neutralising agents, etc.), except one normal fire truck, is available urges immediate action with respect to the Samin problem at the Mingecevir depot.

The oxidizer (nitric acid/nitric oxides) stored at the Mingecevir depot does not pose an immediate risk, even though most tanks already show leaks and larger spills occur from time to time. Except for personal protection equipment, the necessary equipment for emergency response in the case of spills is still available, namely pumps, $10m^3$ of ammonia as a neutralizing agent and backup tanks. However, continuous leakage of smaller quantities of oxidizers is obvious and the storage can not be regarded as safe.

Large quantities of isopropyl nitrate that are stored in sheds separated from Samin and oxidizers by a distance of 50 m and 150 m respectively, pose a larger threat. A strong, isopropyl alcohol like smell inside and outside the sheds indicates that at least some of the barrels lying and standing in the sheds are leaky. As isopropyl nitrate is highly flammable, explosive and forms explosive vapours in ambient air, an acute danger of explosion has to be assumed for the isopropyl nitrate storage sheds. As the quantity of isopropyl nitrate stored in the sheds aggregates to approximately 24 metric tons, an accident would also affect the storage facilities for Samin, oxidizers and conventional fuels (diesel, kerosene, gasoline, etc.), thus resulting in a major accident that would also effect the local population in surrounding villages.

3. Risk Management

All the described dangers need the implementation of safety measures, especially during the time of eliminating the risk. A well-known problem is that effective safety regulations exist but no one controls their compliance. Because all hazards associated with these storage sites can affect life and health, keeping these safety regulations, providing relevant safety equipment and the training in first emergency response and first aid, is necessary. Therefore, a definition of safety areas is needed. The safety area is the border between public areas and the area of danger. Given the possible atmospheric fate of the described chemicals, the safety distances can be defined in kilometres. Within this area a limited number of persons should work and only these who are needed to work there. The population surrounding the sites should be informed how to act in the case of a disaster. But the first point is to eliminate all situations which can lead to this.

The first step, realizing a minimization of the health and environmental risks was started within a NATO “Science for Peace Programme”. Within this program the BTU Cottbus, ANAS and the “Melanj Expert Group” developed a technical solution to destruct the Melanj. The aims for this solution had been environmental friendly, cost-effective and mobile. As result, a plant was inaugurated in July 2006 which produces 0.8 metric tons of calcium nitrate per metric ton of Melanj. This end-product is useable as a fertilizer compound. The processing plant and the associated procedures will ensure an intrinsically safe environment. The plant should be able to neutralize a minimum of 5 metric tons of Melanj in every 24 hour period. The cost per produced ton Melanj is less than 1,000 EURO. The Melanj combustion in Moldova was done at an average of 1,500 EURO per metric ton Melanj in 2002.

Further, the whole plant is transportable in four standard ISO containers (Fig. 14). The assembly and commissioning, or the disassembly, is possible in seven working days. In this way the conversion of Melanj is possible on both depots of Azerbaijan without transporting the hazardous chemicals. Furthermore, this plant is available for

Figure 14. The Melanj pilot plant in Elet constructed within four transportable ISO-Containers.

other countries in the Caucasian region after finishing the conversion of Melanj in Azerbaijan.

4. Results/Conclusion

The recent storage conditions of liquid missile propellants and their components pose an acute risk to life and health. Eliminating the hazards is the right way to reduce this risk and this should be done, consequently, with the proposed technology in the mobile plant. With the destruction of Melanj one risk of the storage sites is reduced, a few more are remaining. With respect to the storage situation of Samin and isopropyl nitrate at the Mingecevir depot, immediate action including the recovery of Samin and TM-185 from tanks in contact with groundwater and the destruction of isopropyl nitrate as soon as possible is still recommended.

Potential solutions for an economic remediation of the storage site after the destruction or removal of Melanj, Samin and isopropyl nitrate should be worked out. After finishing this step, the liquid rocket fuel conversion will be successfully completed and no human or environmental hazards or risks will remain on the storage sites.

I. Apostol et al. (Eds.)

doi:10.3233/978-1-58603-948-6-192

Contaminated Sites – Risk Management in Austria

Heide JOBSTMANN
Umweltbundesamt GmbH, Spittelauer Lände 5, A-1090 Vienna, Austria

Abstract. Groundwater is by far the most frequent receptor at risk from contaminated sites. Between 1989, when the Law for Cleanup of Contaminated Sites (ALSAG) was implemented in Austria, and 2006 the remediation of contaminated sites enabled a qualitative amelioration of groundwater bodies of about 46 million m^3. Remediation targets cannot always be met. Ecologically practicable solutions have to be adapted to the site conditions given.

Keywords. Contaminated sites, groundwater, remediation, chlorinated hydrocarbons

Introduction

Contaminated sites pose a considerable risk to human health and the environment. Generally an impairment of human health takes place via soil and groundwater, e.g. the consumption of contaminated drinking water.

Groundwater represents a valuable source of water in Austria. Overall more than 99% of Austria's drinking water supply comes from groundwater. At about 95% of the registered contaminated sites an impairment of water quality was determined or is to be expected, and 65% of these sites are located in areas with sensitive use of groundwater. About 24% of suspected contaminated sites are situated within groundwater conservation areas and about 10% within groundwater protection zones [1].

The principal sources of groundwater contamination are:

- Industrial and commercial activities;
- Waste disposal activities;
- Agriculture; and
- Mining sites.

The chance that harm will result from consumption of contaminated groundwater is termed the risk, but before a risk can exist, a chain of features must exist, as follows:

- A source of contaminants;
- A pathway for contaminant transport from the source, e.g. groundwater abstraction;
- A receptor or target to which the contaminant may cause harm, e.g. humans, plants, animals; and
- A hazard, i.e. the event or property associated with contaminated groundwater that may potentially cause harm.

Consideration of such a chain, coupled with the chance that harm will result from it, forms the basis for carrying out a risk assessment. By carrying out a structured risk assessment a judgement can be made on whether any potential contamination is likely to cause an unacceptable impact.

In Austria there are about 80.000 sites where substances posing a threat to human health are handled. As of 1st January 2007, 48,678 sites are registered in a database administrated by the Environment Agency. About 1,000 to 2,000 of these sites need to be remediated [2].

1. Contaminated Land Policy

1.1. Legislation

An important approach to dealing with the contaminated site situation in Austria was the implementation of the Law for the Cleanup of Contaminated Sites (ALSAG) in July 1989. The ALSAG defines contaminated sites as "disposal and industrial sites as well as thereby contaminated soil and groundwater bodies which – on the basis of a risk assessment – are believed to pose a considerable threat to public health or the environment" [2]. The definition does not include land polluted by agricultural activities.

The ALSAG provides the legislative structure for the management of contaminated sites, whereby the following key aspects can be summarised:

- A national, uniform structure of registration and assessment for contaminates sites;
- Classification of priorities to identify the urgency for clean-up activities;
- Public funds for investigation and clean-up measures are available (Guidelines for Funding); and
- A description of the responsibilities for different regulatory authorities operating within the above mentioned regulations.

The ALSAG's main focus is on public funding of cleanup measures i.e. the regulation of governmental benefits for site remediation. Only contamination at sites that existed before 1989 is taken into consideration. Funds for the necessary measures are created through charges on disposal, export and temporary storage of waste. The ALSAG hardly comprises any regulations defining targets or criteria for assessment and remediation. There is no existing law for soil protection in Austria. Hence the "Water Act" (1959) serves as the basis for assessment and remediation. As a remediation target groundwater quality has to be returned to drinking water standards.

1.2. Implementation of the Law for Cleanup of Contaminated Sites (ALSAG)

Austria operates under a federal system with the national government supporting the work of nine provincial governments. The provincial governments are responsible for identifying potentially contaminated sites and reporting them to the Federal Ministry for Agriculture, Forestry, Environment and Water Management. The data are then passed on to the Federal Environment Agency (UBA). If a preliminary risk assessment is possible on the basis of the available data, the sites are registered in the register for suspected contaminated sites by the UBA.

Table1. Contaminated sites management

Site class	Available Information	No. of sites recorded	Type of site
Registered sites	Little, reference to former site activities	48,678	10% landfill deposits 90% industrial
Suspected contaminated sites	Result of preliminary assessment	2,100	89% landfill deposits 11% industrial
Contaminated sites	Detailed investigation programme	160	41% landfill deposits 59% industrial

On the basis of a risk assessment, the UBA identifies those potentially contaminated sites which pose a considerable threat to the health of humans or the environment. The risk assessment is based upon adequate investigations and studies. The contaminated sites are then recorded in the register for contaminated sites. The urgency of clean-up measures is expressed by means of a three-stage priority classification.

If a site does not pose a considerable threat to the environment, it is deleted from the register of suspected contaminated sites. Remediated sites are recorded in the register for contaminated sites as either safeguarded or remediated.

2. Contaminated Sites Management

2.1. Scheme for Classification of Sites

Assessment of potentially contaminated sites aims at determining those sites which have already produced an injurious effect or which provide a high risk of environmental pollution and where the need for clean-up is to be established. The assessment procedure can be divided into three phases as follows:

- In the first phase, the preliminary assessment phase, priorities for investigations of the potentially contaminated sites are established. The risk potential of a suspected contaminated site is evaluated;
- The second phase is dedicated to risk assessment, where it is determined if cleaning up is required. Results obtained by site investigations, like groundwater or soil analyses, are essential for the assessment; and
- The assessment procedure is completed with the classification of priorities, documenting the urgency of clean-up measures. (There are three priority classes distinguished.)

Table 1 gives an overview over the contaminated sites management in Austria (status 1.1.2007) [2].

Table 1 shows that initially the percentage of industrial sites reported to the Environment Agency by far exceeds the number of reported landfill deposits. After a preliminary assessment, almost 90% of landfill deposits are recorded as suspected contaminated sites whereas after a detailed investigation programme has been carried out the number of landfill deposits registered as contaminated account for only 41% of the total number of sites.

Figure 1 illustrates the geographical distribution of old landfill deposits and old industrial sites registered in the UBA database (total 48,678) [2].

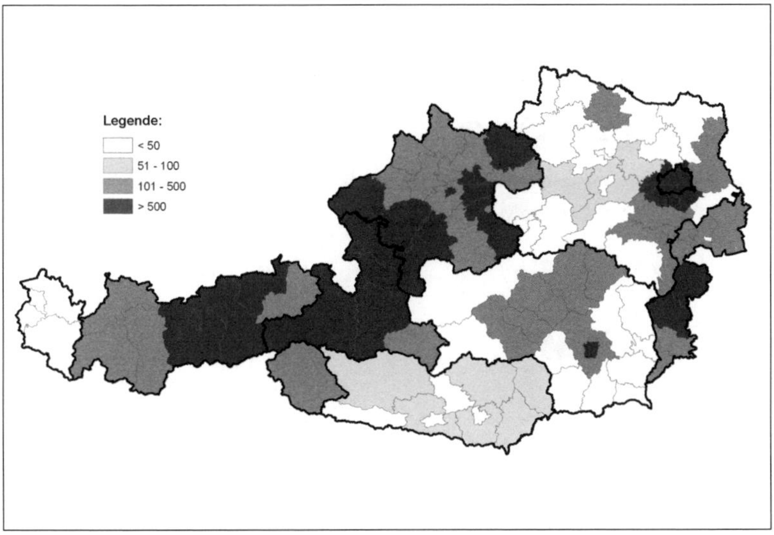

Figure 1. Number of registered old landfill deposits and old industrial sites.

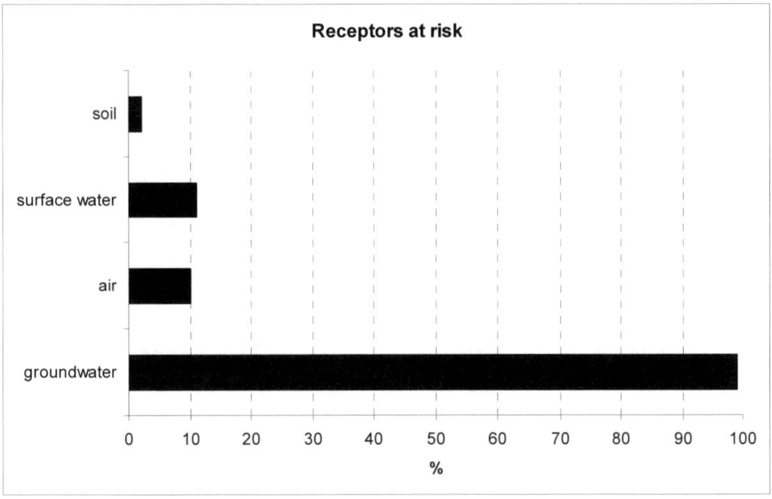

Figure 2. Frequency of receptors at risk from suspected contaminated sites (multiple nominations possible).

2.2. Contamination Sources and Receptors at Risk

Based on a preliminary assessment the UBA determines the receptors at risk from suspected contaminated sites (Fig. 2) [2].

Figure 2 shows that groundwater is by far the receptor most frequently at risk from suspected contaminated sites.

Figure 3 shows the land use of contaminated sites at the time of the risk assessment. The number of industrial sites by far exceeds other forms of land use [2].

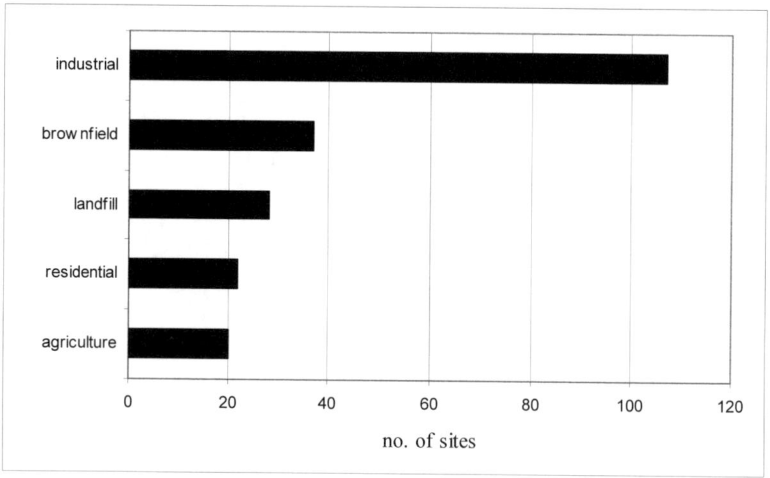

Figure 3. Land use at contaminated sites (multiple nominations possible).

Table 2. Major sources of contamination

Contamination source	Percentage
Metal works	25
Storage sites	25
Dry cleaners	20
Oil refineries	13
Gas works	7
Chemical industry	7
Tar works	3

The major contaminants encountered at a total of 195 contaminated sites were chlorinated hydrocarbons 30%, mineral oils 23%, heavy metals 13%, polycyclic aromatic hydrocarbons 12%, followed by phenols, cyanides, aromatic hydrocarbons (BTEX) and others – all with a frequency of less than 10% [2].

The major sources of contamination are metal works, storage sites and dry cleaners. Table 2 shows the frequency distribution at 123 contaminated sites [2].

3. Remediation

3.1. Funding

Between 1989 and 2006 about €1.1 billion (€136/inhabitant) was spent on the remediation and safeguarding measures of 144 contaminated sites. About 80% of the costs were covered by public means [1]. Public funding is mainly based on a special remediation fund which is fed by a fee on waste treatment. The collected funds are provided by the Ministry for Agriculture, Forestry, Environment and Water Management to the "Österreichische Kommunalkredit AG" (the Bank) which is organised on a private enterprise basis and is responsible for project management. The UBA and the Bank

Table 3. Allocation of priority classes

Class I	Class II	Class III
Groundwater contamination with high emission of contaminants. Actual/potential contamination of public supplies. Rapid spread of contaminants.	Groundwater contamination with low emission of contaminants. No harm to public supply expected. Extent of contamination limited; possibility of further spread is low.	No or irrelevant groundwater contamination with low emission of contaminants. No harm to public supply expected. Extent of contamination limited; possibility of further spread is low.

Table 4. Percentage of maximum available funds for remediation

Priority class	No responsible party found	Non-profit organisation responsible	Profit organisation responsible
I	95%	65%	Max. of €100.000
II	80%	60%	
III	65%	55%	

closely coordinate their activities; the UBA provides the risk assessment and the Bank provides the technical and financial management of the cases. 85% of the funds generated by levies are intended to finance remedial actions with 15% to be used for investigations of those sites where investigations are required urgently and the polluter cannot be made liable.

The aim of the funding is the protection of the environment *via*

- the remediation of contaminated sites with the highest ecological benefit at economically justifiable costs, and
- the safeguarding of contaminated sites if – considering the risk – it is acceptable and remediation at the time is not feasible or can be carried out only at excessive cost.

A requirement for obtaining the necessary funding for remediation of a contaminated site is the allocation to a "priority class", indicating the urgency for remediation. Such an assessment is based on:

- Contaminants' properties;
- Extent/spread of contamination; and
- Vulnerability of the receptor.

Table 3 summarizes how the three priority classes can be defined.

Once a priority class is proposed by the UBA, an application for funding is made to the Bank which is then discussed in a committee, together with the Ministry of Agriculture, Forestry, Environment and Water Management which is the institution allocating the funds.

The priority class influences the sum allocated to carrying out remediation/safeguarding measures. Table 4 shows the maximum available funds related to each priority class [3].

As of 1st of January 2007 a total of 143 contaminated sites have a priority assigned: 36 sites with a Priority I, 54 sites with a Priority II and 53 sites with a Priority III [2].

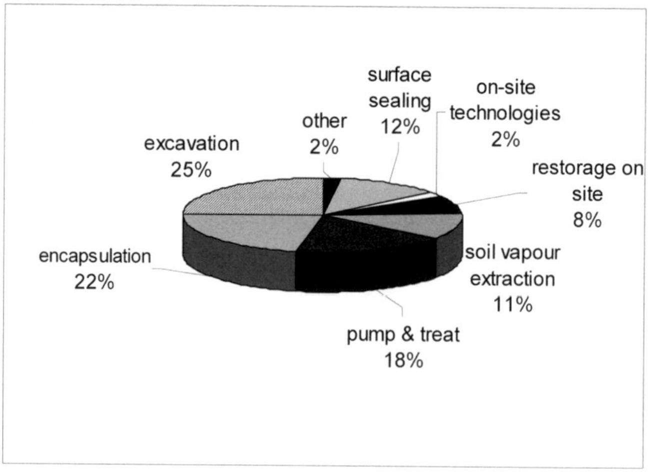

Figure 4a. Techniques used for the remediation of 104 contaminated sites.

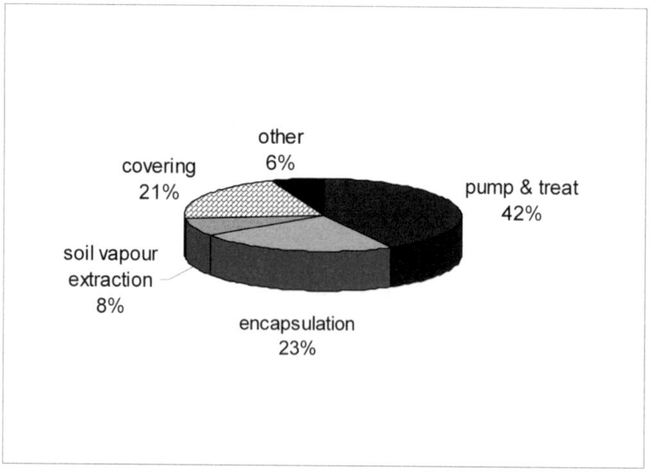

Figure 4b. Safeguarding measures.

3.2. Safeguarding and Remediation Measures

Between 1989 and 2007, 238 contaminated sites were recorded in the register of contaminated sites; 78 of these sites have been remediated or safeguarded, at 67 sites remediation is in progress, and for 93 sites remediation measures are planned [2].

Figure 4a and 4b show the techniques most frequently used to carry out remediation and safeguarding measures.

As for the remediation techniques, excavation, encapsulation and pump-and-treat were most frequently applied. As far as safeguarding measures are concerned pump-and-treat (42%) is by far the most frequently used technique.

3.3. Environmental Effects

- The remediation of contaminated sites enabled a qualitative amelioration of groundwater bodies to the extent of about 46 million m^3. This amount of water corresponds to the annual consumption of about 1.7 million people [1].
- Remediation measures enabled the reintroduction of 145 ha brownfield* land back to the real estate market [1]. (* *'Brownfield' has become the adopted description of any formerly developed land to distinguish it from 'greenfield'.*)
- Through the remediation of contaminated sites the exhaust of greenhouse gases, especially methane, was prevented. In 2004 and 2005 a reduction of annual greenhouse gas emissions by 0.3% (related to the total Austrian greenhouse gas emissions) was achieved [1].

4. Groundwater Remediation at a Site in Linz contaminated with Chlorinated Hydrocarbons (CHC)

Groundwater is by far the most frequent receptor at risk from contaminated sites. In the worst case drinking water supplies are affected as happened in the town Linz where groundwater contamination with CHC occurred in a densely populated residential as well as commercial area.

The Linz sites are in the Danube river valley where the sedimentary basin is filled with well permeable fluvial (more or less) sandy gravel of a hydraulic conductivity of about 5×10^{-3} m/s. The unconfined aquifer of about 20 m thickness is underlain by tertiary marine sands and clays ("Schlier"). The thickness of the groundwater body increases from east to west from 5 m to 12 m. The flow regime is dominated by the river Danube, which provides most of the water flowing into the aquifer by bank infiltration. The groundwater flow is mainly parallel to the river with some local deviations due to a drainage channel and local hydraulic heterogeneities. Groundwater from the shallow aquifer system is used for public water supply and there are two drinking water supplies ("Heilham" and "Plesching") within the catchment of several contaminated sites. The local waterworks "Heilham" is operating at a consent of 120 l/s, "Plesching" at a consent of 300 l/s. In the 1980s contamination with tetrachloroethene (TCE) had been detected in a water supply well at the "Heilham" waterworks where average concentrations of about 500 µg/l TCE were measured. Consequently the waterworks was shut down (1983). Locations of potential sources were identified by a historical survey, focusing on dry-cleaners and other companies using TCE. The impairment of the groundwater quality at "Heilham" originated from several contaminated sites. Massive contamination originated from two dry-cleaners and one plumber (formerly also a site where dry-cleaners were situated) located to the east of the waterworks. Maximum concentrations of > 2000 µg/l TCE were measured. From 1984 a scavenging well was operated about 200 m southeast of the sites in order to prevent a spread of the CHC contamination of the groundwater towards the waterworks "Plesching".

Figure 5 shows the location of the two waterworks and the contaminated sites as well as the direction of groundwater flow.

The site with where the highest TCE concentration in the groundwater was measured is a dry-cleaner located about 2 km northwest of the waterworks Plesching and about 800 m to the southeast of the waterworks Heilham. Since 1992 water has been

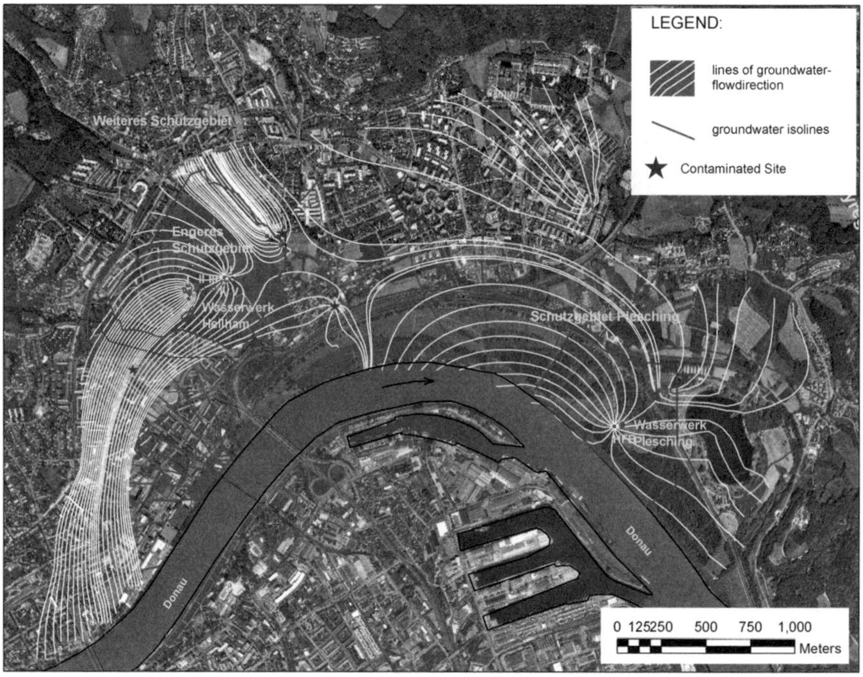

Figure 5. Groundwater flow, location of waterworks and sources of contamination.

pumped continuously (11 l/s) at the site in order to prevent the spread of contamination and an impairment of the waterworks.

In 1997 remediation measures were undertaken at all three sites leading to a marked reduction of the TCE concentration in the groundwater. The clean-up was carried out using the following measures:

- soil vapour extraction and cleaning of the extracted soil vapour; and
- abstraction of groundwater and clean-up of the abstracted water.

Remediation targets were:

1. the decontamination of the vadose zone to prevent the threat of CHC contamination to the groundwater; and
2. the restoration of groundwater quality up to drinking water standards.

Remediation targets for soil-air were set at 10 mg/m^3 for BTEX and the sum of CHC. As for the groundwater, the remediation target for the sum of CHC was set at 18 μg/l, for TCE at 6 μg/l.

From 1998 a distinct reduction of CHC in the groundwater was measured at the "Heilham" waterworks.

Figure 6 shows the remediation measures carried out at the two sites (dry-cleaner, plumber) located to the northeast of the waterworks "Heilham".

To cover the whole cross-section of the contaminant plume downstream of the pollutant source, pumping tests with multiple contaminant concentration measurements were carried out in 2001. Due to the spatial integration of the pumping tests and due to the increasing capture zone with pumping time, both the spatial distribution of the con-

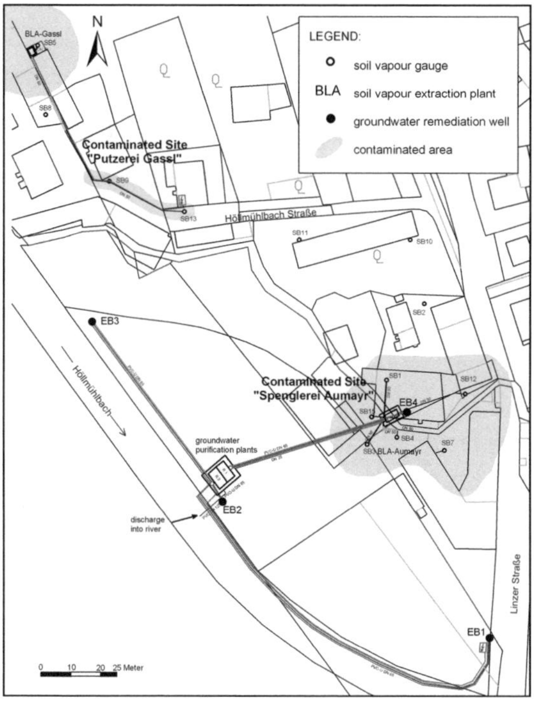

Figure 6. Groundwater remediation – position of the remediation wells.

taminants as well as the total mass flow rate within the contaminant plume could be estimated. A total of 10 pumping wells were placed along three control planes perpendicular to the groundwater flow direction. Pumping rates and pumping times were chosen to yield a full coverage of the control planes. Each of the wells was pumped for a time period of about five days using a constant pumping rate of up to 15 l/s. The concentration time series yielded information on the position and extent of the contaminant plumes as well as on the concentrations of the CHC in the plumes. Using this integral investigation method, further contamination sources could be identified [4].

At the site of the dry-cleaner soil vapour extraction was carried out over a period of 5 years (1997–2002). About 75 kg of CHC, mostly TCE, were removed from the soil.

Between November 1997 and May 2006 about 590 kg of CHC (TCE) were removed from the vadose zone of the soil at the site of the plumber using soil vapour extraction; whereupon more than 200 kg were removed in the first 6 months. Figure 7 shows the total CHC extracted from the soil air at both locations.

For groundwater remediation four remediation wells were constructed. (See Fig. 6.) From 1997 a total of 9 l/s was extracted and cleaned via an air stripper. From 2004 only one of the remediation wells was operated at an abstraction rate of 3 l/s. The groundwater quality was controlled monthly. Figure 8 shows the CHC concentration in the four extraction wells. At the start of the remediation measures the CHC concentrations at the site of the plumber were above 500 μg/l, after about 3 years the CHC contamination was reduced to less than 100 μg/l. Between June 2005 and May 2006 the CHC concentrations ranged between 20 μg/l and 30 μg/l.

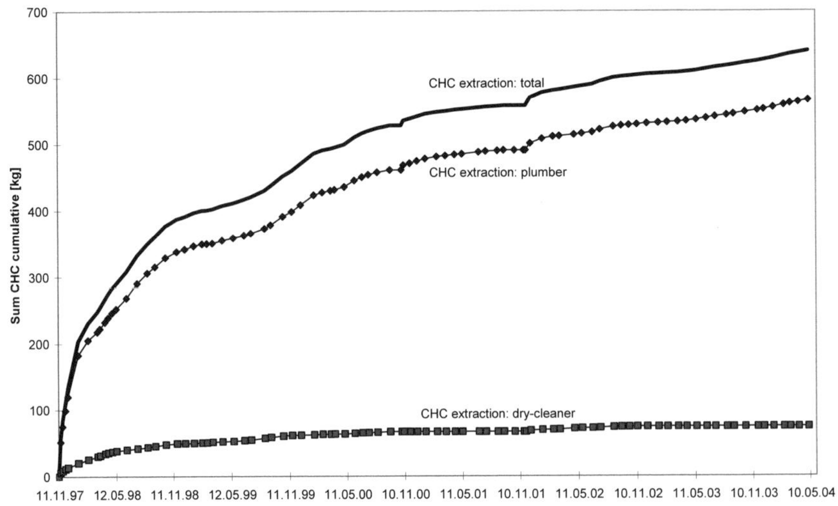

Figure 7. Total CHC extracted from soil air between 1997 and 2004.

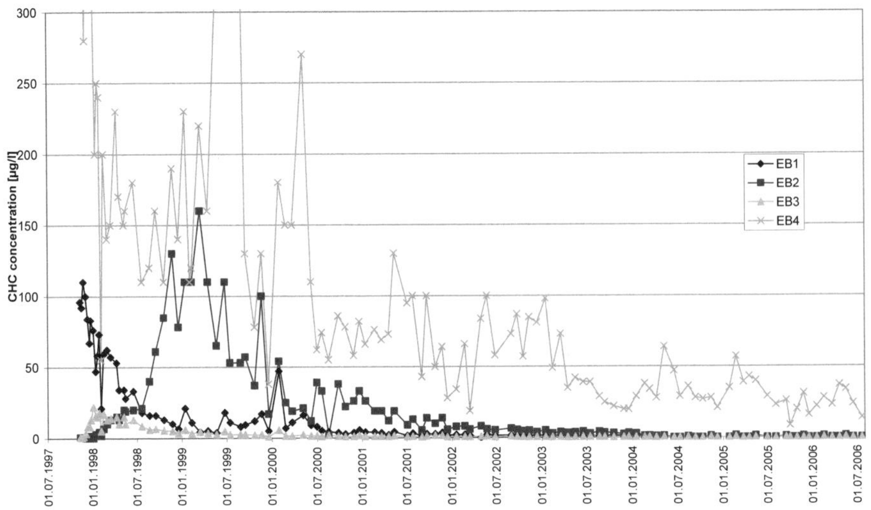

Figure 8. CHC concentrations in the groundwater.

Over a period of 6 years a total of about 65 kg CHC (TCE) were removed from the groundwater. The pollution load in the groundwater could be reduced by 75%. However, the remediation target could not be met, but is unlikely to be achieved within the next 10 to 20 years. Further remediation measures at the site do not seem practicable especially as there is no risk for the waterworks "Heilham" to be expected.

At the same time, two wells at "Heilham" were operated for groundwater remediation, each of them at an abstraction rate between 20 l/s and 25 l/s. From November 1997 to May 2006 about 140 kg CHC were extracted from the groundwater.

Presently remediation measures are planned at two more sites (dry-cleaners) upstream of the waterworks. The "Heilham" waterworks should be back into operation in 2008 or 2009.

References

[1] Altlastensanierung in Österreich. Effekte und Ausblick. Wien, Okt. 2007
[2] Verdachtsflächenkataster und Altlastenatlas. Stand 1. Jänner 2007. Umweltbundesamt.
[3] Altlastensanierung oder Sicherung. Förderungsrichtlinien 2002. Wien, 2002
[4] INCORE – Integrated Concept for Groundwater Remediation. Integral Groundwater Investigation. T. Ptak et al. 2003.

Theme 4

Health/Radiological Hazards/Risks

Optimisation of Disaster Forecasting and Prevention Measures in the Context of Human and Social Dynamics
I. Apostol et al. (Eds.)
IOS Press, 2009.
doi:10.3233/978-1-58603-948-6-207

Radio-Ecological Monitoring in Moldovan Agricultural Industry as a Factor for Forecasting, Evaluating and Mitigating the Impacts of Radiological Pollution of Agricultural Land

Semion NEDEALKOV
Republican Centre of Applied Soil Science, Ministry of Agriculture and Food Industry
6 Cosmonautilor str. MD-2005 Chisinau, Republic of Moldova
E-mail: semned@mail.ru

Abstract. In the article, the assessment of the nuclear environment on the territory of the Republic of Moldova before and after the Chernobyl accident is presented. The results of research on Sr -90, Cs -137 nuclide migration in the soil surface, the factors and parameters causing their accumulation and transfer to the yield of the main agricultural crops are shown. On that basis, the methodical approaches to the assessment, forecasting and regulation levels of contamination of plants and soils are described. The need for a radiological monitoring and its application as a means of nuclear product quality management, and the optimization of protective measures for the diminution of consequences in the agricultural sphere from nuclear contamination of agricultural land, is defined.

Keywords. Nuclear environment, agricultural soil and plant uptake, radiological monitoring, chernozem soils

Introduction

The extensive use of technologies with radioactive substances and nuclear energy in different branches of the economy, and primarily in the development of nuclear energy, is connected to the issue of assurances on safety and protection of the environment and the community in cases of nuclear failures. The acuteness of nuclear energy safety and its impact on the environment, is connected to the deployment and building of nuclear stations in areas with high density of population and intensive agriculture. Some 50% to 90% of the current zones of deployment of functioning nuclear stations occupy agricultural lands [1], while the zone of impact on the environment from any nuclear failures can greatly exceed that limit.

The assessment of the consequences of major nuclear failures shows that in all cases the sphere of agricultural production is the most intensively influenced. The radioactive contamination of agriculture and agricultural products relate to a number of leading factors from which the degree of nuclear hazard, and the scale of measures for recovering the consequences of such failures, depend. That is due to the fact that the main dose-burden on the population is conditioned mainly by the use of agricultural products produced in contaminated areas, while the role of other routes of radiation (on the account of contaminated soil, living dwellings etc.) specifically in later periods, is less important.

1. Radio-Ecological Monitoring

Radio-ecological monitoring in the sphere of agricultural-industrial production is a system of continuous observations, assessment and forecasting of radioactive contamination of natural components and biota elements which are subjects and products of the agricultural activity of humans. The study of the patterns of the behaviour of radionuclides in all parts of the agricultural ecological system, knowledge of the patterns of soil and agricultural plant contamination and methods directed towards its diminution allow for determining the levels of interference, rationally organizing the development of the agricultural industry under conditions of radio-active contamination of areas, and accomplishing scientifically approved systems of measures oriented towards the reception of clean agricultural products [2]. The importance of the accomplishment of the radio-ecological monitoring is conditioned not only by the fact that agricultural products are the main source of access for radio-nuclides entering the human body, but also by the fact that that route of nuclear impact is a more controllablc and possiblc one to regulate.

The beginning of the radio-ecological studies in the Republic of Moldova was set up by the creation of a network of radiological control in the agro-chemical service at the beginning of the 1970s in connection with the testing of nuclear weapons and the development of the nuclear energy. The tasks of those studies were:

- development of a system and accomplishment of control over radioactive contamination of agricultural lands and harvests of the main agricultural crops by the losses of products of nuclear partitioning: radio-nuclides Sr^{90}, Cs^{137}; and
- development of elements of stability/resistance of agriculture under conditions of impact of radiation factors of nuclear bursts, failures at nuclear energy facilities.

2. Historical Background Radioactive Levels

The contamination of the territory of the Republic of Moldova (RM) used to have an overall characteristic of increasing from the south towards the north corresponding, obviously, with the amount of annual precipitation. The increase in levels of contamination was noticed in 1974, 1978, and 1981 [3]. By 1986, the density of contamination of agricultural lands was 0.03–0.05 Ci/km^2. The consequences of the failure at the Chernobyl nuclear power station changed significantly the radiation situation. The beginning of radioactive losses /fall outs/precipitation happened to be from 30 April 1986 to 1 May 1986. The levels of radiation γ-background established on the territory of the RM during the first 20 days of May were 60–430 $\mu R/h$. Taking into account the period of half-disintegration of the mixture of the partitioning products, which at that time was equal to 13 days, the level of radiation in the first days of May on separate sectors of agricultural lands was > 1mR/h. The radioactive precipitation resulted in air contamina-

Table 1. Levels of pollution of a crop of the some plants in 1986 (Bq/kg)

Culture	Basic production		Collateral production	
	Sr^{90}	Cs^{137}	Sr^{90}	Cs^{137}
Winter culture	0.4–19	20–160	60–440	220–3000
Corn	0.1–1.0	0.8–1.3	5.0–13.0	8.0–15.0
Sunflowers	0.3–2.5	0.5–5.9	2.0–5.0	3.0–12.0
Peas	0.7–1.0	4.0–6.0	10.0–40.0	120–165
Soya	0.3–0.7	0.8–1.5	0.9–2.5	7.0–15.0
Vegetables	0.5–3.4	1.5–35.0		
Vegetables (root crops)	0.5–1.5	1.5–5.0		
Apples	1.0–1.5	10–35		
Dried fruits	3.5–9.0	32–130		
Tobacco	18.0–40.0	12.5–36.0		
Lucerne (first hay crop)	1650–3600	3000–10000		
Sugar beet	0.3–0.8	0.5 – 2.3		

tion of green forage, sowings, and later the harvests of agricultural crops. The "Iodine" danger factor manifested itself at that time.

By the time of organizing an operative γ-spectrometric control over the contamination of agricultural products (in the second half of May 1986), the levels of contamination of grasses from I^{131} were up to $2 \cdot 10^4$ Bq/kg at the limit of $1.5 \cdot 10^3$ Bq/kg, which resulted in the contamination of milk, dairy products and the need for measures of interference.

The sowing seed and the harvests from agricultural crops were contaminated by Sr^{90}, Cs^{137} to different extents. As one can see from Table 1, the levels of contamination of different crops depended on the type, morphological peculiarities of plants, and the time of sowing and harvesting.

3. Principal Monitoring Tasks

The determination of real levels of soils contamination is one of the most important tasks, without which it is impossible to assess the radiological status. At the current stage, the radio-ecological situation of agricultural areas is assessed by such a criterion as the density of contamination, i.e. the contents (reserves) of radio-nuclides in the arable or unprocessed layer of soil on an area of 1 m^2. The transition from the content of radio-nuclides in a unit of soil weight to the density of contamination (DC) is calculated following the formula:

$$DC = 2.7 \cdot 10^{-4} \cdot C_s \cdot h \cdot d \text{ Ci/km}^2 = 10^{-2} \cdot C_s \cdot h \cdot d \text{ kBq/km}^2 \quad (1)$$

where C_s is the quantity of the radio-nuclides in soil (in Bq/kg), *h* (in cm) is the depth of an arable layer, and *d* is the specific weight (density) of soil (in g/cm^3).

According to the data of radiological research accomplished in 1987 and 1988 by the radiological subdivisions of the agro-chemical service, the aerogram-pictures and its on-ground connection accomplished by the state industrial association "AGeom" in 1991, the territory of the RM is characterized mainly by a weak level of contamination – up to 1 Ci/km^2 for Cs^{137}, up to 0.15 Ci/km^2 for Sr^{90} and with a consistent predominant contamination of 0.15 Ci/km^2 and 0.06 Ci/km^2, respectively. The sector of local contamination with a density of up to 4.5 Ci/km^2 for Cs^{137} and up to 0.3 Ci/km^2 for Sr-90 on an area of 10 km^2 was identified. The contamination of the territory has an uneven, spot-like pattern with the prevalence on raised elements of landscape and adjacent slopes. To a large extent, the northern part is the most contaminated, while the central and southern parts of the RM are less contaminated.

The expertise of the radiological research accrued during the ante-Chernobyl period, the personnel and the technical staff, allowed for a comparatively short-notice assessment of the radiological status in agriculture, and to organize the radiation control of the quality of agricultural products by periods – according to an integral γ-activity, by isotope I^{131}, radio-nuclides Cs^{137}, Sr^{90} with the use of dozed, radio metrical, γ-spectrometric and radio-chemical methods.

The radio-ecological research in the sphere of the agricultural-industrial complex, after the assessment of the Chernobyl accident consequences, acquired a more distinctly oriented character. In the RM there are no levels of soil contamination needing measures to decrease the contamination of flora products. But even the existing levels allow for the accomplishment of research on the main issues of radio-ecology under certain soil-climate conditions, which is very important for the development under those conditions of practical recommendations for the assessment and forecasting of the radiation situation, and for planting activities under conditions of radio-active contamination of agricultural soils.

The main tasks of monitoring activities are to:

- control the soil and agricultural crop harvest contamination;
- assess the radiological situation in the sphere of agricultural-industrial production by taking into account the deployment around the RM of the nuclear stations network, and the definition of the tendencies and forecasts for its possible modification;
- study the patterns of the radio-nuclides Sr^{90}, Cs^{137} behaviour in the soil-agricultural plant systems, determination of quality parameters of migration of radio-nuclides and the dynamics of their development; and
- submit information regarding the radiation situation and its modification to the state bodies for decision-making.

Large-scale consequences of radiation accidents as well as the overall character of the transition of radioactive substances are connected to the atmosphere, create aerosols, and under the impact of gravity, as well as rain, snow and fog, fall down on the soil surface polluting open water sources, flora and the soil. The control over the radioactive contamination of the next-to-soil layer of atmosphere-aerosols of the air, atmosphere precipitations give the opportunity for the rapid detection of the transition of the most insignificant concentrations of the radio-nuclides, and the assessment of their density of falling and deposition on the soil surface.

This research is done according on a regular basis in Chisinau starting in 1993. The isotope composition and the integral α, – ß – γ-activity is measured by the ß – γ spectro-metric (weekly) and radio-chemical (monthly) definitions of the radio-nuclides Sr^{90}, Cs^{137}. During the whole period of study, in the atmosphere there were no short-life products of nuclear division detected. However, the concentrations of Sr^{90}, Cs^{137}, up to 2003, frequently exceeded 10 times the background figures which, obviously, were connected to their trans-border transition onto the territory of the RM.

4. Soil and Plant Contamination Patterns

The soil, having accumulated the precipitated radio-nuclides on its surface, becomes itself a long-term source of plant contamination. Its physico-chemical properties play a leading role on which the amount of radio-nuclides penetration into the pants depends. Different types of soils have the regular combination of soil and agro-chemical peculiarities. According to the combination of those peculiarities, the chernozem soils (i.e. rich black topsoil) that prevail in the RM have a greater ability of forcing the radio-nuclides to decrease their penetration into the soil.

The physico-chemical peculiarities of soils, precipitations of radioactive substances and the radio-nuclides themselves determine the biological accessibility of the radio-nuclides. The study of the status of chemical forms of the radio-nuclides in soil by means of their consistent extraction with the help of different desorbents (such as an ammonium-acetate buffer solution with the Ph-4.8; IN H1 and 6 GN HCl) allowed for the conclusion that 90% of Cs-137 of its reserve in soil is non-changeable and inaccessible for plant forms. And vice versa, 90% of the Sr-90 reserve in soil is changeable and mobile, and accessible for plants' chemical form. This is obviously characterizing the fact that, as a rule, Sr^{90} is, by approximately an order of magnitude, more accumulated in plants than Cs^{137} is. The migration of radio-nuclides on the vertical profile of soils is apparently explained by the difference of mobility: for Sr^{90} it is equal to 50–60 cm, and for Cs^{137} it is 40–50 cm.

An important role in the formation of the radiation status on agricultural lands is played by the re-distribution of radio-nuclides on landscape elements. This has great importance for the geomorphological conditions of agricultural soils of RM, where 80% of the arable land consists of slope angles of more than 1°. As a result of the water erosion processes resulting from types of rain, there is a wash-out and secondary accumulation of radio-nuclides in the lower parts of the surface. Due to secondary accumulation, the reserves of radio-nuclides in the profile of the deposited soils increased over 50 years by 1.5–4.5 times.

The soil-plant is an active biological system, and the pattern of the radio-nuclide behaviour in this system can be expressed by such parameters as the coefficient of deposition and coefficient of transfer. The coefficient of accumulation (K_a) is an important feature of contamination of plant products as it represents the co-relation between equilibrium concentrations of radio-nuclides in plants (C_p) and in the soil (C_s):

$$K_a = C_p/C_s \tag{2}$$

To asses the penetration of radio-nuclides into the plant, taking into account the density of the soil cover contamination, there is used such a parameter as the coefficient of penetration (K_t), which reflects the deposition of radio-nuclides in the plant on an arable layer unit, and is defined from the co-relation:

$$K_t = C_p/DC \quad (3)$$

Table 2 shows the average values of coefficients of deposition and the penetration of the radio-nuclides Sr^{90}, Cs^{137} into the crops of some plants on chernozem soils. The data were derived from a summary of many years' measurements of the levels of contamination of soils and plants. The selection of samples is effected along with the radiological study of testing areas of agricultural lands on the territory of the RM as well as stationary experiences on fertilizers by the research agencies on field crops, namely, "Selectsia", "Soil Science and Agro-Chemistry", State Agricultural University. According to the data from the radiological study, it was determined that regardless of the low levels of contamination and re-distribution of the radio-nuclides, a close inter-relational link between the volume of the radiation γ-background and the density of soil contamination Cs^{137} (r = 0.89) is:

$$DC = k \cdot P_\gamma \quad (4)$$

In Formula (4) P_γ is the value of the γ-radiation background (m/hour) and *k* is a coefficient of proportionality determined experimentally (by a spectrometric method) on the areas with a wide scale of the values of the γ-radiation background and Cs^{137} concentrations. Such a method of determining the levels of contamination of soils is a cost-efficient one, fast and pretty accurate as the γ-background is being formed from a wider surface and characterizes the average level of its contamination.

As one can see on Table 2, the deposition of radio-nuclides on agricultural crops has substantial comparative differences, depending on their associations with different botanical species. The grain crops are characterized by a lower deposition rate than the beans, and to a lesser extent they are deposited in the grains of corn. The deposition of radio-nuclides in some vegetables is very low and it significantly increases in technical crops, and in the vegetative organs of all the plants. It is obvious that under the conditions in the RM, where the soils with high fertility of the heavy mechanical composition dominate, the decisive factors in deposition and penetration of radio-nuclides into the harvest of agricultural crops are their associations with the species and the biological peculiarities.

5. Forecasting Contamination Levels

Forecasting the levels of contamination of agricultural products by radio-nuclides is being done based on the results of the preliminary assessment of real levels of soils contamination – concentrations of the radio-nuclides in soil (Bq/kg) and density of contamination (kBq/m^2) and coefficients of deposition and transition correspondingly for a real crop (out of inter-relations (2), (3)). With the radiation-hygienic approach to the forecasting of possible contamination of plants products, the basis for decision-making is the hygienic norms of radio-nuclides contents in food products.

Table 2. Average values of factors of accumulation and transition of Sr^{90} and Cs^{137} in plants on chernozem soils

Plant	Body	Sr^{90}		Cs^{137}	
		K_a	K_t	K_a	K_t
GRAIN					
CORN	Grain	0.014	0.04	0.003	0.007
	Stalks	0.39	0.97	0.05	0.13
WINTER WHEAT	Grain	0.043	0.12	0.005	0.013
	Stalks	0.24	0.75	0.03	0.08
BARLEY	Grain	0.05	0.14	0.01	0.035
	Stalks	0.26	0.74	0.06	0.18
GRAIN-BEAN					
PEAS	Grain	0.09	0.24	0.02	0.05
	Stalks	1.0	2.67	0.07	0.19
SOYA	Grain	0.10	0.28	0.03	0.08
	Stalks	0.96	2.64	0.06	0.18
BEANS	Grain	0.11	0.30	0.01	0.02
	Stalks	0.46	1.25	0.03	0.08
TECHNICAL					
SUGAR BEET	root crops	0.07	0.21	0.01	0.03
	Leaves	0.19	0.49	0.15	0.43
SUNFLOWER	sunflower seeds	0.08	0.21	0.03	0.08
	Stalks	0.51	1.42	0.13	0.29
TOBACCO	Leaves	1.80	3.28	0.16	0.44
	Stalks	1.79	4.96	0.12	0.32
FODDER					
CORN	green weight	0.36	0.99	0.037	0.10
LUCERNE	green weight	0.32	0.88	0.03	0.09
	Hay	0.98	2.71	0.07	0.19
BEET FODDER	root crops	0.32	0.85	0.05	0.14
	Leaves	1.17	3.27	0.18	0.49
GRASSES	Hay	0.76	2.89	0.07	0.69
VEGETABLE					
EGGPLANTS		0.013	0.037	0.004	0.010
PEPPER		0.014	0.038	0.004	0.010

Table 2. (Continued.)

Plant	Body	Sr^{90}		Cs^{137}	
		K_a	K_t	K_a	K_t
CUCUMBER		0.017	0.045	0.004	0.012
TOMATOES		0.023	0.062	0.004	0.012
POTATO		0.032	0.086	0.011	0.036
CABBAGE		0.035	0.035	0.004	0.014
BEET		0.06	0.168	0.005	0.014
CARROTS		0.10	0.268	0.006	0.016
ONIONS		0.20	0.541	0.008	0.021
WATER-MELONS		0.003	0.007	0.002	0.005

Based on the limited level of radio-nuclide content (LRNC) in plant products (hygienic normative) and the value of coefficient of penetration (Table 2), one can calculate the limit density of soil contamination (LDSC) for a real crop is:

$$LDSC = LRNC / K_t. \tag{5}$$

The limited density of soil contamination is the density at which the contamination of plant products does not exceed the hygienic norms. Values of limited densities of chernozem land for the harvest of a series of agricultural crops are given on Table 3. As one can see, the range of limited levels of soil contamination for the production of different types of agricultural crops is pretty wide. The level of the limited soil contamination is lower under sunflower and soya for Cs^{137}, and under carrot and food beetroot for Sr^{90}. However, growing agricultural crops on soils with such levels of contamination, regardless of the forecasts of "clean" products, is impossible due to radiation-hygienic aspects. A limiting factor here is the dose burdens (loads) on the population and working staff.

The admissible levels of growing agricultural crops on contaminated areas are: for Cs^{137} – 40 Ci/ km², and for Sr^{90} – 3 Ci/ km².

6. Concluding Comments

Under the RM conditions, the levels of interference when it would be necessary to undertake counter-measures in agriculture under radioactive contamination of agricultural lands seem to be very low. Even where needed, the implemented measures can be limited to those that do not involve significant costs. Potential options in choosing those measures are quite wide and include: variations in agricultural plants; their deployment in fields with consideration of the density of contamination, and the degree of deposition of radio-nuclides; and aims and possibilities of using the resulting products (e.g. food, forage, industrial processing, and seed production). The use of mineral fertilizers and ameliorants in parallel is necessary, both for an increase in soil productivity as well as for a decrease in penetration of the radio-nuclides into the resultant products.

Table 3. Limiting density of pollution of soils for growing of some food plants with levels of maintenance Sr^{90}, Cs^{137} meeting the requirements NFRP-2000

Culture	Density of pollution			
	Sr^{90}		Cs^{137}	
	кBq/m²	Ci/кm²	кBq/m²	Ci/кm²
Corn	3500	95	11430	310
Winter wheat	1150	31	6150	166
Winter barley	1000	27	2285	62
Peas	580	16	1650	45
Soya	500	14	1000	27
Beans	470	13	3720	100
Sunflower	430	12	875	24
Eggplants	1330	36	13000	350
Pepper	1315	35	13000	350
Cucumbers	1100	30	10830	290
Tomatoes	800	23	10830	290
Potato	700	19	8900	240
Cabbage	1000	27	9280	250
Beet	300	8	9280	250
Carrots	186	5	8125	220
Onions	67	2	6200	170
Water-melons	7140	190	26000	700

References

[1] Всероссийский научно-исследовательский институт сельскохозяйственной радиологии и агроэкологии. Организация государственного радиоэкологического мониторинга агроэкосистем в зоне воздействия радиационно-опасных объектов. Методические указания Москва 2000.

[2] Сельскохозяйственная радиоэкология. Москва «Экология», 1991.

[3] С. Недялков, «Радиоэкологические аспекты в агрохимическом обслуживании», Agricultura Moldovei 3/2001, p. 15-17.

[4] S. Nedealcov, I. Bălan, L. Buiuc, I. Burlacu «Forme chimice de existenţa ale Sr^{90} şi Cs^{137} în sol si migrarea lor», Serviciul agrochimic de stat al Republicii Moldova la 35 de ani, Chisinau – 1999, p. 126-143.

[5] Н.А. Лощилов, П.Ф. Бондарь. Ю.А. Иванов, «Научно-экспериментальное обоснование экспресс-методики оценки загрязнения сельскохозяйственных угодий радиоактивными изотопами цезия», «Проблемы сельскохозяйственной радиологии», Сборник научных трудов, Украинский научно-исследовательский институт сельскохозяйственной радиологии, Киев 1991, p. 15-36.

[6] Nedealcov S., Străjescu O., Buiuc L., Rotari E., «Normarea şi Evaluarea Nivelurilor de Poluare a Solurilor cu radionuclizi de Cs^{137} si Sr^{90}». Pedologia modernă în dezvoltarea agriculturii ecologice. Materialele conferinţei ştiinţifico-practice, Chişinău 2006, p. 93-100.

I. Apostol et al. (Eds.)

doi:10.3233/978-1-58603-948-6-216

The Cytogenetic Status of Human Organism as a Diagnostic Parameter in a System of Socio-Ecological Monitoring

Alla GOROVA, Irina KLIMKINA and Yury BUCHAVY
National Mining University of Dnepropetrovsk, Ukraine
K. Marks avenue, 19, Dnepropetrovsk, Ukraine, 49005
E-mail: gorovaa@nmu.org.ua; irina_klimkina@ukr.net; yurique@3g.ua

Abstract. This paper offers a methodology for socio-ecological monitoring. It consists of high-sensitive indicator systems and a unified procedure for estimating the integral environmental condition with regard to the toxic-mutagenic background, and the general and genetic health of the population. It permits the definition of the integral criteria describing the ecological and genetic hazards for biota and human beings from the impact of mutagenic ecological factors. The paper shows a perspective of using a micro-nucleus test in the system of a socio-ecological monitoring in somatic cells of a person. This test also defines the general mutagenic background and the state of a human organism in the cytogenetic status.

Keywords. Socio-ecological monitoring, environment, toxico-mutagen background, population health, bio-indicators, micronucleus test, cytogenetic status

Introduction

The nation's health and environmental quality are significant integral factors of a civilized level of society and its social and economic development. It is not casual that in the developed countries the state of the ecosystems and population health are considered as life quality criteria, and they are one of the main priorities in governmental activity [1,2]. For this reason the estimation and control of the environmental state and health of the nation have increasing value in Ukraine.

The influence of environmental contamination on human health is analyzed in various aspects. Besides, there is a relationship between the emergence or the incidence rate of a specific disease, as well as environmental contamination by this or that substance, or by a sister compound [3,4]. However, in order to project a sustained development of the state it is necessary to make an integral estimation of the anthropogenic influence on human health and on the environment.

A concept of sustained development in the social aspect foresees a gradual decrease in the specific gravity of diseases in the population and the genetic dangers for the next generations from technogenic pollutants circulating in the biosphere. In this connection it is especially necessary to quantitatively assess the influence of ecotoxicants on population health and its genetic fund.

One of the tests in the system of socio-ecological monitoring, and also the most acceptable method for an express-estimation of mutagenic influences, is a calculation of the occurrence frequency of micro-kernels in human cells [6]. It is established that a micro-nucleus test (MN-test) in somatic cells of a human mucous oral cavity in respect of sensitivity can be compared to a test in the study of chromosome aberrations in a culture of leucocytes of peripheral human blood and cells of bone marrow in animals; further, this test is much less labour-intensive. A MN-test is an express method of the analysis for the mutagenic influence of physical, chemical and biological pollutants of the environment [7]. In particular, it should be noted that the MN test is used during an examination of various population segments, so as to detect industrial and ecological mutagenic factors, as well as for a screening of genetically caused instability of the genome.

The purpose of this paper is to show a system development for socio-ecological monitoring considering the factors that describe the cytogenetic status of a human organism. It also deals with the substantiation of priority-oriented managerial decisions directed at a decrease in the technogenic loading, as well as improving of environmental conditions, health and the nation's genofond.

1. Methods of Assessing Environmental Conditions and Population Health

A structural diagram of factors of our complex socio-ecological monitoring is shown in Fig. 1. As can be seen in the diagram, the top (or zero) level characterises an ecological and social state of the integral system of development for the territory at local, regional or national levels. It includes two factors of the lowest (or first) level: the state of population health (a population block), and the environment's ecological state, which can be determined by bio-indicative methods (a bio-indicative block) and by methods of a physico-chemical analysis.

The second structural level is presented by the factors which, in the population block "population health", characterise a natural population movement, physical health of children and adults, and genetic health. The bio-indicative block is characterised by the factors reflecting the ecological state of the atmosphere, hydrosphere and pedosphere in the toxico-mutagenic background.

The third structural level is presented by the factors which fill the blocks of the second structural level, i.e. they represent the factors describing the physical and genetic health of the population. The physical health of children and adult population is characterised by the morbidity rate of the following illnesses: infectious and parasitic, illnesses of the endocrine system, those of blood and blood-forming organs, mental disorders, illnesses of the nervous system and sense organs; blood circulation illnesses, those of respiratory organs, digestion, urogenital system; skin and hypodermic illnesses, those of the musculo-skeletal system, as well as congenital anomalies of development and neoplasm. The genetic health of a population is characterised by such factors as congenital anomalies of development; neoplasms in children and adult population, as well as infant mortality.

The physico-chemical analysis of the state of various objects of the environment permits the definition of the background radiation in the territory, and the concentration of pollutants in the atmospheric air, in the ground, and in natural waters. However, it is impossible to define general toxicity and mutagenicity for all the pollutants present in the environment. In this case is necessary to use sensitive bio-systems.

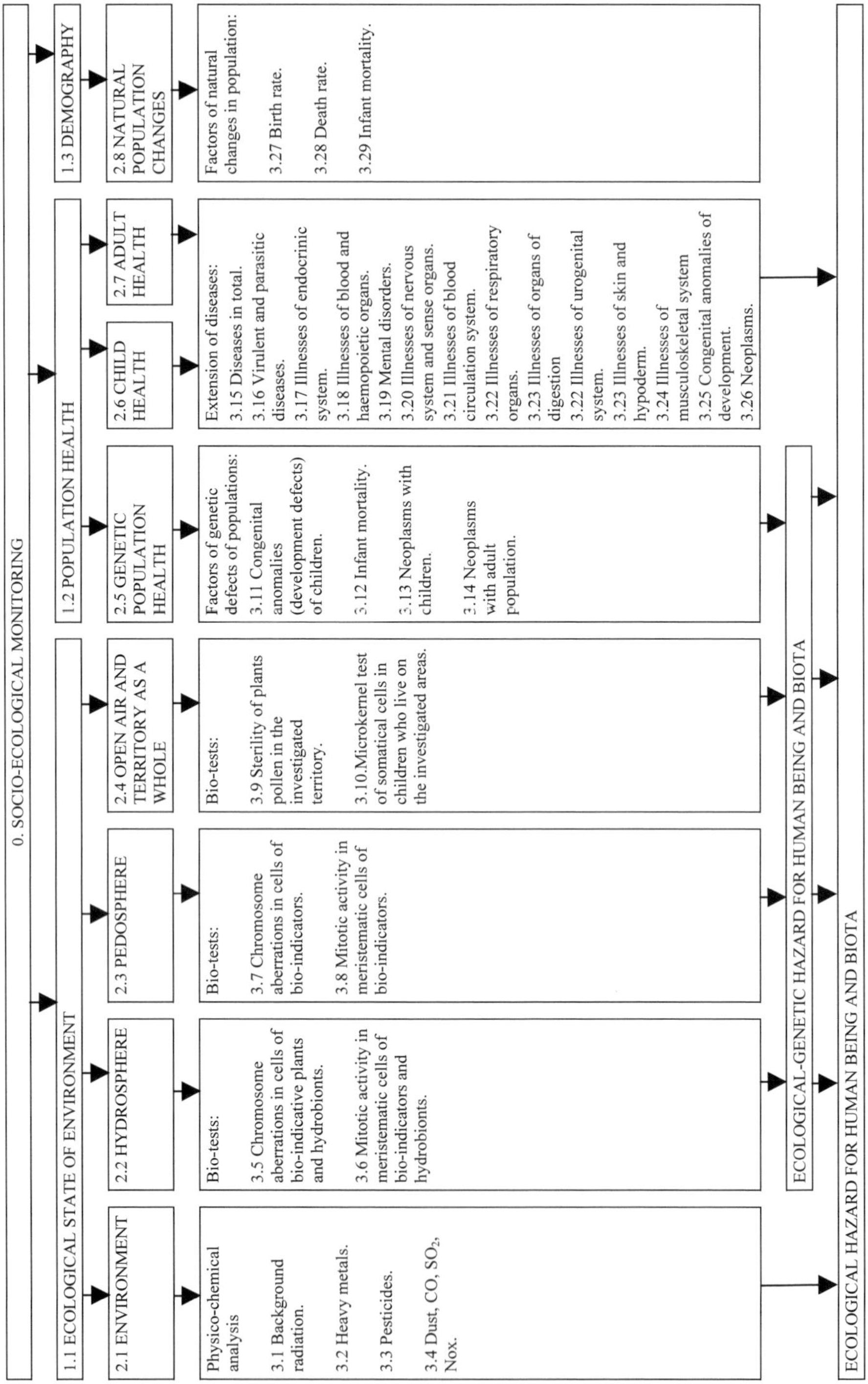

Figure 1. Structural scheme of socio-ecological monitoring.

The ecological state of environment objects in their toxico-mutagenic background is recommended to be estimated by cytogenetic methods of bio-indication which, as it is known, are the most sensitive, informative and sufficient for an adequate assessment of the associated influence on various bio-systems of ecological factors of the environment.

The quality of atmospheric air is identified by the pollen sterility level of the indicator plants growing in the investigated territory.

The quality of hydrosphere in the toxico-mutagenic background can be estimated by the frequency of occurrence of meristematic cells with chromosome aberrations, and at the level of mitotic activity of cells, both exogenic and endogenic bio-indicators, as well as by the MN test in cells of hydrobionts.

The state of pedosphere (grounds) in their mutagenicity reflects levels of genic and chromosome mutations, and their toxic effect is a mitotic index value in the meristematic cells of bio-indicators.

The general mutagenic background of the environment may be characterised by the frequency of microkernel occurrence (i.e. genetic disorders) in the somatic cells of children of the pre-school age living in the territory under investigation. This test characterises the cytogenetic status of a human organism, and is the link between the population health block and that of bioindicative.

Besides, it is necessary to note that the factors of the bio-indicative block, and also the factors describing genetic health at the population level, allow the estimation of the level of genetic danger for humans and biota. The consideration of all factors in the system of socio-ecological monitoring will allow a calculation of the general ecological hazard for all living organisms, including human beings.

As all the factors of bio-indicative and population blocks have their own measurement units, then for integrated assessments it is necessary to reduce them to a one-dimensional form of conditional indices of ‘damageability’ (or vulnerability) or the protection of bio-systems.

The conditional indices of damageability (CID) for separate bio-systems are calculated under Formula 1:

$$\mathrm{CID} = \frac{/\,Ireal - Icomf\,/}{/\,Icrit - Icomf\,/} \qquad (1)$$

where CID is a conditional index of damageability of the bio-parameter caused by the impact of unfavourable factors of the environment. $I_{comf.}$ and $I_{crit.}$ are the values set experimentally (or by experts) for the bio-parameters in comfortable and critical conditions; $I_{real.}$ is the real value of the bio-parameter.

Integral conditional indices of damageability ($ICID_i$) are calculated under Eq. (2):

$$\mathrm{ICID_i} = \frac{1}{n} \times \sum_{i=1}^{n} ICIDi \; \frac{1}{n} \times \sum_{i=1}^{n} \left[\frac{/\,Ireal - Icomf\,/}{/\,Icrit - Icomf\,/} \right]_i \qquad (2)$$

where $ICID_i$ is one of integral conditional indices of damageability of human health or the environment state; $I_{comf.}$, $I_{crit.}$, $I_{real.}$ are, respectively, the comfortable, critical and real values for one of n indices.

The integral parameter describing the environmental state in the general toxico-mutagenic background ($ICID_{bion.}$), assumes the parity of components, and it is calculated under Formula 3:

$$ICID_{biond.} = \frac{1}{n}\,(ICID_1 + ICID_2 + ICID_3 + \ldots \ldots + ICID_n), \quad (3)$$

where $ICID_1$, $ICID_2$, and $ICID_3$ are integral indices of the bio-indication quality of the atmosphere, hydrosphere and pedosphere accordingly; $ICID_n$ is the integral parameter describing the territory as a whole.

The integral parameter describing the general health of the population ($ICID_{popul.}$) is calculated under Formula 4 taking into account the weight factors for the significance of separate parameters set by experts. Thus, greater values of factors are designated to the most sensitive parameters:

$$ICID_{popul.} = 0{,}4 * ICID_1 + 0{,}3 * ICID_2 + 0{,}3 * ICID_3 \quad (4)$$

where $ICID_1$ is the physical health of children, $ICID_2$ is the physical health of adult population; and $ICID_3$ is genetic health.

The integral parameter describing the general ecological hazard (EH) for human beings and biota from the damaging impact of environmental pollutants is calculated under Formula 5:

$$EH = 0{,}60 * ICID_{biond.} + 0{,}40 * ICID_{popul.} \quad (5)$$

Thus, it is necessary to note the priority of the bio-indicative block, parameters of which are defined experimentally.

The integral parameter describing genetic hazard (GH) for humans and biota from the influence of mutagens of the environment is calculated under Formula 6:

$$GH = 0{,}60 * ICID_{biond.} + 0{,}40 * ICID_{genet.\ health.} \quad (6)$$

where $ICID_{genet.\ health}$ is an integral parameter of population genetical health. In this case, as well as in the previous one, the data received experimentally, have priority.

The calculation of parameters in conventional units allows making the comparison and the ranking of the territory of a city, a region, or a country by the state of the environment and population's health. This cannot be performed when these parameters are given in their natural measurement. Besides, it is possible to perform the ranking of territory in the toxico-mutagenic background of different objects of the environment. In addition, it is possible to detect the priority disease incidence of the population in various regions depending on the environmental state.

The values of conditional indices of damageability (CID and ICID) vary within the limits from 0 (comfortable conditions for life activity) to 1 (critical conditions).

To estimate the level of damageability of population health and the environmental state it is recommended to use a single unified scale. See Table 1.

As the normative value is accepted as being CID = 0.300, at this level in our opinion, it is possible to restore the ecological state of the damaged ecosystems [5].

Table 1. Rating scale of the state of human health and the ecological situation

Range of CID values	Damageability level of biosystems and health	State of population health and environment
0.000 ÷ 0.150	Low	**SAFE**
0.151 ÷ 0.300	Lower than average	Alarming
0.301 ÷ 0.450	Average	Conflicting
0.451 ÷ 0.600	Above average	Threatening
0.601 ÷ 0.750	High	Critical
0.751 ÷ 1.000	Maximal	Dangerous

2. A Method of Cytogenetic Population Survey

To detect a harmful influence of technogenic factors on a human organism we have used a method for considering micro-kernels in cells of mucosa of oral cavity [6,7].

To assess the ecological situation in the general mutagenic background we have used the results of a cytogenetic survey of the pre-school age children, because as a result of numerous examinations of population groups of different ages, it has been defined that they are the most sensitive to the unfavourable influence of external factors. The pre-school age children are subjected to their minimal daily migration; in their life history they have no contacts with production factors, no bad habits and, as a rule, are provided with more regular medical surveys [8]. Besides, in a child's organism, in comparison with that of an adult person, metabolic processes and the distribution of cells occur much more intensively. This causes a higher sensitivity of organisms in children to the influence of harmful ecological factors [9]. Therefore, it is possible to use the somatic cells of the pre-school age children as bio-indicators in the ecological-genetic monitoring of the environment.

In a group used for the cytogenetic survey by a special questionnaire there were selected healthy and practically healthy children aged 5–7 years (I and II health groups). The investigated sampling included 25–60 persons with an approximately identical sex ratio [10].

Swabs of the children's mucous oral cavity were taken from the interior of right and left cheeks, and also from the under lip by means of a sterile wadded tampon on an individual chip; then these were applied onto slide plates. The fixation was made in a mixture of spirit and acetic acid (3:1) within 1 hour. Then the swabs were slightly dried in the air till the disappearance of their humectant mist. The specimens were painted with aceto-orcein.

The analysis of specimens was made on microscope MBI-3 with the use of an objective 10 × 60. The MN index was expressed as the occurrence frequency of micro-kernels per cell and was calculated as the absolute spread based upon a relative error value.

The received experimental data have been used for the calculation of conditional indices of damage (CID) to cells of children's organisms (Formula 1). On this basis an assessment was carried out of the ecological situation in the mutagenic background (Table 1).

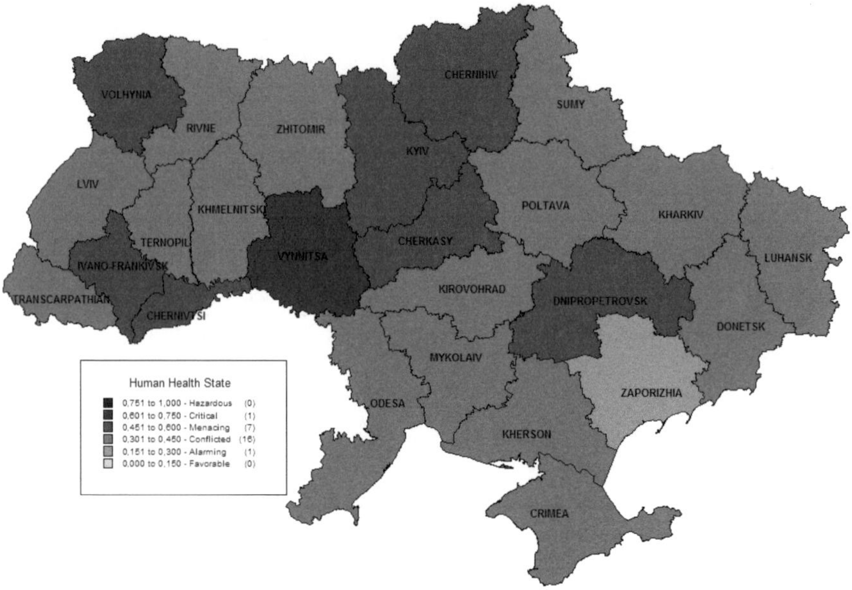

Figure 2. The state of physical population health in Ukraine (2002).

The statistical processing of the received data was carried out according to a method of alternative variation as per Student-Fisher.

3. Results of Assessment of Physical Population Health in Ukraine

The assessment of the state of physical population health in Ukraine was performed on the basis of data presented by the centre of health statistics. According to the methodology we have analysed the factors of morbidity rate for various illnesses characterising the physical health of the adult and child population in the territory of 24 regions and the Autonomous Republic of Crimea. On the basis of the calculated integral parameters we have made maps of integral physical health of the population of Ukraine for 2002 and 2006 (Figs 2 and 3).

These maps testify that in 2002 and 2006 in Ukraine there are no regions with a "dangerous" state of health. The "critical" state of health in 2002 is defined only in one region, Vinnitsa, whereas in 2006 already in 4 regions "high" level and "critical" states of health damage were detected, namely in Vinnitsa, Cherkassk, Dnepropetrovsk and Kharkov regions. In 2002 it was observed that the greatest occurrence (16 regions) was a "conflicting" assessment of population health, and in 2006 it was that of "threatening" assessment (12 regions). As a whole, the integral physical health of children and adult population of Ukraine in 2002 was estimated as "conflicting" (ICID = 0.436), whereas in 2006 it was "threatening" (ICID = 0.520).

The calculated data permit the ranking of the territory according to the state of population health; to define the regions that are most hazardous to human health and which demand the introduction of rehabilitation measures; to estimate the environment state as per these parameters; to specify territories with favourable conditions for the

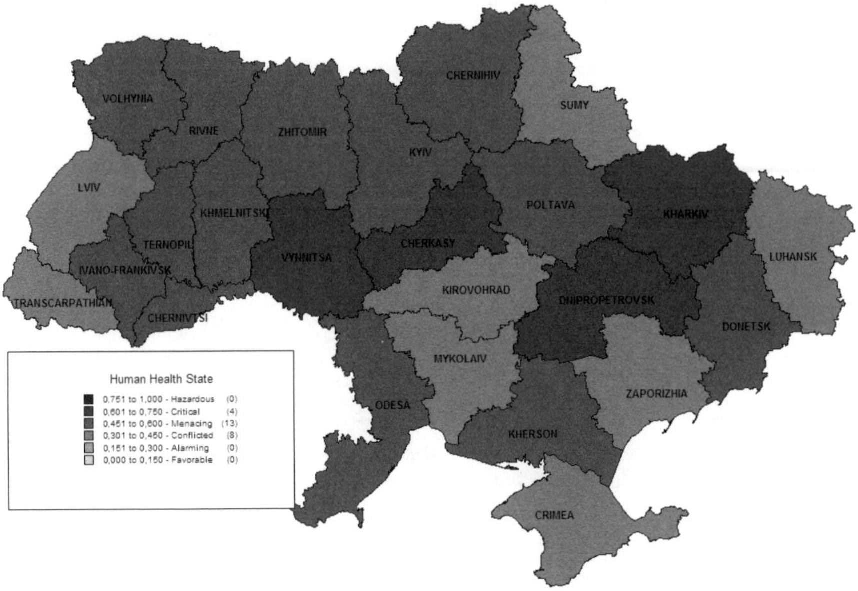

Figure 3. The state of physical population health in Ukraine (2006).

rehabilitation of population health, as well as making environmentally sound administrative decisions.

4. Results of Cytogenetic Population Survey in Various Cities of Ukraine

An investigation of the general mutagenic background of the environment has been carried out on territory with different technogenic loads in the cities of Dnepropetrovsk region (Dnepropetrovsk, Marganets, Zholtyye Vody and Nikopol), the Lviv region (Chervonograd). The investigation has also been carried out in the cities of Chernovtsy and Chernigov, which are intensively developed with metallurgical, mining and processing, chemical and other branches of industry, and there is some deterioration of the chemical composition of the atmospheric air, contamination of soil, and surface and underground waters. The control used was the environment background and the state of population health in "conditionally clear" territories of the settlement of Novotroitskoye in the Dnepropetrovsk region.

There is situated the medical-improving complex "Solyony Liman (Saline Estuary)" and the commitment of Nikita (the Nikitskiy Botanic Garden, the reserve of "Cape Martyan", as well as the Scientific and Experimental Phytocentre of the Ukrainian Academy of Science, AR Crimea).

The results for assessment of the environmental quality in the mutagenic background, using an MN test in human somatic cells, are shown in Fig. 4.

It has been established that the level of micro-kernels in cells of the children who live in the industrial regions have a high technogenic load that is 2.2–4.6 times more than in the "conditionally clear" control located in the territory of the Dnepropetrovsk region (i.e. the medical-improving complex "Solyony Liman") and is 5.2–10.9 times more than the all-Ukrainian control (the commitment of Nikita, AR Crimea).

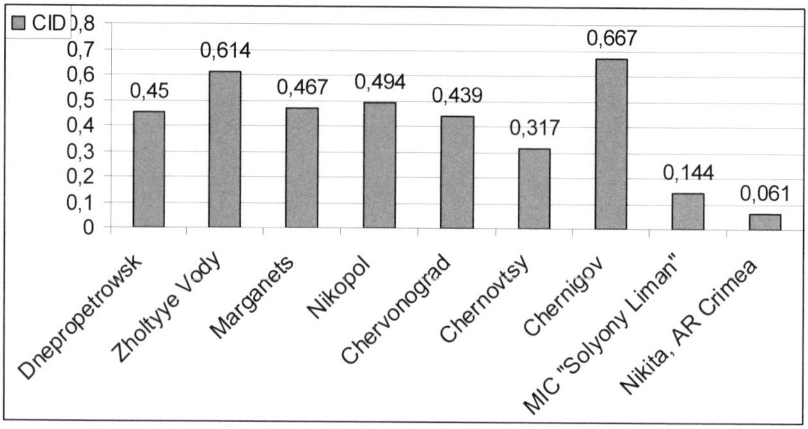

Figure 4. Comparitive characteristics of the rate of cells with micro-kernels (on the basis of their conditional indices of damageability) in children living in different cities of Ukraine, 1999–2006, ($p < 0,05$).

The calculated CID for organisms of children in cytogenetic parameters in view of minimal (P = comfortable) and maximal (P = critical) values of the investigated parameter, testify that in control points there is established a "low" level of genetic damage in epithelial cells, and a "safe" state of organism in the cytogenetic status. It has allowed its consideration as a "reference" in the mutagenic background for the ecological state of control territories of the medical-improving complex "Solyony Liman", as well as the settlement of Nikita in AR Crimea.

In the cities of Dnepropetrovsk, Chervonograd and Chernovtsy was observed an "average" level of damage of cells in children and their "conflicting" state. In the centres of mining and primary processing industries – to which belong the cities of the Dnepropetrovsk region, namely Marganets, Nikopol and Zholtyye Vody (Yellow Waters) – the level of damage of the biosystems is specified as "above average" on the basis of a "threatening" state of children's organism in the cytogenetic status. As a whole, the ecological situation in the general mutagenic background in the territory of investigated technogenic-loaded cities is estimated as "unsatisfactory".

Mostly, genetic disorders were manifest in the cells of the children living in territory which has been subjected to the significant influence of radioactive releases owing to the disaster in the Chernobyl Atomic Power Station, namely in the city of Chernigov. That area has been specified as having a "high" level of genetic disorders in the cells of children, and a "critical" state of children's organisms in the cytogenetic status. The ecological situation in the mutagenic background is defined here as "critical".

Thus, the above-mentioned methodology permits not only the ranking of the territories in the toxico-mutagenic background, but also the definition of the levels of ecogenetic danger for all live organisms, including humans. This is imperative for the development and realisation of programmes for the rehabilitation of the state of environmental objects and the population's health.

The offered methodology of the socio-ecological monitoring can be used for similar research in the whole of Ukraine and in other states.

References

[1] The European Policy in Health Reaching for Everybody in the Twenty First Century. // A Working Document for Consultations. The World Health Organisation. 1997. Page 216.
[2] Tasks for Reaching Health for All: the European Policy of Public Health Services. Copenhagen: the World Health Organisation. The European Regional Bureau. 1993. Page 322.
[3] K. V. Pikul. Abnormal Development in Children from the Nitrate Polluted Territory. / The Environment and Health. 2003. No 2 (25). Page 18-20.
[4] E. A. Derkachev, L. B. Ogip, K. Yu. Ogip, I. O. Gubar. The Influence of Ground Pollution with Heavy Metals on the State of Population Health and the Forecast of its Possible Changes. // The Hygiene of Populous Cities. Edition 45. Kiev. 2005. Page 159-165.
[5] A. I. Gorovaya, I. I. Klimkina. The Methodology of Socio-Ecological Monitoring with the Use of Cytogenetic Methods. // C. Mothersill et al. (eds.), Multiple Stressors: A Challenge for the Future. – S. 91-102. 2007 Springer. Printed in the Netherlands.
[6] A. I. Gorovaya, I. I. Klimkina. The Use of a Cytogenetic Testing for the Assessment of the Ecological Situation and Effectiveness of Sanitation for children and adults with Natural Adaptogenes. // Cytology and Genetics. 2002. No. 5. Page 21-25.
[7] R. M. Arutunyan, E. R. Tumanyan, G. S. Shiriyan. The Analysis of Microkernels in Mucous Oral Cavity for the Assessment of Cytogenetic Effect of Environment Pollutants. // Cytology and Genetics. 1990. – 24, No. 2. Page 57-60.
[8] Bottom-Zoological Diagnostics of Population Health State in Connection with the Influence of Environmental Factors. // Methodical Recommendations. MR 2.2.12.068. 2000. Page 42.
[9] O. V. Berdnik. The Sensitivity of an Organism to Environmental Factors. // The Environment and Health. 2000. – No. 1 (12). Page 38-42.

I. Apostol et al. (Eds.)
IOS Press, 2009.
doi:10.3233/978-1-58603-948-6-226

Medical and Biological Aspects of the Chernobyl Nuclear Accident: Influence on the Population of the Republic of Moldova

Liubov CORETCHI and Ion BAHNAREL
National Scientific and Applied Centre of Preventive Medicine, Gh. Asachi 67 A, Chisinau, Republic of Moldova

Abstract. During the 1996–2004 period 850 patients, who were participants in diminishing the consequences of the Chernobyl nuclear accident (PDCCNA), and their children, were investigated in terms of clinical, immunological and cytogenetic analyses. The clinical investigations indicate that the PDCCNA patients, when compared with patients of a control group, were more susceptible to infectious and non-infectious diseases, with the prevalence of large polymorphism of nervous, heart-vascular and gastric-intestinal system, which were accompanied by circulatory disorder of the vegetative nervous system. The immunological analysis revealed alterations in the immune system of the PDCCNA. Cytogenetic research of the lymphocyte cultures of peripheral blood of PDCCNA members living in the Republic of Moldova in the last 15–20 years after the accident, and their children, revealed the deterioration of the hereditary system, being expressed through a high level of genomic, chromosomal, and chromatid type aberration. Chromosomal type of aberrations prevailed in the adults and chromatid type in the children.

Keywords. Chernobyl nuclear accident, clinical investigations, cytogenetic analysis, immunology

Introduction

Stress factor effects on population health evaluation [9], especially on emergency workers, remains one of the most important problems of contemporary medicine [2] and in this regard the Chernobyl nuclear accident (CNA) that took place on the 26th April 1986 is an eloquent example. Radioactive substances produced as a result of the CNA fell out over a significant part of Europe, including the Republic of Moldova, affecting more than 5,000,000 persons. In the clearance and abatement of the CNA consequences there was participation by a lot of military staff including a great number of reservists. Lack of previous experience in the field (since it was the first large-scale nuclear accident) made it impossible to prepare specially trained personnel for such control and clearance tasks. Consequently, many military staff, even from the first days, were presented to medical authorities with a range of symptoms which were characterized as somatic diseases after detailed investigations [7].

The ionizing radiation influence on the health status of the participants attempting to diminish the consequences of the Chernobyl nuclear accident (PDCCNA) were diffi-

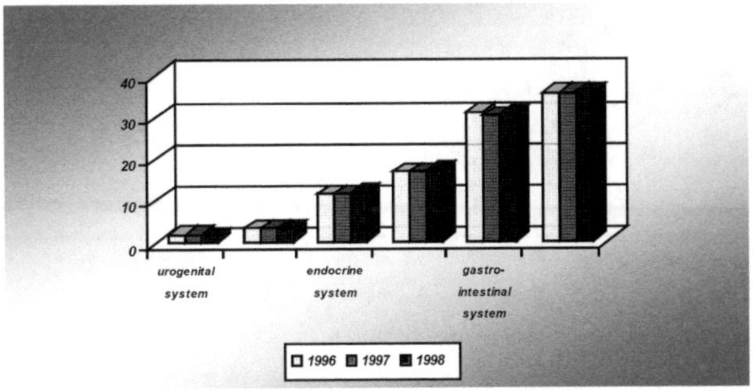

Figure 1. Diseases spectrum in the CNA workers (%).

cult enough to evaluate, and so called for an adequate multi-lateral study applying modern diagnostic techniques. Large studies were conducted in the Russian Federation, the Ukraine and the Republic of Belarus. Acquired data suggested the existence of a noticeable deteriorating effect of ionizing radiation, produced as a secondary effect of the CNA, with an increased incidence of health-status disturbances in the affected population [3,5,6]. Approximately 3,500 inhabitants from the Republic of Moldova took part in the clearance of the CNA consequences.

This study objective comprises the determination of clinical, immunological and cytogenetic features in the PDCCNA from the population of the Republic of Moldova and their descendants.

1. Clinical Aspects of Therapeutic Pathology Manifestation in Cohort of PDCCNA

The group comprised patients within the age range 32–54 with an exposure period to ionizing radiation ranging from 15 to 180 days during the clearance of the CNA effects, which during 1986–1987 comprised 90% of participants and during 1988–1989 it was 10% of participants. The control group included 62 persons, also within the age range 32–54, that was a relatively healthy group which was not previously exposed to ionizing radiation.

A detailed study of the ionizing radiation exposure influence on the health status of persons situated in the increased radiation activity zone due to the CNA, determined that general morbidity of these patients has its peculiarities. It showed that 308 patients were concomitantly supervised by more than one specialist, i.e. they suffered from multi-systemic pathology. It must be mentioned that psycho-neurological pathologies prevailed in the determined diseases spectrum and in 1998 these were encountered in about 36% of all diagnosed pathologies, thus being the dominant effect. The second most dominant effect was gastro-intestinal system pathology (30.55%), followed by cardiovascular diseases (17.0%). Endocrine system pathology, which was in a state of continuing increase, was the fourth most dominant with 11.65% of cases. Osteo-articular pathology was rather rarely met, being diagnosed in 3.25%, 3.4% and 3.2% respectively in 1996, 1997, and 1998 (Fig. 1). Urogenital diseases were determined in 1.85%, 1.9% and 1.6% of cases respectively in 1996, 1997, and 1998. The analysis carried out allowed the possibility of revealing a dynamic increase overall and in all systems of pathology incidence in the PDCCNA.

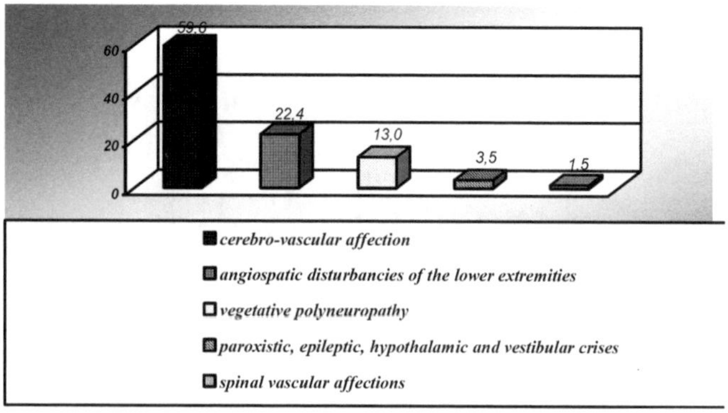

Figure 2. Nervous system diseases in the PDCCNA (%).

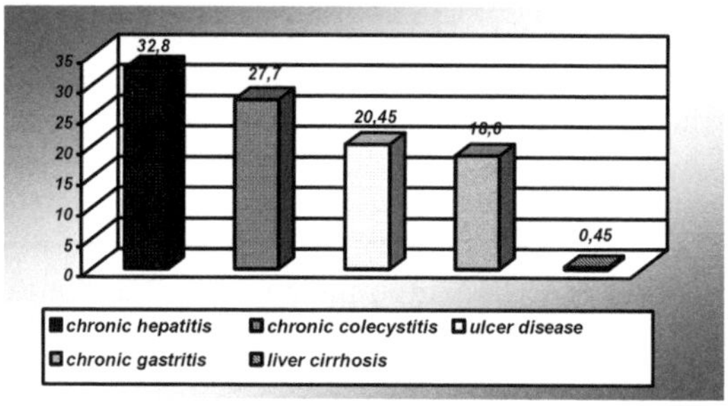

Figure 3. Gastrointestinal diseases in the PDCCNA presentation (%).

Cerebrovascular pathology comprised up to 59.6% of the overall nervous system diseases. Patients who suffered from neuro-circulatory dystonia and various degrees discirculatory encephalopathies were predominant. Angiospastic and angiodystonic disturbances in superior and inferior extremity vessels were found in 22.4% cases. Vegetative polyneuropathy was diagnosed in 13.0% cases, mostly being associated with hypothalamic paroxysmal epileptiform crises (3.5%). Spinal vascular affections were registered in only 1.5% of patients with nervous system pathology (Fig. 2.). The most frequent complaints of the patients with predominating nervous system pathology included permanent intensive headache, particularly in the second half of the day, vertigo, marked asthenia, nervousness, insomnia, both superior and inferior extremities numbness, working ability diminution and precoma stages.

Chronic liver diseases are the most frequent manifestations of gastro-intestinal system diseases among the PDCCNA (32.8%). Ulcer diseases, chronic gastritis and liver cirrhoses comprised, respectively, 20.45%, 18.6% and 0.45% of the overall number of patients with this group of diseases (Fig. 3).

Figure 4 shows the PDCCNA morbidity structure 20 years after the CNA.

2. Cellular Immunity Peculiarities in the PDCCNA

Many authors appreciate the major importance of the lymphoid system in the processes of regeneration, rehabilitation, and creation of an increased resistance of the organ-

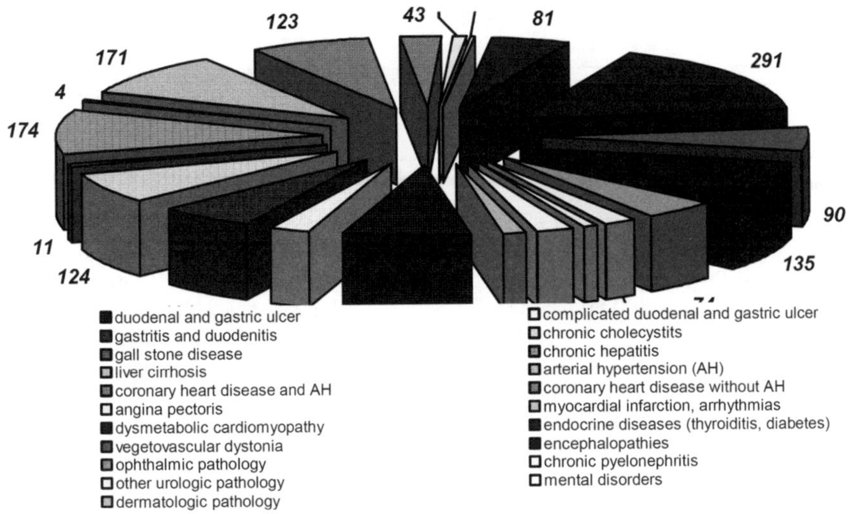

Figure 4. Morbidity structure of the PDCCNA 20 years after the CAN.

ism [1,8]. An important role in this direction is attributed to monocytes. Nowadays, it is considered that granule cytopenia, relative lymphocytosis, hyper segmentation, and fragmentation of the neutrophiles nucleus, the presence of lymphocytosis with modifications within the nuclei and cytoplasm, represent certain specific markers of chronic ionizing irradiation. As a result, the ionizing radiation, together with chemical, physical and psychological factors "give birth" to the "Chernobyl syndrome".

Table 1 includes comparative results for the immunological indicators in the PDCCNA and the patients who did not belong to the control group. As can be seen, we analyzed the state of the persons that did not take part in the decrease of consequences of the CNA and were not subject of radiological investigations (n = 62). The analysis of the populations and sub-populations structures of the lymphocytes of the peripheral blood was applied to 100 of the PDCCNA that, for the clinical study, had been divided into three groups. The results of the immunological study showed an authentic decrease ($P < 0.05$) of the leucocytes number in the PDCCNA (5.51 ± 0.23), in comparison with the patients from the control group (6.6 ± 0.26); the data are included in Table 1.

In the PDCCNA the percentage of the T-total lymphocytes was lower (44.33 ± 1.23) than in the control group, i.e. 44.33 ± 1.23 Vs. 47.55 ± 1.07 ($P < 0.05$). Concerning the T-teophylline-resistant, teophylline-sensitive, and thermo stable lymphocytes, we did not notice an essential difference between the values of these indicators in the PDCCNA and of the control group patients.

Further, we established that in the PDCCNA, the B-complementary lymphocytes number diminished under the influence of stressful factors of the CNA. In this context, their percentage constituted 22.39 ± 1.15%, in comparison with the patients from the control group – 21.79 ± 0.76% ($P < 0.05$), their portion being equal to 0.4 ± 0.15%, (CEW) and for the patients from the control group – 0.57 ± 0.02% ($P < 0.05$) (Table 1).

The above-mentioned results confirm the data from the specialized literature which claims that, under the influence of the ionizing radiations, some changes appear in the system of the cellular immunity.

Table 1. Characterization of the cellular immunity indicators in the PDCCNA and of patients from the control group (%)

Nr	Immunological indicators	PDCCNA group (n=100)$x \pm m_x$	Control group (n = 62) $x \pm m_x$
1.	Leucocytes, g/l	5.51 ± 0.23	6.60 ± 0.26*
2.	Lymphocytes, %	34.31 ± 1.56	29.60 ± 1.19*
3.	Lymphocytes, g/l	1.92 ± 0.12	1.84 ± 0.07
4.	T-active Lymphocytes, %	23.15 ± 2.17	24.60 ± 1.06
5.	T-active Lymphocytes 10^9/l	0.44 ± 0.05	0.44 ± 0.03
6.	T-total Lymphocytes, %	44.33 ± 1.23	47.55 ± 1.07*
7.	T-total Lymphocytes,10^9/l	0.82 ± 0.05	0.85 ± 0.04
8.	T-teophylline-resistant Lymphocytes, %	30.08 ± 1.20	31.14 ± 1.08
9.	T-teophylline-resistant Lymphocytes, 10^9/l	0.56 ± 0.05	0.57 ± 0.03
10.	T-teophylline-sensitive Lymphocytes, %	14.41 ± 0.77	15.51 ± 0.58
11.	T-teophylline-sensitive Lymphocytes, 10^9/l	0.27 ± 0.02	0.28 ± 0.01
12.	T-thermostable Lymphocytes, %	9.45 ± 1.45	8.07 ± 0.89
13.	T-thermostable Lymphocytes, 10^9/l	0.17 ± 0.13	0.15 ± 0.02
14.	B-complementary Lymphocytes, %	22.39 ± 1.15	21.79 ± 0.76*
15.	B-complementary Lymphocytes, 10^9/l	0.40 ± 0.15	0.57 ± 0.02*

Notes: * – The difference between the cellular immunity indicators in the PDCCNA and patients from the control group is genuine after the t-Criterion Student ($P<0.05$); $x \pm m_x$ – the average value with error.

A rather decisive indicator concerning the evaluation of the functional state of the T-lymphocytes subpopulations constitutes the co-report between the T-helpers and T-suppressors populations. The decrease in number of T-suppressors is explained by some authors [10] through their amplified radio sensitivity, and the diminishing of their function can lead to the appearance of self-aggression and worsening of patients' chronic diseases. Of great importance is the simultaneous reduction of the number of T-lymphocytes and T-suppressors, considered as an initiation in the development of secondary immunologic disturbances. In the case of the PDCCNA investigations, we established that the value of the co-report T-helpers/T-suppressors was higher than the same value for the patients from the control group, which indicates an augmentation of the number of T-helpers in the PDCCNA, in comparison with the patients from the control group.

From the total number of stressful factors that influenced the PDCCNA, radiation was found to be the principal one, and so showing that the cellular marker was more sensitive to ionizing irradiation was of great interest. For this purpose, we made the co-relational and regressive analysis for each immunological indicator, for the exposure time of the PDCCNA and for their age, counting the coefficients of correlation with the dose of ionizing irradiation. We traced out that the most sensitive indicators of the chronic effect in the majority of doses of ionizing radiations were both the percentage and the absolute number of the B-complementary lymphocytes ($r = -0.54 \pm 0.01$ and $r = -0.66 \pm 0.01$). For the rest of the cases, the dependence dose-effect was of a negative nature too, yet the values of the coefficients were lower and we established an unimportant dependence. For example, in the case of the relation between the dose and

Table 2. Characterization of the immunoglobulin in the PDCCNA (g/l)

	Name of the studied groups	IgA		IgM		IgG	
		1	2	1	2	1	2
1.	The group with dysmetabolic cardiomiopatitis; n = 21	2.58	1.1–10.9	1.74	0.29–15.6	13.44	0.8–18.4
2.	The group with psycho neurological disorders; n = 22	2.27	1.6–2.4	2.41	0.5–12.4	15.68	0.64–23.0
3.	The group with gastrointestinal disorders; n = 19	3.99	1–35	1.09	0.41–2.1	14.3	9.8–23.2
4.	Optimal norm of the immune rank in men aged 21–40 years old		1.75–2.43		1.12–1.68	8	8.38–10.07

Notes: 1 – average value; 2 – the interval.

the lymphocytes, r = –0.28 ± 0.2; the system T-active lymphocytes – dose, r = –0.33 ± 0.1; and for the system T-total lymphocytes – dose r = –0.06 ± 0.77.

3. Humoral Immunity Indices Modifications in the PDCCNA

We examined the IgA, IgM and IgG (i.e. immunoglobulins A, M and G) levels in 62 of the PDCCNA in order to study the influence of the ionizing radiation factor upon the values of humoral immunity indicators in PDCCNA. These persons were distributed into three groups according to the clinical results: (i) having dysmetabolic cardiomiopatitis; (ii) disturbances of gastrointestinal tract; and (iii) psycho-neurological disorders.

The results of the investigation showed an increase of IgA levels in every studied group, the values being 2.58 g/l, 2.27 g/l, and 3.99 g/l, respectively, for the patients having dysmetabolic cardiomiopatitis, disturbances of gastrointestinal tract, and psycho-neurological disorders. The optimal norm of the IgA values was between 1.75–2.43 g/l. A noticeable increase (3.99 g/l) was traced out in patients with gastrointestinal tract disturbances, the interval of variability of that indicator being larger, i.e. 1–35 g/l.

Concerning IgM, we established a slight increase of this indicator (2.41 g/l) in the group with psycho-neurological disorders. The minimal value was 0.5 g/l, and that maximum was 12.4 g/l. As for the other two groups of patients, the average IgM values corresponded to the norm and constituted 1.74 g/l (minimum value – 0.29 g/l, maximum – 15.6 g/l) for the group with dysmetabolic cardiomiopatitis, and 1.09 g/l (minimum value – 0.4 g/l, maximum – 2.1 g/l) for the group with gastrointestinal tract disturbances.

For men aged 21–40 years old (which age range corresponds with the age of men from the PDCCNA group) the normal IgA value constituted 1.12–1.68 g/l.

Regarding IgG there could be observed a growth of 30–50% in the amount in all investigated groups when compared with the maximum value of the optimal norm (8.38–10.07 g/l), especially in the group with psycho-neurological disorders where the IgG average value represented 15.68 g/l, with limit intervals of 0.64 g/l – minimum value and 23.0 g/l – maximum value (Table 2).

The increases in IgM and IgA values were associated with a high level of lymphocytes proliferations, which presumes the presence of antigen stimulation of the immune system.

4. Cytogenetic Side Effects in the PDCCNA Inhabitants of the Republic of Moldova

The negative influence of the ionizing radiation on the cells depends on the direct transmission of the energy of charged particles or secondary electrons towards the atoms and the molecules of the cellular material. The main results of this action are that the ions and the atoms are excited by the particle that crosses the cell. Then follows the IInd stage when the ions interact with the molecules from the cell and the last, being excited, can dissociate forming free radicals. The IIIrd stage (chemical) is the result of the interaction between free radicals and intracellular macromolecules: DNA, proteins and macromolecules [6].

The biological effects of the ionizing radiation depend on the quantity of energy absorbed by the tissues or cells and that is why measuring the dose is of great importance, i.e. the severity of the effect depends on the dose absorbed by the tissues. One of the characteristic manifestations of the influence of the ionizing radiation on the cell is the inhibition of the mitotic activity. Substantial doses of ionizing radiation, if applied, can affect large cytoplasmatic structures, as well as processes of synthesis, that occur within the cytoplasm.

The investigation of cytogenetic effects, as an element of retro-biodosimetry consisted of the estimation of the genetic risk associated with the ionizing irradiation, with the study of the CNA influence on the PDCCNA.

In order to achieve this purpose, we followed some objectives:

- The detection of the influence of ionizing radiation actions on chromosomal apparatus in the PDCCNA and their descendants, that resulted after the CNA;
- Output of the biological dosimeter based on the cytogenetic analysis, within the PDCCNA group, according to the dicentric analysis scheme;

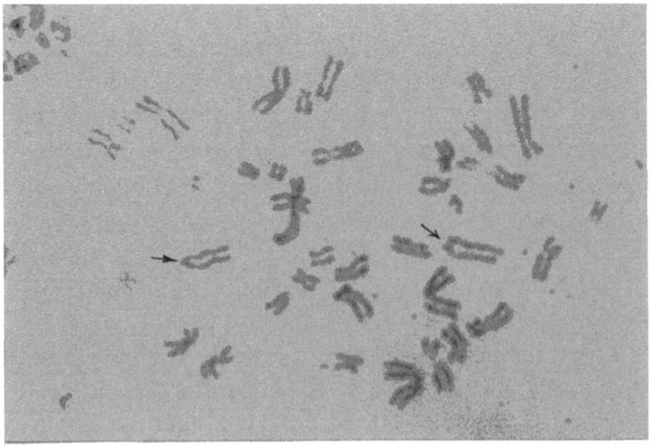

Figure 5. Chromosomes aberration detected in the PDCCNA.

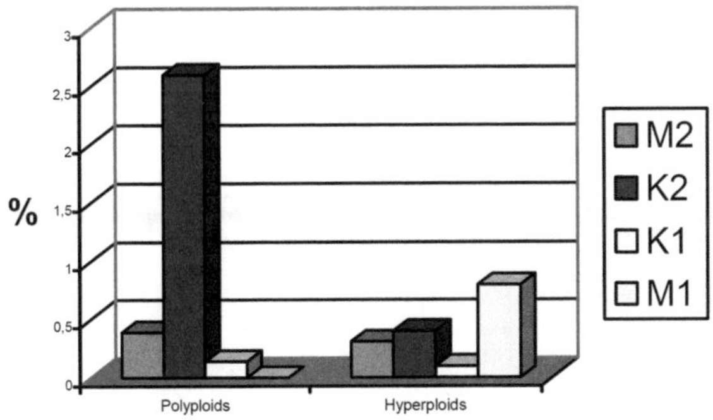

Figure 6. Genomic mutations detected in the PDCCNA and in the patients from the control lot during the first (M_1, K_1) and second (M_2, K_2) mitosis, ($X \pm m_X$, %).

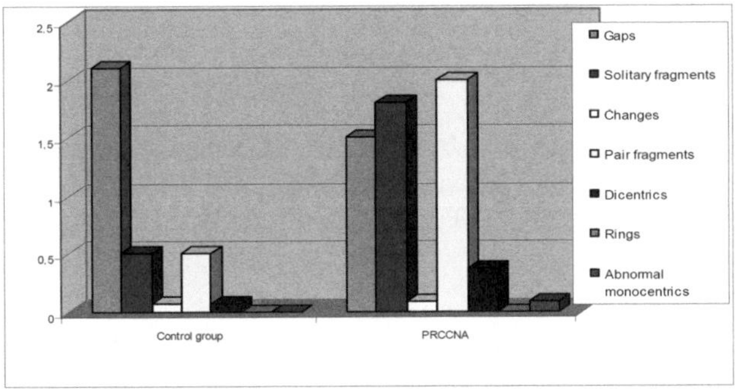

Figure 7. The frequency of chromosomes aberration detected in the PDCCNA and in patients from the control group.

- To establish the ionizing irradiation dose and the incidence of the chromosomal aberrations correlation; and
- To detect the PDCCNA children's hereditary diseases frequency.

The results denote that in the PDCCNA, the frequency of genomic mutations and chromosomal aberrations was higher, in comparison with that of the control group (Figs 6 and 7).

The significance of the study of the mechanism of chromosomal aberrations is determined by the fact that they play a major role in the development of the deterioration because of the ionizing radiation action. Biodosimetric research shows that during the cytogenetic study of the blood lymphocytes, the quantitative analysis of the dicentrics and ring chromosomes has an essential role. The results of the investigations show that there were detected both cells with one dicentric and two dicentrics, the former predominating.

Summarizing the above mentioned factors, we can consider that the results obtained on the cytogenetic analysis in the PDCCNA who were living in the Republic of Moldova, demonstrate the increase of chromosomal mutagenesis in somatic cells of the investigated persons, which include those patients in the group with a high risk of pathologies with a genetic component.

Table 3. Frequency and the spectrum of chromosomal aberrations in the PDCCNA's children

The examined parameters	Detected aberrations	$X \pm m_{X, \%}$
Chromosomes general state in metaphase	1. Endoreduplication	0.04 ± 0.004
	2. Incomplete separation of chromatids	0.13 ± 0.08
	3. Complete separation of chromatids	0.70 ± 0.2
Genomic anomalies	Polyploidy	0.39 ± 0.1
	Hyperploids	0.31 ± 0.1
Chromatid-type aberrations	1. Gaps	3.9 ± 0.4
	2. Solitary fragments	2.35 ± 0.3
	3. Changes	0.17 ± 0.09
Chromosome-type aberrations	Pair fragments:	
	1. Changes	0.57 ± 0.2
	2. Dicentrics	0.09 ± 0.06
	3. Rings	0.04 ± 0.004
	4. Abnormal monocentrics	0.04 ± 0.004

During the investigations of the PDCCNA descendants' karyota, there were involved 23 boys and girls, born between 1989 and 1992. The total number of investigated metaphases amounted to 2,300 and the results obtained are presented in Table 3.

5. Congenital Malformations Study in Children from the Republic of Moldova

In order to assess and monitor the influence of mutagenic factors within the populations, radio-ecological and genetic monitoring is applied, as well as the use of the most efficient methods, namely systems of supervision and assessment of the frequency of congenital malformations. The morphogenetic disturbances present a major risk (morphological congenital anomalies without functional disturbances), and are considered minor anomalies of development. They can be caused by both endogenous and exogenous factors. The presence of the minor anomalies within the population can serve as a criterion for a hostile ecology. The detection of three or more minor anomalies allows us to suppose the appearance of a hereditary polygene disease, but the presence of five or even more anomalies, presents a major factor of risk.

The purpose of our research was to study the frequency of congenital anomalies in children from different geographical zones of the Republic of Moldova. The study included 863 children aged 1–15 years old from 29 areas of the republic: from the North, 236 children, from the Center, 354, and from the South, 273 children.

We assessed the following indicators of the hereditary pathologies:

1. Minor anomalies (MA) caused by the disturbances of the processes of morphogenesis, during the intra-uterine period. We classified them according to the number of MA detected in a child.
2. Congenital malformations (CM) of polygene etiology have been described according to the universal classification of diseases.
3. Multiple syndromes of hereditary etiology (chromosomal and genes syndromes) were described.
4. Polygene hereditary diseases are mainly caused by external factors.

Table 4. The frequency of congenital malformations and hereditary diseases in different zones of the Republic of Moldova (10,000 children)

Pathologies	North	Centre	South	RM	P
	1	2	3	4	
Total number of minor anomalies	3927.9 ± 317.9	3607.0 ± 255.2	4234.4 ± 299.0	4017.3 ± 166.8	P > 0.05
Minor anomalies 1–2	118.6 ± 21.0	251.4 ± 23.0	157.5 ± 22.0	185.4 ± 13.2	P1.2 < 0.01 P1.4 < 0.05 P2.3 < 0.05 P2.4 < 0.05
Minor anomalies 3–5	622.8 ± 31.5	327.6 ± 24.9	274.7 ± 27.0	391.6 ± 16.6	P1.2 < 0.01 P1.3 < 0.01 P1.4 < 0.01 P3.4 < 0.01
Minor anomalies 6–8	194.9 ± 25.7	288.1 ± 24.0	424.9 ± 29.9	305.9 ± 15.6	P1.2 < 0.05 P1.3 < 0.01 P2.3 < 0.01
Minor anomalies 9–11	–	25.4 ± 8.3	69.6 ± 15.4	33.6 ± 6.1	P2.3 < 0.05
Congenital malformations	266.9 ± 22.8	248.5 ± 36.8	358.9 ± 40.8	288.5 ± 22.9	P2.3 < 0.01 P3.4 < 0.05
Total number of hereditary diseases	38.1 ± 12.4	11.3 ± 5.6	25.6 ± 9.5	23.1 ± 5.1	p > 0.05
Monogenic pathologies	29.6 ± 11.0	8.4 ± 4.8	3.6 ± 3.6	11.5 ± 3.6	P1.3<0.05
Pathologies with chromosomal etiology	8.7 ± 5.9	2.8 ± 2.8	21.9 ± 8.8	10.4 ± 3.4	P2.3 < 0.05
Polygenic pathologies	258.4 ± 28.5	355.9 ± 25.4	212.4 ± 24.7	283.8 ± 15.3	P1.2 < 0.05 P2.3 < 0.01

The analysis of the results shows that, according to the spread frequency of all MA between the zones of the Republic of Moldova, we did not notice an authentic difference. Concurrently, we established that MA frequency, (1–2 anomalies in a child/multiple MA (6–8 in a child), was lower in children from the northern part of the country, constituting 194.0 ± 25.79 cases, and higher in those from the southern part, namely. 424.91 ± 29.92 cases ($p < 0.01$). In this region, 9–11 MA in one child, were also very frequent, representing 69.6 ± 15.4 cases in 10,000 children, the average in the republic being 33.6 ± 6.1 ($p < 0.05$) cases (Table 4).

The analysis of CM-spreading frequency observed an increase in children from the southern parts – 358.97 ± 40.89 cases ($p < 0.05$). The anomalies of the following systems were prevalent: digestive – 150.18 ± 21.62 cases; urinary – 146.52 ± 21.4 cases;

Table 5. The frequency of congenital malformations according to genetic monitoring of newborn children between 2000 and 2001 in the Republic of Moldova (10,000 newborn)

Type	North	Centre	South	Republic	P
	1	2	3	4	
Minor anomalies	28.58 ± 1.8	45.1 ± 1.8	58.11 ± 2.71	43.64 ± 1.01	P1.4 < 0.01 P1.3 < 0.01 P1.2 < 0.01 P3.4 < 0.01 P2.3 < 0.01
Congenital malformation	154.11 ± 4.1	195.33 ± 3.8	217.03 ± 5.2	188.86 ± 2.49	P1.4 < 0.01 P3.4 < 0.01 P1.2 < 0.01 P1.3 < 0.01 P2.3 < 0.01

musculoskeletal – 36.63 ± 11.37 cases; and genital – 14.65 ± 7.27 cases (p > 0.05). MA and CM frequencies were compared with the results obtained via the monitoring of congenital anomalies in newborns, which has been applied in the Republic of Moldova since 1989. The results, presented in Table 5, show a major frequency both of CM (217.03 ± 5.2), and MA (58.11 ± 2.71) in newborn children from the southern part (p < 0.01). Concerning the northern part, we established a lower frequency of CM – 154.11 ± 4.15 cases and MA – 28.58 ± 1.8 cases, in comparison with the central part of the republic (p < 0,01).

6. Conclusions

1. The clinical study of the general morbidity structures of the PDCCNA allowed the highlighting of the fact that diseases of psycho-neurological, gastro-intestinal and cardiovascular systems prevail. We noticed an increase of 3–4 times the frequency of disturbances of the above-mentioned systems' pathologies in the PDCCNA, in comparison with the pre-CNA period.
2. Classical rosette formation tests explained the following disturbances in the PDCCNA immune status: diminution of the T-total and B-complementary lymphocytes absolute number, and net increases in IgG and IgA immunoglobulin levels. The correlative and regressive analysis of immunological parameters' dependence on the ionizing radiation dose determined the negative linear correlation between the B-complementary lymphocytes number and the absorbed dose (r = –0.54).
3. The cytogenetic examinations of the lymphocyte population have traced the effect on the reproductive system in the PDCCNA and manifest through the increase in chromosomal aberrations frequency on the genomic, chromosomal and chromatid levels. The chromosomal aberrations predominated. Thus, the average frequency of the hyperploids cells at the participants was 8.0 times higher in comparison with the control group. The level of solitary and pair

fragments was 3.6 and 5.0 times higher in comparison with the control group. The analysis of the dicentric distribution allowed us to carry out the retro-biodosimetry in the PDCCNA.

4. The determination of the type of immune response according to the level and the co-report of determinants expressing on the surface of the immunology regulatory T-lymphocytes makes the method more informative and it can serve as a marker in clinical forecasting. One cannot exclude the fact that the PDCCNA disturbances of the immune system, according to the IInd type of reactions, presents the group at high risk to further development of the lymph proliferate diseases and so requires regular immunological monitoring.

References

[1] Н. Опополь, Р. Коробов, Эколого-гигиенический Мониторинг: проблемы и решения. Кишинев: Центр. Типогр, 2001, 238.

[2] I. Bahnarel, The prevention of the local nuclear accidents in the Republic of Modova, IAEA-CN-70/88, Contributed papers "Safety of radiation sources and security of radioactive materials", Conference held in Dijon, France, 1998.

[3] А.А. Ильин и др., Экологические особенности и медико-биологические последствия аварии на ЧАЭС, Медицинская радиология (1989), № 11, 59-81.

[4] V.G. Bebeshko, V.I. Klimenko, L.N. Yukhimuk, et al., The hematopoietic system and bone marrow microenvironment state of the persones which were heavily irradiated as a result of the Chernobyl accident, Proccedings of the International round table "Chernobyl: Never again". Italy (1994), 87-89.

[5] E. Botezatu, O. Iacob, Contribution of Chernobyl Accident to human contamination with Strontium-90, Long-term health consequences of the Chernobyl disaster. 2nd International Conference. Kiev (1998), 25.

[6] G.G. Boroday, Zh.V. Usatenko, Indices pathological affection among children included into clinical and epidemiological register, Long-term consequences of the Cernobyl disaster, 2nd International Conference. Kiev (1998), 23.

[7] L. Andrieş, G. Pădure, L. Rusu et al., Metode unificate de cercetare ale statusului imun. Chişinău (1993), 30.

[8] В.Ю. Нугис, А.А. Чирков, Способ оценки дозы и величины облученного объема тела при частичном радиационном поражении по результатам цитогенетического анализа культур лимфоцитов периферической крови, Радиобиология 29 (1986), № 6, 838-840.

[9] N. Dubinin, Genetica moleculară şi acțiunea radiațiilor asupra eredității, Bucureşti: Ed. Ştiințifică (1963), 286.

[10] О.И. Потетня, Сравнительная оценка структурных повреждений хромосом лимфоцитов человека в различных стадиях митотического цикла при облучении источниками ^{60}Co и ^{232}To с разной мощности дозы. Автореф. дис. на соиск. уч. степ. канд. биол. наук. Обнинск (1990).

Theme 5

Hazard/Risk Communication/Public Participation

Optimisation of Disaster Forecasting and Prevention Measures in the Context of Human and Social Dynamics
I. Apostol et al. (Eds.)
IOS Press, 2009.
doi:10.3233/978-1-58603-948-6-241

PIMS as a Communication Tool Between PfP Nations in Support of Civil Emergency Preparedness

Prepared by PIMS Program
e-mail:support@pims.org
http://www.pims.org

Abstract. PIMS support for PfP nations has been evolving for more than ten years. Originally, the efforts were primarily focused on providing connectivity and computer equipment, as well as supporting various events. But over the past three years PIMS has additionally been developing and implementing tools for better information sharing, employing Internet based concepts and technologies

Keywords: PIMS, communication, innovative technologies, communities of interest.

Introduction

PIMS, which stands for Partnership for Peace Information Management System, designs, integrates, and provides innovative technologies and services to facilitate collaboration and strengthen relationships in the Euro-Atlantic and Partnership for Peace (PfP) community. PIMS is people-centric and strives to increase participant interaction and interoperability so that the community can cooperate fully on priority areas such a Civil Military Emergency Preparedness (CMEP), Global War on Terrorism (GWOT), Civil Emergency Planning, Peacekeeping Operations and others.

1. PIMS Role in the PfP Community

For over a decade, PIMS has provided a secure distributed Intranet to link PfP Partners with US and NATO Colleagues. A specific focus is to build partner capacity related to technology while focusing on priority topics such as Defence Institution Building, Defence Reform, and Response to Terrorism.

PIMS designs and delivers technical solutions with the aim of increasing partner interoperability, integration, efficiency and transparency in order to prepare PfP partners for future coalition operations. This capability reaches "beyond desktop" to include hardware, software, peripherals, satellite bandwidth/local ISP rental, as well as the logistics, IT support, network administration to keep the network functioning. Additionally, PIMS extends these capabilities to the site of interoperability exercises and conferences so that they function efficiently and participants reap the most benefit.

In November 2004 the European Office of SPAWAR System Charleston was selected by OSD to take over the management of PIMS due to their technical competency and experience in such projects. Since then, the technical branch upgraded the Internet-based collaborative and communications capabilities available to PfP Partners and Exercises. Now, through the PIMS Members, the PfP Training Centres, PfP Consortium and other powered-by-PIMS websites, communities of interest can easily engage in discussion on priority topics, harmonize policies and procedures, and search for expertise in password protected environments

PIMS seeks cost-effective solutions wherever possible. As local connectivity becomes more reliable, PIMS Network Technicians and In-Country Coordinators adjust the network configurations to maintain a good level of connectivity. These collaborative environments assist participants to share and capture knowledge that can be tagged and searched in ways not previously possible. Through this collaborative platform participants have access to integrated instant-messaging clients, voice-calling and can collaborate on projects through blogs and threaded discussions.

Essentially, PIMS is a multinational social network, a community of practice, for exchanging information on the security cooperation programmes of the members involved, as the aim of PIMS is "to coordinate the efforts" offering assistance and with the countries implementing their IPAPs and potential MAPs.

PIMS acts as a communication tool between PfP nations as well through the following methods:

1. To meet in person at different events (workshops, conferences, exercises)
2. To keep in touch by other means : PIMS Members Website, exchanging email (we offer the benefits of effective email, cheap, fast with ability to reach many people almost instantly), sending and receiving each other documents.
3. Working groups where members can consolidate and keep track of the conversations. With these working groups, there is an opportunity to build upon the knowledge of previous meetings, instead of reinventing the wheel each time.

2. "Many-to-Many" Communication versus to "One-to-One" and/or "One-to-Many" Communication

PIMS websites act as a "Marketplace" or "Common Virtual Office" for the PfP community. Participants keep in touch with others through the contact directory, find upcoming events, and consult reference pages on NATO/PfP Cooperative Topics and a Glossary of terms. Instead of communication "One-to-One" and/or "One-to-Many" PIMS sites offer the option of "Many-to-Many "communication.

The early Internet applications of e-mail, FTP and Telnet are characterized as "one-to-one," because they are primarily communication means from one individual (or computer) to another. The benefits are that this kind communication is relatively cheap and it's easy to reach a number of people quickly. But, these methods have limitations. People are inundated and overwhelmed by email. Email chains create "inbox spam", and also with the entry and exit of members of the group, it's difficult to know who should be part of the email announcement. When People exit the group their background knowledge (mostly stored in their Inbox) goes with them. Books, printing press, TV tower are examples of "One-to-Many" communication.

In "Many-to-Many" communication environments people are able to both contribute and receive information. As a member of PIMS supported sites, one can start an instant messaging session, use a multi-lingual chat aid, and post a comment, share information on the topic of interest. Users are there to talk with each other. These tools facilitate communication and interoperability.

For its NATO/PfP audience, PIMS provides technology that enables communication and collaboration in new ways. PIMS capabilities are manifested primarily in the form of Internet-based applications that enable groups of people to communicate, coordinate, and collaborate. These password-protected environments differ from static website where one webmaster is responsible for posting content. Instead, on a PIMS instance, the participants themselves are the primary content creators. The members are enabled, through the tools, to publish their conversations and communications among small teams or entire communities.

The overall goal of the PIMS Members site is to be the electronic manifestation of the human social network and assist in day-to-day working life, which is making these bilateral/multilateral processes work more smoothly and effectively.

3. Communities of Interest (COI) and/or Online Working Groups

PIMS Members Website hosts **Communities of Interest (COIs),** mostly called in IT terminology "Online Working Groups", on priority topics critical to NATO-PfP Interoperability and Partner Capacity Building. PIMS sites integrate the best of "Web 2.0" technology, to provide opportunities for communication online in a "need to share" environment.

PfP Partner and NATO Practitioners and topic Subject-Matter Experts are encouraged to take part in interactive discussions about the NATO Cooperative topics. One may browse the listing of communities and subscribe to groups of interest. Participants can subscribe to topics of interest relating to the PfP Community, NATO, CMEP papers and events.

PIMS online environments like members.pims.org, the PfP Consortium Portal, or exercise instances such as Combined Endeavor and RESCUER, are tailored for the specific group of people, yet all these sites enable the participants of each group to post and "pull" data from the site that makes it useful to their everyday lives. For example, working groups that may have traditionally communicated only on email, now have an opportunity to post presentations, reports, audio and video files to the site for sharing. Then, others can begin conversations in a "threaded discussion" that displays the discussion for future reference.

The instances offer improved asynchronous communication, i.e. exchanges that do not require the parties to be present at the same time. But, these online sites also have real-time communication tools embedded into the software. For example, if you see a colleague present online, you can start a web-based instant-message conversation. PIMS sites have a new Multi-Lingual Chat capability that does basic translations that can assist with conversations in an international environment (Figure 1).

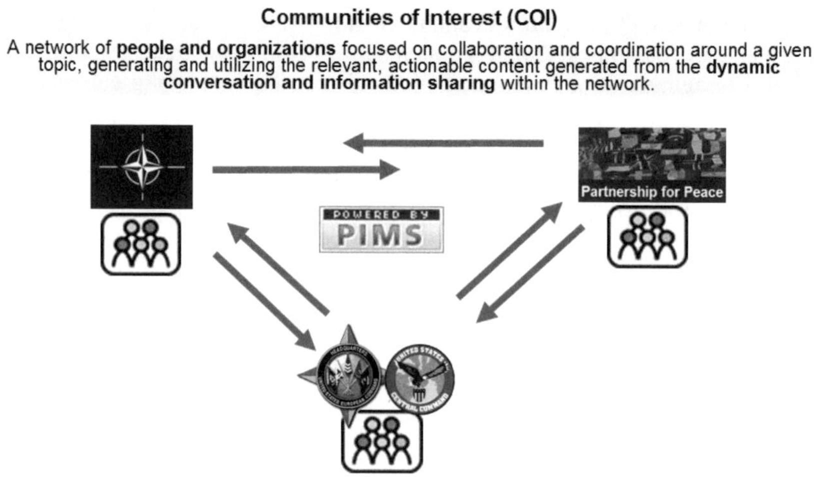

Figure1: Multi-lingual Chat concept

The online working groups are set up to be permissions-based, meaning that what you see on the website is in accord with the groups to which you belong. The groups are "closed", meaning that new members must be approved by a designated administrator. These groups reside among other PIMS Members working groups, but are more controlled than regular communities of interest. PIMS technicians assist the Chairperson and members with user questions, and account management issues.

This site is a "two way street." It does "broadcast' information as many websites do, but it should be thought of as a messaging platform, discussion area as well. An individual can also start their own discussion, make a comment, or pose a question in response to someone else. These online conversations are saved for others to see and take part in.

4. In Support of Civil Emergency Preparedness

PIMS is being used extensively for information management, civil military emergency planning, and support for the Euro-Atlantic Disaster Response Coordination Center, a NATO activity conducted in collaboration with donor nations. PIMS supports the Civil Military Emergency Preparedness (CMEP) initiative for several years, through the technical and personnel support of the CMEP events, and, recently with the new, powered by PIMS, CMEP website (https://cmep.pims.org). Civil Military Emergency Preparedness (CMEP) is the US Department of Defense term for a program to encourage the agencies responsible for Civil Emergency Planning (CEP) in each Partner nation, and their military counterparts, to cooperate in sustained information exchange for emergency preparedness and crisis management.

At the moment, the PIMS staff is building the CMEP online library (archive), where the content related to all these events can be found; as one of the CMEP event participants once said: "Here you can find the whole wisdom of CMEP through the years".

Before creating the software, the design team developed a point of view of knowledge that provided a framework to work:

- Knowledge is both a "good" and a "flow";
- Knowledge is dynamic and unpredictable;
- People are the source of Knowledge;
- Information transactions form basis of understanding Knowledge;
- The higher the number of transactions and participants in exchange the more valuable the model (Moore's law);
- Intelligence about transactions informs where Knowledge lives;
- Focus on derivative models for Knowledge trend identification; and
- Interoperability and Extensibility of Knowledge interactions are framework.

The goal is to develop a modular transactional framework... a "Network of Networks" (Figure 2).

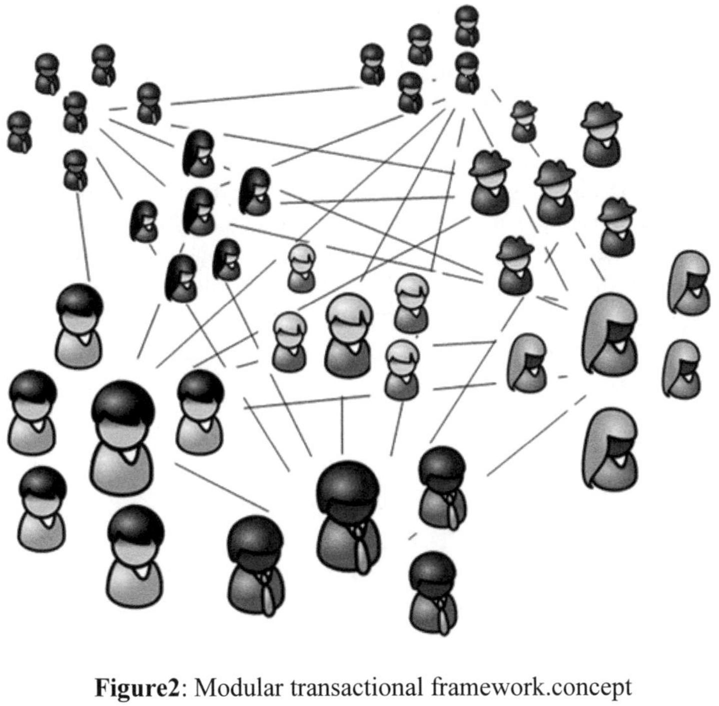

Figure2: Modular transactional framework.concept

The software developers kept six design rules in mind when designing the platform architecture:

- Be Modular - Evolving and Emerging Modular design
- Simplify!
- Embrace Diversity
- Be Standards Driven & Self Descriptive
- Embrace and Extend
- Post and Smart Pull

PIMS end-to-end event-support capabilities are designed to increase participant interaction so that groups can work together before, during and after a simulated disaster, in the case of Civil Emergency Preparedness. End-to-end means that PIMS provides a full range of capabilities and services including comprehensive connectivity, event websites, and IT experts to train and assist planners and participants. PIMS capabilities support the entire event life cycle - from concept development to after-action support. Most websites are used year after year, building an online knowledge and lessons learned repository on the recurring event.

5. Conclusions

PIMS websites provide a rich and collaborative online experience that is designed to easily integrate with the way people work already. PIMS-powered environments utilize innovative technologies that function behind a user-friendly interface. These collaborative websites strive to increase the interactivity of the users and allow them to find and create information ready to be shared among them. With instant messaging, chat, voice, and video conferencing, PIMS users can search, find, and contact potential partners and colleagues. With blogs and distributed web publishing, PIMS users can share their information with others, add their expertise to topics, and collaborate with working group colleagues – all in real time.

PIMS collaborative sites offer a network platform for disaster management collaborative messaging and communications capability to support nations in their efforts to coordinate the disaster management and humanitarian emergency response, civil emergency planning, and to provide GIS tools and other resources that aid in a nation's overall capability to effectively respond to natural and man-made environmental disasters.

References

[1] PIMS Members website https://members.pims.org
[2] CMEP PIMS website https://cmep.pims.org
[3] Public PIMS site http://www.pims.org
[4] SPAWARMNIS collaborative environment https://mnis.spawaeurope.net
[5] WIKIPEDIA http://www.wikipedia.org

Optimisation of Disaster Forecasting and Prevention Measures in the Context of Human and Social Dynamics
I. Apostol et al. (Eds.)

doi:10.3233/978-1-58603-948-6-247

Licensing of Hazardous Industries and Public Participation in the Ukraine

Tetyana BODNARCHUK
State Ecological Inspection, Lviv Region

Abstract. In Ukraine goals of industrial development should correspond to the strategy of technological and ecological safety on the basis of sustainable development. This article deals with the mechanisms of technological emergency management and licensing of environmentally hazardous branches of industry in Ukraine, and the public impact on environmental policy.

Keywords. Technological safety, licensing, regulation, emergencies

Background

In times of establishing new types of economic relations, it is important to find out a model of industrial development that provides economic growth and takes into account environmental needs. Insofar as there is no integral system for economic and ecological assessment that is suitable for modern industrial relations, there is a need for methodological developments and concrete proposals, especially for chemical and oil industries.

Due to a number of negative and positive factors, social and economic consequences of transformation towards market relations are very contradictory. For example, structural changes in Ukrainian industry go together with a sharp decline in production and a high rate of reduction of hazardous discharges, especially in chemical and oil industries. Some large enterprises are technically outdated, and are therefore especially dangerous for the environment, or have totally stopped production. There is trend to move towards less polluting industries, although this is not due to technical, technological and ecological transitions in the industry. At the same time economic crises led to incredible reductions in expenditure to compensate the economic damage. So the issue of environmentally-friendly industry became especially urgent. In times of structural changes such transition should be based on the main principles of sustainable development.

Scientific research in Ukraine, and outside, concerns forms and methods of management for environmentally hazardous companies. However, the issues of environmental impact assessment of hazardous enterprises in Ukraine and its regions at the stage of transformational structural changes are not studied to the full extent.

1. Research Goals and Tasks

The goal of the research is to assess the impact of institutional, structural, regional and market changes on development of environmentally dangerous industries and to analyse the conditions of licensing in Ukraine.

According to this goal the tasks are as follows:

- To analyse structural changes in Ukrainian industry, the state and tendencies of development of the chemical and oil industry;
- To identify modern environmental and economic problems related to the impact of industry on the environment and their regional features;
- To identify external impacts and ecological risks from environmentally dangerous industries;
- To generalise the main trends of dangerous impacts on the environment from ecologically dangerous industries;
- To analyse the system of licensing dangerous types of industry in Ukraine; and
- To analyse the influence of public opinion on environmental decision-making in Ukraine.

2. Economic and Environmental Problems of Dangerous Operations in Ukraine

Ukrainian industry, as currently located, has severe economic and ecological problems, leading to significant losses due to the deterioration of environmental quality, when compared with normal values, due to external impacts. Therefore, according to the adopted classifications, the majority of Ukrainian regions are considered as zones of environmental catastrophes. This is the case for Kyiv, Dnipropetrovsk, Zaporizzhya, Donetsk and Odessa regions.

However, the territorial structure of the chemical industry of Ukraine has been changed. First of all this is the case for regions where environmentally hazardous industries were the most developed. The share of chemical industry in the Dnipropetrovsk region almost doubled (from 9.5% in 1990 to 17.4% in 2007), even more so in the Kyiv region (including Kyiv city) where it was 2.4 times (from 8.9 to 21.2%). From 1994 the chemical industry in the Kyiv region stabilised; starting from 1996, oil and pharmaceutical industries started to grow. It is proven that the high concentration of chemical industry in the Kyiv region due to the increased growth of chemical and oil chemical potential of Ukraine in 1960–1980 was one of the most important causes of the regional environmental crisis. Its features are the effect of synergy of small but very toxic doses of chemical substances with radioactive pollution, which causes irreversible harm to the environment and people's health.

Detailed analysis of primary statistical documents for the majority of environmental hazardous chemical enterprises of the region showed that the total decrease in production due to the transformation of Ukrainian economy was 51%, the reduction of hazardous substance emissions reduced by 56%. For the Kyiv region, these figures are 37% and 51% respectively or, in absolute figures, 140,000 tons, out of which 9,000 tons (or 80%) is due to chemical and oil industry. However, despite the significant reduction in emissions due to the closure of the especially hazardous operations (Table 1), their technological impact, according to national and international norms, can be characterised as an internal regional environmental catastrophe.

Since 1996 the chemical industry has grown annually by 3.5%, and so there is a threat for further deterioration of the environment. There is a need to develop a methodology for assessing the regional environmental state reflecting the process of restructuring of industry as a whole, and especially the chemical and oil industry. The state should control the process of structural change and economic growth in the chemical industry using environmental and economic assessment of such processes and mechanisms for compensation of the costs of environmental protection. Firstly, the methodology of such assessment should be improved; secondly, complex method of cost compensation should be developed. This mechanism should reflect peculiarities of the transition period, the establishment of new market relations and the need for strengthening the regulating role of the state in environmental protection. Using such an approach to economic development, Ukraine will be capable of implementing the main principles of the concept of sustainable development.

On the basis of conducted research, there is a need to improve the methodology for the assessment of hazardous impacts of the chemical and oil industry on the environment because:

- There is no assessment of complex (integrated) hazardous impacts of industrial enterprises and a methodological approach to their synergistic impact – when the integrated impact is more environmentally hazardous than the simple sum of the individual values of the impacts of all the enterprises, which for each individual enterprise do not exceed significantly the maximum allowable concentrations;
- A system of statistical reporting (ecological monitoring) in Ukraine does not correspond to international requirements for complex analysis and the integrated assessment of hazardous impact on the environment.

One of the main tasks of state policy in the field of technological safety and civic protection for the future, described in "Strategy of Economic and Social Development for 2000–2004" and Action Plan of the government, approved by the Verkhovna Rada of Ukraine, is to establish: reliable guarantees for the safe life of the people; technological safety; to prevent emergencies at specially hazardous enterprises; and to reach high norms and standards of protection of population and area from natural and technical emergencies. To implement these tasks, the country should improve the mechanism of technological emergency management and develop relevant legislation.

3. Natural and Technological Safety of Ukraine

The present level of natural and technological safety of Ukraine is characterised by technological overloads on the environment. Regions with an excessive industrial load are the zones with very high risks of emergencies. These risks are constantly growing due to an increase in the percentage of outdated technologies and equipment, and a reduction in the speed of restoration and modernization of production processes. Depreciation of plant in all sectors of the Ukrainian economy is around 50%. Potentially hazardous industries form a large part in the structure of the national economy – they produce around 1/3 of all production.

In conditions of the market economy, in order to ensure proper safety of citizens and the state in case of emergencies, the country should use economic instruments of emergency management at all levels. However, at present, such mechanisms are not systematised and their efficiency and integrated impact on risk levels in the region is not assessed.

Studies of national and international experience of development and the use of economic instruments to prevent natural and technological hazards, show that there are

Table 1. Economic mechanisms for technological emergency management

	Type of mechanism	Characteristics of the mechanism
1	Economic responsibility	Fines for exceeding allowable levels of risks (quotas) or payments for the levels of risk
2	Environmental fund and mechanism for budget funding	Direct funding of activities to reduce risks from state budget or environmental or other funds. As a rule they are mechanisms of fund distribution according to priorities. Some cases when enterprises are not able to cover the expenditure to reduce the level of technological risk are envisaged
3	Mechanism to establish the reserve of financial, labour and material resources	Establishment and use of the reserves of material resources to cover the consequences of the emergency (that is, reserves of the state, local authorities, organizations and enterprises). Volumes of reserves are established by taking into account a prognosis of the type and scale of emergency, and the scope of work to cover its potential consequences
4	Mechanism for stimulating an increase in the level of safety (preferential taxation, credits)	Preferential credits or preferential taxation if environmentally safe technologies are used. Stimuli are created if the level of taxes depends on the level of risk of emergency and if it increased in case of exceeding of the norms
5	Mechanism for redistribution of the risk and a mechanism of insurance	Redistribution of losses due to an emergency between insurers, in case of significant losses when it is hard for one enterprise to compensate them

different economic mechanisms for emergency management (Table 1). They include first of all:

- **State standards:** the main goal of these is to identify norms to prevent emergencies and to avoid losses. Here different types of standards (all-state or per branch of industry) can be used as well as construction norms and rules (Fig. 1). Safety standards are based on the norms of an allowable level of risk for people and the environment; the level of possibility for losing unique natural features; prohibition on living in dangerous zones and prohibition of dangerous works and activities in the region; temporary limitation or stopping of operations of potentially hazardous enterprises; ensuring population safety in case of emergency; economic sanctions towards industrial activities in case of emergencies and improper operations; preparation and further training of staff.
- **Regulation:** power to develop and approve safety norms, rules and requirements. The regulation includes technological criteria and norms for pollution to ensure environmental safety by main factors as well as requirements concerning obtaining licences for all types of activities using hazardous technologies;
- **Safety management for potentially hazardous enterprises:** includes the issuing of permits (licensing) for production activities; the establishment of requirements and limits, frameworks and conditions of operation for such enterprises (by regulation); state control on compliance with requirements and conditions stated in the licenses of the hazardous enterprises.

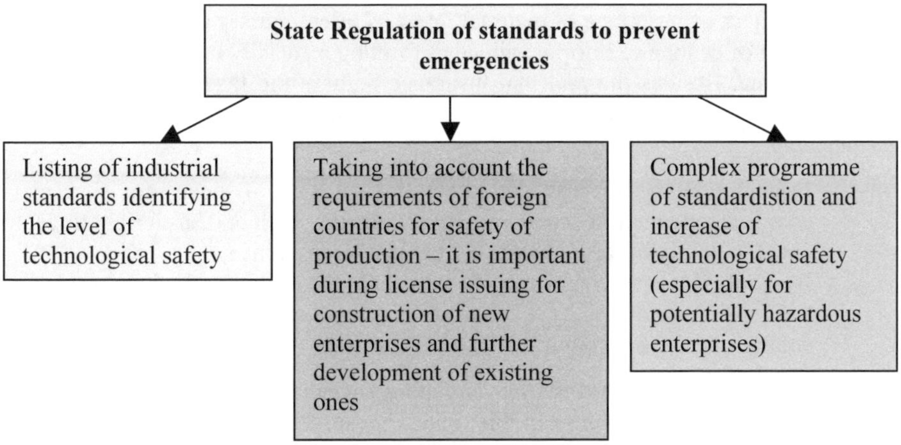

Figure 1. Directions of state regulation of standards.

Table 2. Functions of Insurance

Function of insurance	Characteristics of the function of insurance
Risk compensation	Compensation by money to affected bodies and citizens. Possibility to finance by legal entities for potential losses by means of establishment of their own funds in the form of insurance companies. The Ministry of Emergencies is the main coordinator in the field of risk insurance services for all enterprises.
Prevention	Introduction of obligatory insurance for responsibility by owners of especially dangerous objects for losses caused, as well as property insurance of such enterprises using special conditions
Savings	Related to income, because enterprises can use additionally earned money for activities to prevent emergency
Control	Right of the founder of the insurance fund to control use of the money according to the goals.

From an institutional point of view in the field of emergency prevention there are the following means of regulations:

- ecological requirements for several types of economic activities (industrial production, agriculture, land reclamation, energy sector, construction, military sector);
- requirements to protect and restore the environment (protection of air, water, flora and fauna, soils from pollution);
- listing of potentially dangerous enterprises and hazardous activities, approved by the Cabinet of Ministers of Ukraine;
- insurance, with efficiency based on quality of preparation of external data about the sources of danger, quality of assessment and risk analysis and, most of all, methods and mechanisms of risk management for potential hazardous enterprises (Table 2).

Development of insurance of material interests of enterprises and citizens is an integral element of economic reforms, aimed at forming a socially oriented market economy in Ukraine. Besides the fact that insurance reduces the load on the expenditure part of the budget, and reduces losses in case of emergencies, it plays two more important functions:

1. Insurance allows successful solutions for the issue of social provision, as far as it is the element of social system of the state. Each one has a right to get income, not only in form of state aid but also as insurance;
2. Insurance is the mechanism of investment in the economy. In developed countries, insurance due to its peculiarities and functions in the society belongs to a strategic sector of economy.

One of the important means of overcoming financial difficulties is to introduce state privileges for the companies in this field. For potentially hazardous operating enterprises, there is the system of discounts from this sum of insurance for protective measures. Foreign experience shows that such discounts due to savings for insurance premiums pay off such measures over 6 years, reducing the possibility of danger to a minimum [3]. So it brings profit for the enterprise as well as for the insurer. If to stimulate activities of insurance companies in the field of potentially dangerous bodies, significant savings can be made which can be used to cover losses from technological emergencies, so as a result the general level of social safety is increased. It is worth of mentioning that the large scale introduction of insurance of potentially hazardous enterprises in Ukraine is mostly a task for the future.

At the same time, such economic mechanisms cannot work without proper legislation. It is important to study the impact of a system of such mechanisms at the level of technological and environmental safety in the region. To increase it, the country should activate a system of ecological certification. Further, it is important to develop it together with the licensing of potentially hazardous enterprises, ecological audits, and obligatory insurance.

Emergency management should cover all problems related to emergencies, including the stages of their prognosis, prevention and preparation for operation in case of emergencies, as well as elimination of their consequences. Special means of direct emergency management should be considered from the point of view of the operational regimes for the management system:

- everyday activities (stationary functioning);
- increased readiness (active preparation and implementation of preventative activities);
- post-emergency – elimination of consequences of emergency.

For the first regime, namely “everyday activities”, it is typical not to have information about the clear threats of emergencies.

Management systems start to react in an emergency by adopting emergency and radical activities. In case of emergencies the following actions are taken:

- identification of the situation, preparation of the necessary maps, study of the causes of emergencies, and safety ensuring;
- prognosis of emergency development, modelling the dynamics of its development and assessment of the resources (materials, financial, labour etc.) for its elimination, and assessment of the need to evacuate the population;

- development and analysis of the strategies to liquidate an emergency and its consequences, split of the area into zones and assignment of the staff for each zone, identification of the necessary operative brigades and their composition, distribution of the brigades, organization of the patrol, and evacuation;
- planning and operational management of the works organisations by identified directions, identification of the priority works, assignment of those responsible for their implementation, and distribution of the limited resources;
- rescuing people, emergency construction works and other urgent works by such directions: reconnaissance, identification of victims, provision of urgent medical aid, conduct fire extinguishing, chemical and other activities, organisation of accommodation and temporary infrastructure, organisation of consumer services, means of transport etc.

The following procedures of emergency management are used most recently: environmental impact assessment, plan of liquidation of emergencies, safety standards in case of emergencies (state standard ДСТУ 3891-99 "*Safety in Case of emergencies. Terms and Definitions of the Main Notions*", state standard ДСТУ 3900-99 "*Safety in Case of Emergencies, Main Notions*" state standard ДСТУ 3970-2000 "*Safety in case of Emergencies. Emergencies on areas of water. Terms and Notions*").

Generally speaking, environmental impact assessment (EIA) includes assessment of possible consequences of any types of activities on the environment. Practice shows that the standard procedure of EIA is expertise of large enterprises, using approved "*Methodology of Ecological and Economical Assessment of the Projects*", developed by the Council of Productive Forces of Ukraine.

Public evaluation of the project is an important part of EIA. Therefore, the process of EIA becomes a procedure for seeking a compromise, where the actors are the entrepreneur (initiator of the project), the investor (if a bank gives credit), the project implementer, the local administration, and population.

Practice shows the increased role of ecological expertise in projects for the construction and operation of enterprises. Here the expertise is done in case of large enterprises from the point of view of emergency prevention and possible ways of their operation. Namely, such decisions and projects can include plans to reconstruct branches of the economy, projects for utilisation and processing of radioactive and toxic waste, programmes for the reliable function of basic assets of industry and construction, and environmental protection in the case of emergency for pipe transport.

Therefore, the growth of exploitation of natural resources in industry and lack of funding, together with an increase in the share of outdated technologies and equipment, increase the risk of technological catastrophes. Due to this, there is a need to develop directions to improve the system for managing the protection of the population and the environment in the region.

It is important to study economic mechanisms for emergency management (such as economic responsibility, funds and mechanism of budget funding, reserves of resources, stimulating increases in the level of safety, redistribution of risk and insurance, and situational management), especially for the regions with a large industrial potential.

Assessment of the risks of technological and natural emergencies includes first of all cause-effect analysis, identification of the time of their activation and the amount of resultant losses. In order to implement these tasks, scientific analysis of economic, socio-economic and demographic factors, defining social development and their inter-relations, is conducted. In this case the criteria are GDP, quality of life of people, life expectancy (or losses in terms of life duration).

Social protection of the population from the consequences of emergencies covers many issues at all management levels. The mechanism for this protection is integrated and dynamic, and it should take into account scales of consequences of the emergency, the material state of the affected population and other components.

4. Licensing and Regulation

The law of Ukraine "on licensing of some types of economic activities" defines types of economic activities which should be licensed, establishes the state control in the field of licensing, responsibility of management agents and licensing bodies for violation of legislation in the field of licensing.

A licence is a state document giving the right to the licence holder to conduct the type of economic activity mentioned in the licence during the stated period in case of compliance with licensing conditions, A licence is the only permit-type of document which gives the right to do a specific type of economic activity, which should be limited according the legislation.

The licensing authority is the executive authority, approved by the Cabinet of Ministers of Ukraine or specially authorized executive authority of Rada for licensing of some types of economic activity.

The main principles of state policy in the field of licensing is ensuring; the equity of rights, legal interests of all management agents; protection of rights, legal interests, life and health of the population; environmental protection and state safety; introduction of the common order of licensing of economic activity of the territory of Ukraine; establishment of the general list of the types of economic activities, which should be licensed.

Licensing cannot be used to limit competition in the field of economics.

The agents of the state policy in the field of licensing are the Cabinet of Ministers of Ukraine, specially authorized body for licensing, executive authorities, appointed by the Cabinet of Ministers of Ukraine, empowered to license some types of economic activities.

Licence conditions form a legal act, where qualification, organizational, technological and other requirements on how to conduct a type of economic activity are stated. The management agent should organize his activity, if it should be licensed, according to the established conditions for this type of activity licence.

Licence conditions for types of economic activity, where special knowledge is needed, include qualification requirements to staff – legal entities and (or) people – private entrepreneurs. In cases where special requirements for premises, equipment and other technical means are needed to conduct this type of economic activity, such requirements are included in the licensing conditions.

Licence conditions and changes to licence conditions should be made public in an order, which is established by the law, and they come into force 10 days after the date of state registration of the law, unless it includes a later date.

In order to get the licence, the management agent who plans to conduct some type of economic activity that needs to be licensed, should appeal personally, or via authorized organization or person, to the relevant licensing body with a statement written in accord with the provided sample concerning the issuing of a licence. In the statement for issuance of a licence the following data should be mentioned:

1. information about the management agent;
2. title, location, bank information, ID code – for legal entity;
3. surname, name, passport data (passport number, who and when issued, place of living), ID of tax payer and other payments – for physical agent;
4. type of economic activity (fully or partly) for which the applicant wants to get license.

If the applicant has branches, or other separate sub-divisions, which will arrange economic activity on the basis of obtained license, he should mention their location in the statement.

Out of all types of economic activity which should be licensed, only operations in the field of hazardous waste treatment are environmentally dangerous. The company which has such types of waste should have a separate licence for each type of waste.

The applicant should attach a copy of the state registration of the entrepreneur or a copy of the document about inclusion in the General state register of enterprises and organisations of Ukraine, certified by notary or the body which issued the original document.

The licensing authority decides whether or not to issue the licence with terms, not more than 10 working days after the date of the statement about issuing of the licence and documents, attached to the statement, that is if another term of licence issuance for some types of activities is not envisaged by a special law regulating relations in some fuels of economic management.

Notification about the decision concerning the licence is issued to the applicant in written form within 3 working days of the date of the relevant decision-making. In the case of a negative decision, the reasons for this should be stated.

For the types of economic activities related to use of limited resources, in order to use those limited resources effectively and to promote the use of modern technologies and equipment, open competition between applicants is conducted and license is issued based only on results of the competition. Licensing authorities in Ukraine use standard licences based on a template approved by the Cabinet of Ministers. The licence templates are strictly reported and have a serial number.

A licence includes:

- title of the licensing authority issuing the licence;
- type of economic activity for which the licence is issued;
- title of legal entity, surname and name of the person/individual entrepreneur;
- ID code of legal entity or ID of the person – tax payer;
- location of the legal entity or place of residence of the person/ individual entrepreneur;
- date of licence issue and number of the decision about the issuance of the license;
- duration of the licence;
- position, surname and initials of the person, who signed the licence;
- date of issuance of the license.

The licence is signed by the head of the licensing authority or his deputy and certified by the stamp of this authority.

The applicant should pay the amount of one untaxed minimum income of a citizen for the issuance of the copy of licence. This money goes to the state budget of Ukraine.

The specially authorized authority for licensing keeps a General Licensing Register, containing the data of licence registers and ID of licensing authorities. Information in General Licensing Register and licensing registers is open. One has to pay for their use, and the money goes to the state budget of Ukraine.

Management agents, who act without licence, must pay fines in amounts stated by law. The fines go to the state budget of Ukraine. A decision about fines is taken by the authority which, according to the current legislation, should control the presence of licence.

5. Ecological Audit, Ecological Programmes and Insurance

Ecological audit in Ukraine is a type of scientific and practical activity of specially authorized state authorities, ecological expert organizations and public groups, based on inter-institutional ecological study, analysis and assessment of future and existing enterprises, implementation of which can negatively affect the state of environment. It is aimed to prepare a conclusion on whether planned or conducted activities conform to norms and the requirements of environmental legislation, rational use and restoration of natural resources, thus ensuring ecological safety.

According to the Article 4 of Law of Ukraine "On Ecological expertise" from 09.02.95, the goal of environmental expertise is to prevent negative impacts from human activities on the state of the environment and people's health, as well as assessment of the level of ecological safety for economic activity and the ecological situation at some areas.

Ecological programmes are developed in order to organise and co-ordinate activities to protect the environment, provide ecological safety, and rational use and restoration of natural resources. The examples of ecological programmes include the Programme of Establishment of a national ecological network of Ukraine in 2000–2015, approved by the Law of Ukraine on 21.09.00; Programme of prevention and reaction to technological and natural emergencies for 2000–2005, approved by the decree of the Cabinet of Ministers of Ukraine on 22.08.00; All-state programme of toxic waste treatment, approved by the Law of Ukraine on 14.09.00; Programme of search and sterilization of remaining chemical weapons sunk in the marine economic zone of Ukraine for 1997–2002, approved by the Decree of the Cabinet of Ministers of Ukraine on 25.11.96.

It is worth of mentioning that the total negative impact of technological emergencies and catastrophes in Ukraine at present is clearly growing. This can be explained first of all by the active growth in the volumes of production, unclear economic instrument of state regulation in the field of prevention and elimination of negative technological impacts on environment and the absence of market insurance ecological services.

One of the ways to solve the above-mentioned problem is to introduce environmental insurance. This is the component of a financial mechanism that is aimed at compensating damage done to the lives and health of people, as well as losses to the property of a legal entity exclusively on a market basis using private capital. The Cabinet of Ministers of Ukraine developed the draft law of Ukraine "On Ecological insurance". This draft regulates relations in the field of ecological insurance and is aimed at compensating for damage done to people and legal entities due to emergency pollution of environment. Ecological insurance will become obligatory. Activities of enterprises, institutional bodies and organizations, which can be environmentally hazardous and are related to emergency pollution of environment, should be insured.

6. Public Participation in the Resolution of Environmental Problems

Public participation in environmental decision-making is required for realising citizens' rights for participation in state governance, stated in Article 38 of Ukrainian Constitution and for protecting against unlimited power of authorities and for avoiding their abuse of the ecological and attached rights of the population.

National legislation of Ukraine envisages functions of environmental management by citizens and their unions. In many cases it regulates public participation in the process of environmental management in a manner wider than it is required by international norms, namely, Article 9 of the law of Ukraine "On Environmental Protection" which states the ecological rights of Ukrainian citizens. Chapter IV of this law "Competence of the Environmental Authorities" has Article 21 "Competence of public organizations in the field of environmental protection", where the rights of public environmental organizations are described.

On the basis of the analysis of legislation concerning public participation in environmental decision-making, one can identify several types of public participation:

- Influence on the formation of ecological policy at different levels and participation in decision-making;
- Public environmental monitoring;
- Initiation and implementation of ecological audit;
- Implementation of public ecological control.

Partly these rights are envisaged directly in environmental legislation, partly in other legal acts and it is a form of implementation of the general rights of citizens. Ecological rights of citizens are protected by control over the state of environment. State control is done by local radas and their executive bodies, Ministry of Environmental Protection of Ukraine, their bodies and other specially authorized state authorities.

Public control is done by public inspections according to Regulations on public environmental inspectors, approved by the decree of the Ministry of Environment and Natural Resources of Ukraine on 27.02.02, according to which public control in the field of environment is done by public environmental inspectors. Authorities, belonging to the Ministry of Environment organize and co-ordinate activities of public inspectors.

Modern practice shows that solving environmental problems at the local level by means of mobilization of public sources is the most effective instrument for improving the ecological situation in general. At the same time, public representatives are often not environmental specialists, so a lot of different factors can influence their position. Such situations should be solved by legislation.

State officials, due to objective factors, have often limited capacities to react in time on any violation of environmental legislation. Citizens can fully compensate for this when either at work or during holidays they find out any violation of environmental legislation, namely sources of pollution or their consequences (such as oil spots or garbage), illegal actions of some people (such as the unauthorized use of nature or violation of the rules of behaviour in nature). As a rule, this is done by making observations concerning the state of environment as well as by some indicator of flora and fauna.

Public ecological monitoring can be also organised. In the U.S. there is a movement of so-called "river pedestrians", namely, people who voluntarily walk along rivers and streams in order to identify any companies discharging wastewater into rivers. They have helped to identify all company-pollutants and to react to this from the point of view of the state authorities. Similar activities are done by NGOs in Ukraine. The main use of such information is to transfer it to environmental authorities to conduct actions to stop violations of environmental legislation.

The law of Ukraine "On Ecological Expertise" (or audit) (Articles 10 and 11) states that ecological audit client should inform, via the media, about conducting ecological audits in a special Statement about the ecological consequences of activities of enterprises, which can possibly have a negative impact on the environment in case of construction or operation. After completion of the ecological audit, environmental expert bodies publish their conclusions via the media.

In order to take into account of public opinion, bodies of ecological audits conduct public hearings or open meetings. Public participation in the process of ecological audits can be carried out by means of presentations for media, written comments, proposals and recommendations, inclusion of NGOs representatives into the expert commissions and groups to conduct ecological audits. Preparation of conclusions of ecological audits and decision-making concerning further implementation (use, operation etc.) of the object of the ecological audit is done taking into account public opinion.

At the same time, it is worth mentioning that time places new requirements on the role of ecological audit as a function of ecological management. Especially the law "On Ecological Expertise" states terms of implementation of state ecological audit, particularly the right to get the information needed for implementing ecological expertise, on request. Experts of public ecological audits dos not have such rights but according to "On Ecological Expertise" (article 29) he has all the obligations of an expert to conduct ecological audits. They include the obligation to provide complex, objective, good quality and effective implementation of ecological audits, which is impossible if there are none of the needed materials

Unfortunately, there are no financial means in Ukraine to stimulate people whose activity helped to stop ecological violations, to prosecute the guilty and provide damage compensation due to breaches of the law.

Activating a public ecological movement caused some changes in citizen's attitude and authorities to environmental problems. So citizens and their unions more actively become agents of ecological management and, most often, ecological audits concern environmental aspects of economic activities, because such activities can cause negative impacts on the environment. In conditions of democratisation and 'ecologisation' of social processes, the public gets more and more possibilities to participate in ecological management and actively realises them, both alone and in partnership with state environmental and industrial authorities. Of course, this effective direction of public activities should have relevant legal background.

7. Conclusions

The present structure of the Ukrainian economy is such that it constantly leads to pollution of the environment. According to the adopted classification, the majority of Ukrainian regions are considered as zones of ecological disaster. The urgent problem is to decrease negative impacts on the environment, first of all from the chemical and oil industry. In order to solve it effectively, it is important to conduct environmental and ecological assessments of their impacts, which should take into account the damage done by operators' activities.

Analysis of the methodology and methodological approaches to the assessment of the impacts of the chemical and oil industries on the environment show that they do not reflect properly the modern state because environmental problems worsen in relation to the transition in the economy and the consequences of Chornobyl catastrophe. Of course, they should be improved, taking into account the present situation.

The introduction of an improved methodology for research, taking into account structural changes due to market transformations in industry, will show significant changes in the dynamics of the development and regional location of environmentally hazardous industries. The main fact is that during 1990–2007 the structure of industrial production has been changed and the share of oil-energy and natural resources increased correspondingly from 3.4% to 22.8% and from 13.3% to 21.3% in Kyiv, and from 10.2% to 31.7% and from 24.7% till 35% in the Kyiv region (in Ukraine the increases are from 8.9% to 28% and from 23.3% to 34.3%).

In the strategy for reforming, the economy and ecology should be considered in close interaction. So while solving problems of environmental protection, one should more often use relevant market relations with economic and legal instruments as well as administrative ones, such as ecological certification, audits, licensing, tax on products that pollute the environment in one of the periods of their life cycles, market permits, ecological insurance, subsidies etc. The use of such instruments will significantly stimulate a decrease in environmental damage, the use of new environmentally-clean technologies and the production of environmentally clean production. So it stimulates innovative processes of development in the field of the chemical industry.

The system of environmental and legal relations should be oriented towards international standards as much as possible, especially the system of ecological management and audit of industrial activities. The introduction of this system will take Ukraine closer to world standards of ecological safety, which is one the main factors for further development in the world community.

References

[1] ЗУ "Про ліцензування певних видів господарської діяльності" N 1775-III від 1 червня 2000 року м. Київ (1775-14).

[2] Постанова Кабінету Міністрів України "Про затвердження Порядку формування, ведення і користування відомостями ліцензійного реєстру та подання їх до Єдиного ліцензійного реєстру" від 8 листопада 2000 р. N 1658 м.Київ.

[3] Выморкова Н. Возможности решения экологических проблем в странах содружества // Экономист. – 2001. – № 4. – С. 78-81.

[4] Дорогунцов С., Федорищева А. Государственное регулирование техногенно-екологической безопасности в регионах Украины // Экономика Украины. – 2002. – № 4. – С. 70-77.

[5] Ларионов Г. А. Общественный экологический контроль // Государство и право. – 1996. – № 2. – С. 65.

[6] Екологічні права громадян: як їх захистити за допомогою закону / Благод. фонд "Екоправо-Львів": Центр громадської екологічної адвокатури "Правнича ініціатива" для центр. та схід. Європи; Регіон. екол. центр для центр. та схід. Європи. – К.: Інформ. агентство "Эхо-Восток", 1997. – с. 14.
[7] Участие общественности в правовом регулировании воздействия на окружающую среду / Вильям Фатрел, Гровер Рен, Ан Пауерз и др.: пер. с англ. – Вашингтон: Ин-т законодательства окружающей среды, 1991. – с. 18.

I. Apostol et al. (Eds.)
IOS Press, 2009.
doi:10.3233/978-1-58603-948-6-261

Man-Made Disaster Prevention: The Role of Risk Assessment in Development Control

D.L. BARRY, BE, FICE, MCIWM
Director, DLB Environmental, Surrey UK
dlbarry@btinternet.com

Abstract. The risks to human and the environment from natural hazards and man-made sources of pollution can be minimised by effective land-use planning controls that takes due account of hazards sources and their links with potential receptors. These controls can be facilitated in the first instance by the use of vulnerability maps that relate to different environmental, social or physical aspects. The effective use of risk assessment techniques is a logical part of the development control process.

Keywords. Risk assessment; vulnerability maps; development control

Introduction

Controlling the creation of new development areas and features, whether they are for residential, commercial or infrastructural purposes, can help mitigate the potentially significant consequences from placing such developments too close to either natural and man-made hazard sources. While history cannot be changed in terms of the proximity of, for example, existing population centres and facilities to earthquake, mining or natural flooding zones, the use of basic risk assessment processes in land-use planning can help reduce the relevant risks to future development areas. These controls are becoming even more important in developing countries where there can be significant on-going changes in terms of the types of industries being developed, and in the range of new facilities for meeting increased socio-economic needs. These changes can, in many cases, also result in greater pressure to recycle old industrial sites and/or greatly expand urban zones. Some of these changes can exacerbate old problems and/or create new ones and, therefore, the resultant changes need to be managed effectively so that future potential disasters can be avoided, or at least minimised.

Controls on new development should be applied in a positive way from national level down to local levels, depending on the strategic importance of the particular type of development. However, it can be difficult to create an integrated control process that properly recognises the balance that should be drawn between national and local needs in terms of environmental, social and economic factors. On the other hand, the increasing and effective use of environmental impact assessment techniques (which techniques embody the essential spirit of risk assessments) can enable the relevant regulators to minimise future risks by ensuring a sound understanding of the relevant hazards and the potential links with potentially sensitive receptors.

1. Historical Risks

In addition to the well-established risks associated with many natural disaster sources (such as earthquakes, landslides or floods), there can also be other significant risks from former (and existing) industrial areas. Typical polluting industries, where the extent of consequential ground and groundwater contamination can often extend well beyond the property boundaries include:

- Old military areas;
- Chemical/petrochemical plants;
- Metallurgical industries;
- Oil manufacturing industries;
- Waste disposal sites; and
- Mining facilities.

On the other hand, critically contaminating industries are not always large since some relatively small facilities can have a disproportionate polluting effects; for example, dry cleaning workshops and fuel service areas/garages in urban areas; and domestic fuel tanks in the countryside where they can overlie critical groundwater supplies.

A particular additional risk concerns landfill gas generated from, for example, old municipal waste sites, the critical effects of which can be quite extensive spatially and can exist for many decades after waste disposal operations have ceased. Landfill gas, which principally comprises methane and carbon dioxide, differs in many ways from 'normal' ground contamination because of the following characteristics in particular:

- Acute risks of explosions;
- Omni-directional migration potential;
- Airborne, ground-borne and water-borne pathways;
- High perceived risk levels (usually >> actual risk level).
- Greater dynamics in on-going risk levels, temporal and spatial.

Thus, as with mines gas, landfill gas can migrate considerable distances from a source, aided by a back-pressure at source or high ground permeability, as well as atmospheric pressure variations. The gas can also be carried in groundwater in the dissolved phase and be released a result of reduced atmospheric pressure.

2. Hazard and Risk Assessments

There is frequent confusion between 'hazards' and 'risks', even among professionals. In contrast, there is usually little or no confusion about the definition of a hazard. However, a common definition of risk is that it is a combination of:

(i) the probability, or frequency, of occurrence of a defined hazard (e.g. exposure or instability); and
(ii) the magnitude of the consequences to a specified 'receptor'.

Example Conceptual Site Model

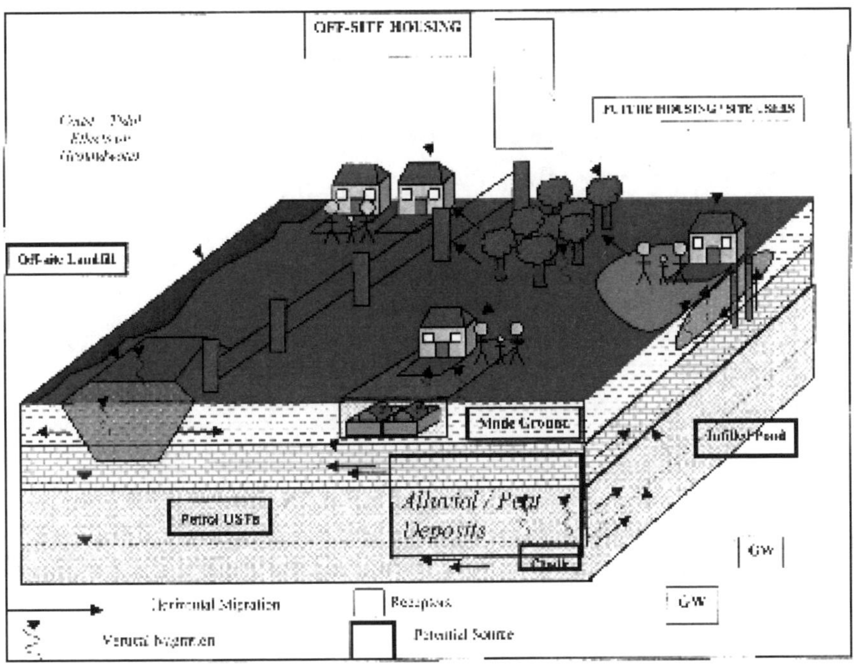

Figure 1. Typical Conceptual site Model.

Therefore, when we say "*It's a risk*" we mean (or should mean) that "*Something specific could happen*" such as a certain intensity of earthquake.

In order to assess any risk it is first essential to understand:

(i) the hazard and its characteristics;
(ii) the sensitivity (or vulnerability) of the actual or potential 'receptor' or 'target'; and
(iii) the potential linkages between the hazard and receptor.

Thus, a risk cannot be managed cost-effectively unless it is understood sufficiently.

A preliminary risk assessment process usually has the following elements:

- a desk study (including a review of existing information and data, and a site visit);
- development of a Conceptual Site Model (CSM) showing potential 'pollutant linkages' (see Fig. 1);
- making qualitative/semi-quantitative assessment of risks for particular scenarios; and
- defining further information and data needs for enhancing the assessment process.

Figure 2 shows the typical elements that feature in a Conceptual Risk Model.

A CSM is a simple model that places (a) known (or potential) hazards in (b) generalised physical contexts that show (c) actual or potential links with any type of 'recep-

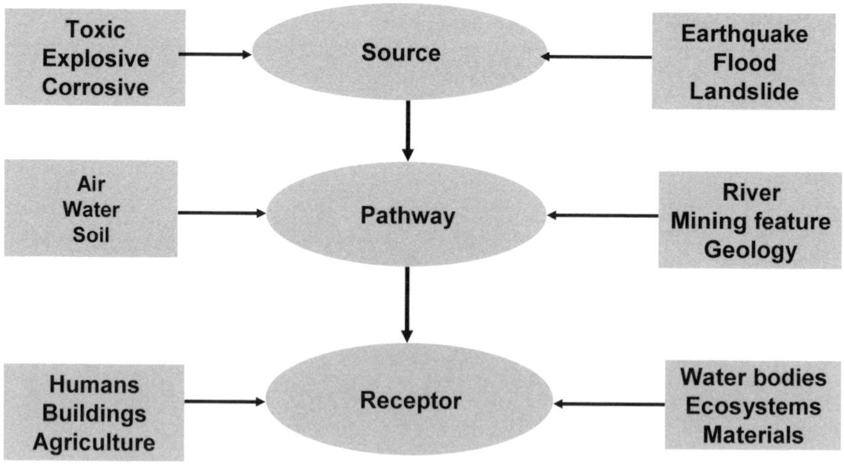

Figure 2. Conceptual Risk Model – Typical Elements.

tor', such as humans, water bodies, ecosystems, buildings, infrastructure, or land, for example. The key to managing the identified risks is to break the 'linkages' by one of the following:

- modify/remove the Source (e.g. make it non- hazardous or less hazardous); or
- modify/break the Pathway (e.g. create a 'barrier' – physical or chemical); or
- modify/replace the Receptor (e.g. select a less sensitive use or location); or
- any combination of above elements.

3. New Development Areas

In controlling the location of new vulnerable developments, such as residential property, then it is essential to consider the potential effects of not just the major natural disaster sources but also the man-made ones mentioned earlier, i.e. ground and groundwater contaminated by industrial usage which can have significant health and other effects on new development users and features. Indeed, those potential effects can sometimes be dominated more by the public's perception of risk scales rather than by the levels of actual risk. This is because residents can be particularly fearful of health effects on, for example, children playing outside the dwelling, or the growing of vegetables for domestic consumption. Such perceptions are usually much lower in the case of less sensitive developments, such as commercial buildings or apartment blocks, where there can be a negligible association between the soil quality and human interactions. In any case, these risk perceptions by the public can lead to 'economic disasters' if the value of their property is compromised such that they can no longer live in it, cannot sell it, or they may even have inherited some liability to remediate the conditions.

Similar perceptions can also be created when sensitive developments are in close proximity to explicitly polluting sources such as municipal landfill sites which, as outlined earlier, can continue to generate hazardous gases for many decades after completion of the waste disposal activities. Sterile development zones might well be created

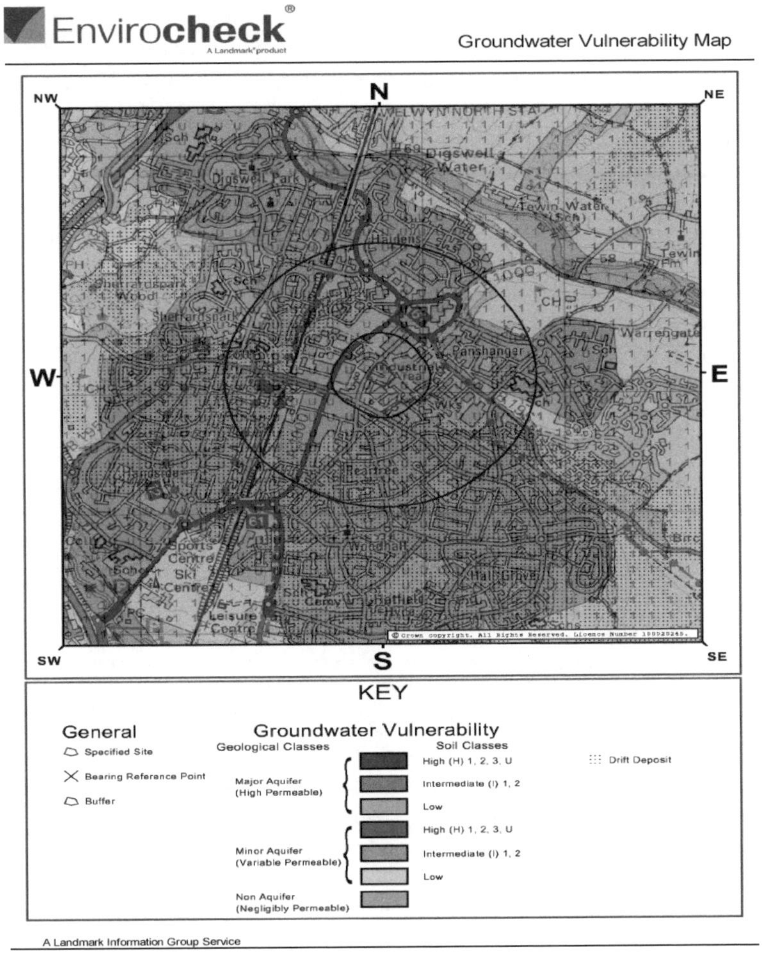

Figure 3. Typical Vulnerability Map for Groundwater (UK).

around such pollution sources. Overall, new land development should therefore take due account of the sensitivity of development type and its features, being particularly aware of the importance of investment security (e.g. land values), as well as the perceived effects of any post-construction remediation works.

4. Vulnerability Maps and Land-Use Zoning

A key technique for helping prevent potential disasters would be the creation of '*Vulnerability maps*'; these would be analogous to maps of earthquake, landslip and flooding zones, for example. Indeed, key elements of such maps may already exist in many countries where vulnerable hydrogeological regimes have been identified (see Fig. 3). A more basic map would relate to the vulnerability of surface water bodies since this is usually more easily defined because of the topographic data and fact that the pollution dynamics are essentially two-dimensional. In contrast, groundwater dynamics are three-dimensional and actual risks can be dominated by the permeability of superficial layers.

Vulnerability maps are also relevant to defining existing risks from either operational or closed facilities/industries, whether waste disposal sites or industries that have/had a high pollution migration potential, for example. In some cases, the critical pollution might not yet be manifest in terms of water quality, but when it becomes manifest it could constitute a pollution disaster. Thus, while it might be too late to prevent the worst effects, there might still be an opportunity for applying some mitigation, not least of which could be the removal or treatment of the primary polluting source.

In all cases vulnerability maps should be seen as giving preliminary indications of risks only; actual risks should be calculated for specific scenarios and features. Thus, the maps would represent 'early warnings' at a strategic development planning level and would need to be supplemented with more detailed factors, including technical and economic ones. Such development planning controls are probably best addressed at the local level rather than at a national level because there can be greater familiarity with local vulnerability factors with respect to either the proposed development or the relevant natural resources, or both. The local regulating authority can, based on an effective understanding of these local sensitivities require the site developer to carry out appropriate risk assessments. These will then enable the developer to include essential safeguards in the facility design, thereby reducing the potential for future disasters. This is not a novel approach, but it is sometimes wrongly considered to be relevant only for major potentially-polluting industries where there is a more obvious need for close evaluation of environmental risks.

With increased knowledge of all the relevant environmental, social and economic factors, local regulators can begin to generate effective land-use maps that help the more appropriate siting of future installations and so reduce the risks from and to future land uses. In generating such types of general development control it is vital that all concerned, not least the regulators, fully understand the profound distinction that must always be made between 'hazard' and 'risk'. All too often these terms are confused and so imply that it is the hazard that must be managed. In many cases this is not possible (especially with natural hazards) and so the emphasis should be placed on risk management, which will include management of the hazard where possible.

5. Concluding Comments

The management of risks through the land-use planning process should occur at the strategic level and requires active liaison between the key regulators, whether at the planning level or the pollution control level. Risk assessment tools can play a crucial role but they are simply tools to aid decision-making and must be used intelligently. The concept of 'zero risk' is generally unaffordable and not usually sustainable and so compromises are necessary, involving technical, environmental, practical, economic and political factors. In all events, it is essential that information on risks is communicated effectively to all interested parties, whether professional or community representatives.

Optimisation of Disaster Forecasting and Prevention Measures in the Context of Human and Social Dynamics
I. Apostol et al. (Eds.)
IOS Press, 2009.
doi:10.3233/978-1-58603-948-6-267

International Cooperation for Emergency Warning and Prevention of Catastrophes in Kura River Basin

Kristine Sahakyan
JINJ Ltd., Armenia:National expert of project "Development of the trans-boundary cooperation for hazard prevention in the Kura river basin"
Tel: (+374 10) 54 01 02 (off.) E-mail: kristine_sahakyan@jinjconsult.com

Abstract. This paper is devoted to the activities of the regional project on "Development of Cross-National Cooperation for Incident Planning in the Kura River Basin Watershed" implemented in 2003-2006. The project was funded by the German Federal Ministry for the Environment, Nature Conservation, and Reactor Safety. South Caucasus countries Armenia, Georgia and Azerbaijan participated in the project. The main goal of the project was the development of international cooperation and exchange of international experience aimed at protection of water basins against industrial pollution and increasing industrial safety in the Kura river basin. Prior to the implementation of this project in South Caucasus, similar projects have been implemented for 3 international major transboundary rivers - the rivers Rhine, Danube and Elba, for which International Commissions for protection have been established. With the help of the checklists developed by them (covering organizational and technical measures related to industrial safety) pilot investigations were carried out in the industrial enterprises with high potential hazards to water bodies and corresponding recommendations were provided. As a result of the project an international warning and alarm plan was developed in the Kura River Basin Watershed, the purpose of which was providing information exchange among the countries if needed. Based upon the information and warning systems existing in the countries at the national level, the corresponding departments in the participant countries were charged with the responsibilities of the International main warning centres, based on the justifications represented on the allocation of warning and international communication centres in the Kura basin

Introduction

The Kura river basin is of significant political and economic importance for the basin countries. The total area of the basin is 188,000 sq. km, of which 102,000 sq. km is the Araks river basin flowing mainly through Armenia (54%). The Kura and Araks basin mostly coincides with Georgia, Azerbaijan and Armenia. The Kura is also of great importance for the economy of the basin countries, and being a most vital artery of the region, it is a determining factor for ecological stability. The Kura basin protection has appeared at the centre of attention relating to the increasing importance of the rational use of water bodies and providing drinking water for the population.

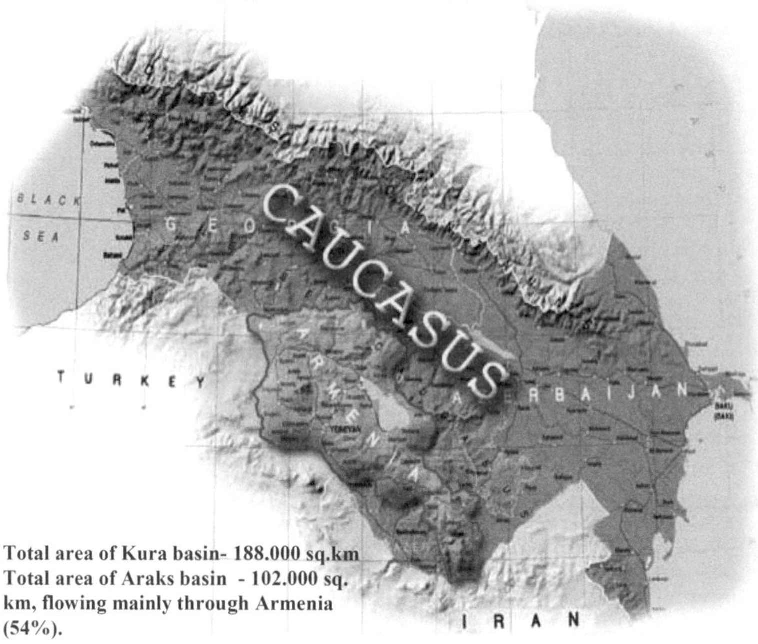

In the process of "Environment for Europe" Georgia, Azerbaijan and Armenia have occupied a firm place in recent years within the framework of European cooperation in the field of environmental protection. The German Federal Ministry for the Environment, Nature Conservation, and Reactor Safety, having a certain input within the framework of this process, has funded the project on "Development of Cross-National Cooperation for Incident Planning in the Kura River Basin Watershed", using the means of Consulting support fund for CACENA countries. An expression of this support is the transfer of information and technologies. Besides, support is also rendered in the sphere of introducing European standards related to industrial safety of dangerous facilities.

This project, which was implemented in 2003-2006, was based upon the necessity to develop basin cooperation between Armenia, Georgia and Azerbaijan in the sphere of emergency warning in the Kura River basin. It promoted the transfer of Western European experience to international river basins of the countries of CACENA. Therewith, a full series of requirements for water body protection and providing water quality in surface reservoirs, was taken into account (e.g. Framework Directive on Water Policy, requirements in accordance with ecological management on EMAS).

1. Project objectives

The project supports exchange and transfer of Western European experience in trans-boundary river basins in Eastern Europe and is aimed at the development of international cooperation, in particular in the field of increasing ecological safety and efficient use of water bodies. The main goal of the project is the transfer of technological experience aimed at development of international cooperation in the field of protection of reservoirs against industrial pollution and increasing industrial safety in the Kura river basin in the following directions, according to the UN Convention of EEC:

- protection of reservoirs against the impact of industrial pollution; increasing industrial safety level
- emergency warning trans-boundary management.

This project differs from those in the region, since it takes into account the importance of trans-boundary impact. The project uses the experience of International Commissions for protection of the Rhine, Danube and Elba (ICRP, ICDP, and ICEP), which have developed recommendations for measures for providing safety of industrial and warehouse facilities. One example of this is the checklists developed by the FEA for evaluating the hazard potential in industrial handling of materials hazardous to water, which are a proven practical instrument for collecting, managing, and monitoring required data under the UNECE industrial convention.

This instrument is recommended for approbation to South Caucasian partners and representatives of responsible authorities. The developed organizational and technical measures for industrial safety promote a gradual increase in the safety levels of industrial facilities, taking into account peculiarities of local conditions.

The German Federal Ministry for the Environment, Nature Conservation, and Reactor Safety attached a special importance to the project, which was included in National strategy of sustainable development of Federal Republic of Germany (NHS) in the part of “Global responsibility”, as well as in the report of FME to the Parliament of FRG. Besides, the project was included in the Programme of Partners projects on the implementation of ecological strategies for countries of CACENA within the framework of “Environment for Europe” process of UN European Economic Commission.

The Ministries of Environment of the participant countries assigned national responsible persons for the project activities, through whom coordination of tasks, methods and possible forms of task implementation, as well as all of the planned project measures in each country, was implemented. In accordance with the nature of the tasks, specialists and agencies engaged in the project activities were determined.

Representatives of different levels were engaged in the project – Ministries, various state services of the region, specialists and national coordinators, engaged for provision of quality and competent implementation of the project activities. During the project all of the participants were regularly and purposefully informed in accordance with their responsibility level in the implementation of activities.

2. Main project activities

The following six main activities were carried out during 2003-2006 within the framework of the project.

- *Identification of industrial plants with potential hazards to water, and their evaluation (2003 – 2004)*

The local experts of the participant countries carried out the collection of basic information on prevention of major accidents in the South-Caucasus countries and preliminary determination of accident-potential in industrial enterprises in the Kura river basin. Based upon the obtained data, the hazard potential of 15 industrial enterprises and 15 former industrial areas was assessed in each of the participating countries. The potential for an incident was evaluated using the WRI (water risk index) method developed in the Danube river basin watershed, which was also recommended for use by the UNECE countries. To derive the WRI, the materials hazardous to water that are permanently present at the plant were surveyed. The materials are classified by class of hazard to water bodies (http://www.umweltbundesamt.de/wgs/wgs-index.htm).

- *Selection and Investigation of Relevant Plants with High Potential Hazards to Water (2003 - 2004)*

Based on the identification of the plants with potential hazards, three model-enterprises with high potential hazards were determined in each country, to investigate within the project (according to the assessment results, these enterprises were with high incident potential - WRI >=6). The model investigations were done using the "Checklists for Investigation and Evaluation of Systems with Materials Hazardous to Water" developed by the FEA . The studies in one of the selected enterprises in each country were done with the participation of an independent expert from Germany (who, according to German law, has the right of access for the expert examination of facilities with water hazardous materials). Local specialists got acquainted with the approach and method of research and two other model enterprises in each country were studied independently by the specialists of the participant-countries. In addition to international requirements, the economic and other conditions of the individual countries were given great consideration in the development of short, medium, and long-term actions, in order to be able to recommend measures that would have a sustainable effect on the region.

The enterprises were from different industrial sectors: chemical, mining, textile manufacture, oil-processing, and power supply.

Works were carried out in the following enterprises:

Armenia

- Chemicals plant (r. Hrazdan-Araks),
- Chemicals plant (r. Debed-Kura),
- Ore mining and processing enterprise (r. Debed-Kura).

Georgia

- Chemicals plant (r. Kura)
- Railway-carriage repair works (r. Kura)
- Ore mining and processing enterprise (r. Mashavera– Kura)

Azerbaijan

- Textile mill (r. Alazani/Kura)
- Petroleum refinery (at Caspian Sea)
- Hydropower plant (r. Kura)

On the basis of the investigation results, the necessary organizational and technical measures for preventive protection of water bodies were recommended.

- Development of checklists for investigation and evaluation of industrial plants with substances and preparations hazardous to water (2004 – 2006)

It is to be noted that as a result of the inspection works new checklists were developed for the investigation of mine tailings, as well as checklists for closing of hazardous enterprises, and confirms the cooperation and experience exchanges among analogous projects, and emphasizes the importance of such cooperation.

A checklist for investigation of tailings is applicable for mining extraction and processing sites, if handling with waste rock or tailings is necessary, including facilities with direct impact on the tailings. The checklist was developed on the basis of BREF (Best Available Techniques Reference Document) for management of tailings and waste rock - MTWR. A checklist for control over the closing process of hazardous industrial facilities and plants, as a result of discussions with local experts, was developed in the form of two independent checklists. Such an approach was necessary for taking into account the situation in countries with economies that are in transition. The economic situation develops in a way that the plants closed for an indefinite period must be subject to conservation. According to the conditions, there should be checklists for both temporary (conservation) and permanent (liquidation) plant closings.

- *Development of an International Warning and Alarm Plan in the Kura River Basin Watershed (IWAK), (2005 – 2006)*

The goal was the development of a functional, secure alarm system between Azerbaijan, Georgia, and Armenia that would alert the environmental agencies in all three countries in case of incident-related water pollution, and provide them with critical information about the accident. International main warning centres were implemented, within the existing early warning structures, to exchange information among the countries. The main task of the International Warning and Alarm Plan is the provision of a functional system of information transfer in the Kura River Watershed with respect to the place, time and scale of the pollution of water resources as a result of accident.

The IWAK plan was developed taking into account the practical experience of International Commissions for the Elbe, Rhine, and Danube, as well as current results from the Neman project. During the plan development, also the warning systems functioning in the Republics of South Caucasus were taken into account, along with the corresponding warning system on industrial emergencies, UNECE, as proposed in October 2004 in Budapest.

Plan IWAK is considered a «living document». Its proposed actual version has been discussed within the project, modified by local experts in accordance with the regional conditions and confirmed by the project participant countries. An important component of IKWA are the alarm criteria, jointly developed and concretely defined by the project experts. The International Kura Warning and Alarm Plan (IKWA), developed by German and South Caucasus experts during the project, is workable and was tested successfully during alarm exercises.

For alarm propagation, the reporting model of “Yerevan – Tbilisi – Baku – Tbilisi – Yerevan” was

used, until direct “Yerevan – Baku – Yerevan” and “Baku – Yerevan – Baku” communication becomes possible. Transmission of information during test exercises showed a few deficiencies at first, the causes of which were investigated by the project team, but still confirmed the principal functional suitability of the model.

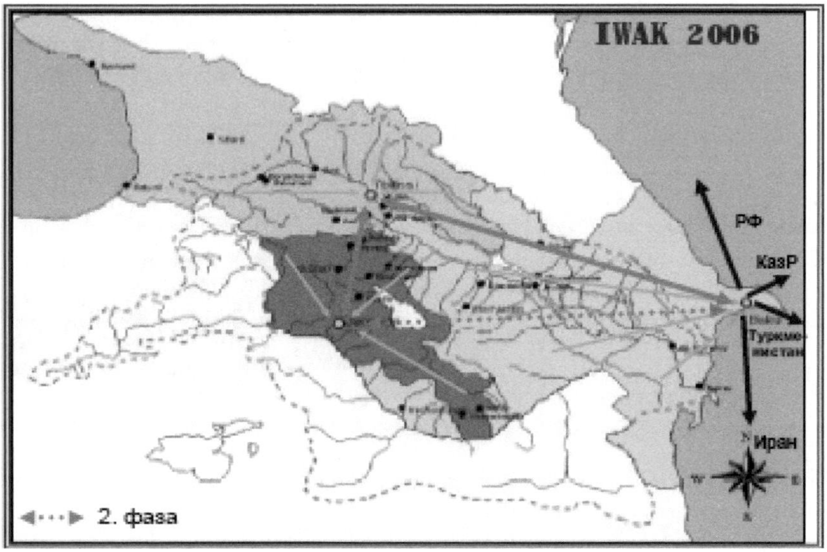

Fig. 3 Information transmission scheme

According to the IWAK plan, the Kura river basin is subdivided into 3 warning areas and one International Main Warning Centre (IMWC) is functioning in each area. According to the warning plan, information is transferred by baton model, therewith, information is always transferred through communication node in Tbilisi.

For the first communication the IMWC is responsible, in the territory of which (area of responsibility) the accident has taken place. This means that in case of emergency situation in the Kura river basin in the territory of Armenia, the information is to be transferred by the IMWC in Yerevan to the address of the IMWC in Tbilisi and from Tbilisi – to Baku. If an accident has taken place in the territory of Azerbaijan, the above-mentioned countries must be informed.

Information transfer by IMWCs is carried out in accordance with the existing regional and national warning plans. Further adaptation and development of the International Warning and Alarm Plan (after project completion) will be carried out by the Expert group of the project.

- *Implementation of International Main Warning Centres and information transfer (IHWZ), (2005-2006)*

Introducing International Main Warning Centres in the existing early warning systems was carried out in accordance with plan of IWAK, taking into consideration the optimal use of the existing communication base (Baku, Yerevan, Tbilisi). Supplementing and improving the required technical equipping of the Centres (communication techniques) was carried out from the project funds. For implementation of plan IWAK, and introducing International Main Warning Centres, the responsible contact persons of the participant countries presented to the project management team the actual information on the existing national warning systems, as well as brief justifications and recommendations on the allocation of warning and international communication centres in the Kura basin.

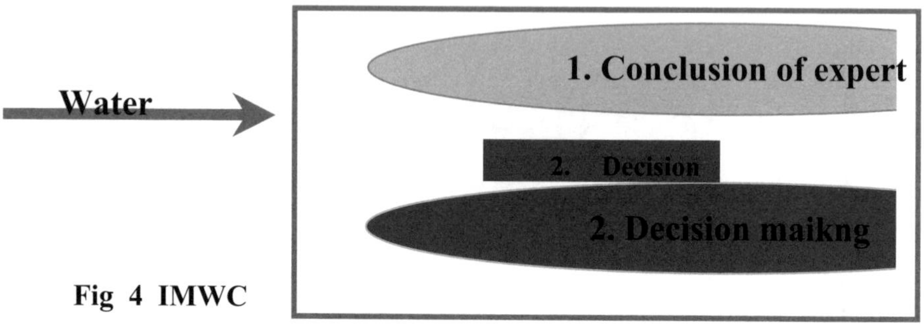

Fig 4 IMWC

The IMWCs combine various functions, which are designated as follows:

- Expert assessment
- Decision making
- Information transfer - international communications (Fig. 4).

The IMWC function in Armenia is carried out by the Emergency response centre ("Crisis Management Centre") of the Rescue Service of the Ministry for Territorial Governance of the Republic of Armenia in Yerevan. The IMWC function in Azerbaijan is carried out by Caspian Complex Environmental Monitoring Administration of the Ministry of Ecology and Natural Resources of the Republic of Azerbaijan in Baku. The IMWC function in Georgia is carried out by IHWZ-Centre of Monitoring and Forecasting of the Ministry of Environment Protection and Natural Resources of Georgia in Tbilisi. In close cooperation with the IMWCs also other state agencies are working, for example State technical supervision, ministries of environment in the three countries, etc.

Expansion and improvement of the necessary technical equipment (communications technology) for the IMWC was supported by project funds. The information paths and internal national legal prerequisites for the notification procedure were developed under consideration of the existing notification procedures in these countries, and tested with alarm exercises. The regional expert group will take over the completion of the work of the IMWC and the notification paths.

- *Establishment of Permanent Expert Group for Emergency situations (PEGAS)*

The jointly-developed cross-border warning and alarm system is to be defined as a first step in the direction of an international river basin commission for the protection of the Kura.

3. Conclusions

Above and beyond the agreed-upon project assignments, the results of the cooperation of the South Caucasus countries and the Federal Republic of Germany, a suggestion was made by the international project steering group to found a standing expert working group to protect the Kura (PEGAS), based on the working group formed within the project. The suggestion and the necessity to establish PEGAS was supported by the experts of the project "Development and introduction of measures for preventing accidents in Kura river basin". Within the framework of analysis and evaluation of the project results, at the meeting of the International group of the project coordination, the representatives of the participant countries came to a general agreement that establishment of the standing expert group for emergency situation (PEGAS) in the Kura river basin was a fundamental necessity welcomed by them. Therewith, the project experts who will continue the initiated work and are the primary chain for future establishment of PEGAS are of special importance. The principal benefits and conclusions from the project are:

- Under the project, activities were implemented and specialists trained for the development of professional and international cooperation in the field of early warning system in South Caucasus.
- A real possibility for quality improvement of prerequisites for practical introduction of international requirements related to emergency management in the Kura river basin was created by the project..
- The project proved that the presence of a clear professional work plan, based on the professional experience of local responsible persons, provides efficient and tangible results which will promote sustainable development of the region.

Also, the solution of problems in the field of trans-boundary river basin protection at international level has a direct impact on gradual development of mutual trust among the adjacent countries, and thus, makes a certain input in the sustainable development of the states' economics and ecology. The carried out work confirmed once again that nature does not recognize any borders.

The project results are available at
http://www.umweltbundesamt.de/anlagen/index.html,
as well as www.kura.iabg.de.

References

Final Report of project "Development of Cross-National Cooperation for Incident Planning in the Kura River Basin Watershed"

Progress Reports
www.kura.iabg.de
http://www.umweltbundesamt.de/anlagen/index.htm

I. Apostol et al. (Eds.)
IOS Press, 2009.
doi:10.3233/978-1-58603-948-6-278

Communication Problems during an Emergency and Lessons Learned

Aysen TURKMAN[a], Ayla UYSAL[b]
[a]European University of Lefke, Gemikonagi Mersin 10, Turkey
[b]Suleyman Demirel University, Department of Environmental Engineering Isparta, Turkey

Abstract. Communication during an emergency and preparedness are very important since they can cause fatalities if not properly managed. In many countries, including Turkey, there are many examples of bad communication in emergency situations. In this study, communication problems encountered during an emergency are discussed. The harmful effects that occurred due to the miscommunication are also discussed.

Keywords. Communication problems, emergency, miscommunication

Introduction

A disaster is defined as a serious disruption to the functioning of society, causing widespread human, material or environmental losses which exceed the ability of an affected society to cope using only its own resources [1]. The extent of a disaster depends on both the intensity of the hazard event and the degree of vulnerability of the society.

Every year many people are affected by natural disasters or technological accidents world-wide. More than 60,000 people are killed and material damage accounts for €69 billion a year in the last decade. While the number of geophysical disasters reported over the last decade has remained fairly steady, there has been a steep increase of hydro-meteorological disaster events (such as floods, tropical storms, and droughts) since 1996. Many scientists assume that this trend will continue and could even be reinforced as a result of global climate change. Together with increasing population pressure and changing habitation patterns in the coming 35 years, this scenario suggests that, a few years down the road, the number of people affected by natural disasters could increase massively. On top of that, some scientists suggest that climate change may cause large scale migration of populations and trigger new or exacerbate existing conflicts about scarce resources like arable land or water [2].

During the past four decades, natural hazards such as earthquakes, volcanic activity, landslides, tropical cyclones, floods, drought, and other hazards have caused major loss of human lives and livelihoods. They equally destroyed economic and social infrastructure and created environmental damage [3]. Economic losses have increased almost 10 times during this period. In recent years, floods in Bangladesh, Mozambique and elsewhere, volcanic eruptions in Ecuador, DRC, Indonesia and the Philippines, and earthquakes in Afghanistan, El Salvador, Indonesia, Peru and Turkey have created widespread social, economic and environmental destruction.

The government is the dominant actor in moving towards sustainable development and disaster risk management, but also the private sector and civil society are playing an ever more active role in successful disaster risk reduction. It is being increasingly recognized that disaster risk management at the local level is a key element in any viable national strategy to reduce disaster risk [4].

1. Disaster Risk Management and its Components

Due to the increasing frequency of disasters worldwide, a lot of international organizations, governments and NGOs are upgrading the priority of disaster risk management policies, and are developing techniques and tools for disaster mitigation, rehabilitation and reconstruction. According to ISDR Secretariat disaster risk management means the systematic process of using administrative decisions, organization, operational skills and capacities to implement policies, strategies and coping capacities of the society and communities to lessen the impacts of natural hazards and related environmental and technological disasters. This comprises all forms of activities, including structural and non-structural measures to avoid (i.e. prevention) or to limit (i.e. mitigation and preparedness) adverse effects of hazards.

Generally, the disaster risk management process (cycle) is composed of the following main elements [5]:

- **Risk identification and assessment** (determining and analyzing the potential, origin, characteristics and behaviour of the hazard – e.g. frequency of occurrence/magnitude of consequences);
- **Knowledge management** (information programmes and systems, public awareness policy, education and training, research in disaster reduction);
- **Political commitment** and **institutional development** (good governance to elevate disaster risk reduction as a policy priority, integration in development planning and sectoral policies, implementing organizational structures, legal and regulatory framework);
- **Application of risk reduction measures** (planning and implementation of structural interventions (e.g. dams, dykes) or non-structural measures like disaster legislation);
- **Early warning** (provision of timely and effective information, through identified institutions, that allow individuals exposed to a hazard, to take action to avoid or reduce their risk and prepare for effective response);
- **Disaster preparedness** and **emergency management** (activities and measures taken in advance to ensure effective response to the impact of a hazard, including measures related to timely and effective warnings as well as evacuation and emergency planning); and
- **Recovery/Reconstruction** (decisions and actions taken in the post-disaster phase with a view to restoring the living conditions of the affected population)

Based on the above specified components, disaster risk management includes measures *before* (risk analysis, prevention, preparedness), *during* (emergency aid) and *after* a disaster (reconstruction). Sometimes disaster risk management includes only a part of disaster management, focusing on the *before* of the extreme natural event.

2. Disaster Preparedness

Disaster preparedness is seen as that action taken when the occurrence of a tropical cyclone, flood or storm surge threatens to become a disaster. Preparedness activities are designed to reduce social disruption and losses to existing property and are an essential component of overall disaster planning. They can serve in the absence of more permanent measures to reduce the threat to loss of life and property. The main types of disaster preparedness include [6]:

- forecasting and warning systems
- evacuation from affected areas
- flood fighting
- flood relief
- cyclone shelters

Depending on the size of the drainage basin, the length of river and the time of concentration of floodwater in the main channel, flood forecasts and warnings may be issued well in advance of the arrival of the flood crest on large rivers. Flash floods originating on small catchments present special problems and usually require some form of forecasting based on rainfall estimates. Although the forecasts for cyclones and floods may be accurate and timely they may have little or no effects on the intended recipients if the warning system for dissemination of the forecast is inadequate. Each agency responsible for emergency operations should receive prompt forecasts and warnings of the changing circumstances so that action needed to meet the emergency can be achieved. Dissemination of forecasts requires an effective communications system based on radio broadcasts, television, newspapers, telephone and special warning systems. The evacuation of people from a potential or actual disaster area is one of the most important elements of disaster mitigation. Careful planning is necessary for the efficient evacuation and relief of flood victims. To be effective the plan should define hazardous areas and potential dangers. However, the difficulty in evacuating victims and property can be increased if escape routes cannot cope with the traffic volume, if evacuation services cannot be contacted or suitable evacuation equipment such as trucks, boats and helicopters are not available [6].

3. Problems in the Risk Communication

There are many examples of emergency miscommunication in the world. These problems not only create negative human health and environmental effects, but also psychological effects on the community as explained below.

Lack of credibility: Lack of credibility alters the communication process by adding distrust and acrimony. The most important factors detracting from the credibility of a risk message relate to the accuracy of the message and the legitimacy of the process by which its contents were determined, as perceived by the recipients.

The perceived legitimacy of the process by which the contents of the message are determined depends on the following: legal standing, justification for the communication programme, access of affected parties to the decision-making process, and a fair review of conflicting claims. Perhaps the most difficult problem for credibility is a past record of deceit, misrepresentation, or coercion. When the responsible government organizations have been proven to have lied, it is not surprising that people want independent verification.

Establishing and defending credibility is difficult when the message represents a departure from previous positions. In large part credibility derives from the demonstration over time of consistent competence and fairness. Both scientific incertitude and changes in policy can serve to undermine credibility to the lay public. The necessity of correcting mistaken statements or positions can undermine credibility with the public. Care must be taken to demonstrate why the interpretation of scientific or policy conclusions has changed [7].

Incomprehensible messages: For those who are not familiar with it, the technical terminology of risk assessment is very difficult to understand.

Preparing messages with few data and no time: Sometimes the risk communicator must disseminate messages when there are not enough relevant data to draw satisfactory conclusions and there is no time to obtain better information. This usually occurs in the following situations:

1. An emergency requires immediate action, or
2. Events lead to requests for information prior to the completion of study or analysis.

The problem is most extreme when external events take control and require action when no preparation has been made in advance of the event. For example, the Nuclear Regulatory Commission was almost totally unprepared for an accident at the time of Three Mile Island. There was no effective management structure to support emergency decision-making and time was lost in figuring out who should do what [8].

4. Bad Communication Examples

There are many bad communication examples in the world and several of them are mentioned below [9]:

Porto Marghera Chemical Plant, Italy

An accident occurred at the Porto Marghera chemical plant on November 28, 2002 and toxic chemicals were released to the environment. The Porto Marghera plant caused a serious health risk for people exposed to its fumes and several workers have died after being exposed to carcinogenic substances produced in the chemical plant.

The real mortality rate from vinyl chloride is different from the 'official' one, as asserted in a report published by an Italian magazine, *Medicina Democratica*, in 1994. The data on deaths from angiosarcoma (an obscure cancer of the liver tied to vinyl chloride exposure) given by public bodies and companies are not reliable. Among those who died at Montedison because of vinyl chloride, only three have been officially recognized.

The situation at Porto Marghera is a genuine emergency, but for the past 25 years it has been dealt with only by adopting inadequate temporary measures, said Fabrizio Fabbri of Greenpeace Italy. Not even the helpful action of the Venice judges, who pointed out several environmental crimes committed by chemical companies in the area, could persuade the politicians to take concrete action [10].

Bhopal India

On the night of 2/3 December 1984, a major accident occurred in Bhopal at a pesticide plant owned by the Union Carbide Corporation [11]. This accident triggered a long-term industrial *crisis* for the entire population of Bhopal, for government agencies in India, and for the Union Carbide Corporation (UCC) [12].

The Bhopal crisis was triggered by a technological accident: 45 tons of methyl isocyanate (MIC) gas escaped from two underground storage tanks at the plant. The accident occurred between 10pm (2 December) and 1.30am (3 December) when the plant was on second shift and the surrounding population was asleep in slum "hutments" that are densely packed together in this part of Bhopal.

Leaked gases were trapped under a nocturnal temperature inversion in a shallow bubble that blanketed the city within five miles of the plant. Next morning, over 2,000 people were dead and 300,000 were injured. Another 1,500 people died in subsequent months owing to injuries caused by the accident. At least 7,000 animals perished but damage to the natural environment remains largely unassessed [13].

Emergency services were completely overwhelmed and confusion was rampant in the affected neighbourhood. Police instructed people to run away from the area, but many of those who did so inhaled large amounts of toxic MIC and succumbed to its effects. Residents were unaware that the simple act of covering their faces with wet cloths and lying indoors on the floor would provide effective protection against the gas. That night, and in the days that followed, nearly 400,000 people fled the city in a haphazard and uncontrolled evacuation. Two weeks later, during government attempts to neutralize the plant's remaining MIC, another wave of mass flight involved 200,000 people [14, 15, 16].

An important issue also raised by the Bhopal accident is the location of industrial plants. Legislation is required to ensure that dangerous sites do not exist in close proximity to heavily populated areas. All industrial plants should be built with the potential for a disaster in mind, thereby minimizing the risk to the population if a spill, fire or leak occurs. In the case of Bhopal, it appears that the dense population distribution around the plant occurred as a result of people needing to live close to work [17].

Initially, the state government tried to place all the blame squarely on UCC and sued them for damages on behalf of victims. In a largely symbolic gesture against the company, UCC's Chief Executive, Warren Anderson, was arrested on his arrival in Bhopal. The government thwarted several efforts by UCC to provide relief to victims, in an attempt to prevent the company from earning goodwill among the public. This early political management was very effective. In nationwide elections that took place four weeks after the accident, the Congress Party won both the state legislative assembly and the national parliament seats from Madhya Pradesh by wide margins [12].

Chernobyl, Ukraine

The accident of 26 April 1986 at the Chernobyl nuclear power plant, located about 20 km south of the border with Belarus, was the most serious ever to have occurred in the nuclear industry. It caused the deaths, within a few days or weeks, of 30 power plant employees and firemen (including 28 with acute radiation syndrome) and brought about the evacuation, in 1986, of about 116,000 people from areas surrounding the reactor and the relocation, after 1986, of about 220,000 people from Belarus, the Russian Federation and Ukraine. Vast territories of those three countries (at that time republics of the Soviet Union) were contaminated, and trace deposition of released radio-nuclides was measurable in all countries of the northern hemisphere [18].

The explosion released 400 times more radioactive material into the atmosphere than the US nuclear bombing of Hiroshima, Japan in 1945. This resulted in tons of contaminated food being consumed by millions of people. Officials, including ministers and scientists, systematically suppressed information about Turkish areas and food contaminated by the radioactive fallout from Chernobyl.

A large number of babies were showing deformities, especially those born to mothers who were in their second month of pregnancy when the accident occurred. Also, many miscarriages and abnormal births had been observed. One of the places the magazine mentioned was the village of Düzce, on the western coast of the Black Sea where, in November 1986, an extremely uncommon concentration of babies - numbering 10 in that year alone - were born with their brains outside their skulls. In the city of Trabzon, also near the Black Sea, the number of abnormal births has quadrupled since 1986 [19].

The Turkish Atomic Energy Agency (TAEK) knew about the contamination in the Black Sea region. But the agency did not warn people who grow tea in this region or factories processing the tea. Between May and December 1986, when tea was harvested, people were left without a warning. Contaminated tea processed and packed during this period of eight months was sold on the market. Best quality tea was sold most likely in Germany and consumed by Turkish workers living there. Turkish scientist Dr. Yuksel Atakan, who lives in Germany, published in 1990 a study showing that tea from Turkey was heavily contaminated. Results of measurements in Germany of tea bought in Turkey in June 1987 varied from dangerous levels of 6,000 to 30,000 Bq/kg.

By the end of 1992, Mr. Aral confessed: "The government has indeed hidden the facts and figures on the impact of Chernobyl in Turkey." "I am sorry", the former Minister of Industry and Trade, Cahit Aral, remarked recently, "but we couldn't protect the Turkish nation." After Chernobyl, the Turkish people were burdened with highly contaminated food [19].

The Turkey earthquake of August 17, 1999

The Turkey earthquake of August 17, 1999 was one of the strongest earthquakes ever to hit an industrialized region. The earthquake had a magnitude of Mw 7.4 and caused over 15,000 deaths and 40,000 injuries. It is estimated that 214,000 residential units and 30,500 business units either collapsed or were lightly to heavily damaged leaving more than 250,000 people homeless (USGS, 2000). The earthquake not only cost thousands of lives, but also caused direct and indirect economic losses estimated as $16 billion USD [20].

This earthquake offered a unique opportunity to study risk management practices and emergency response to accidental releases of hazardous materials triggered by seismic movement. While there has been some attention devoted to releases from pipeline breaks during earthquakes, there has been little consideration of earthquake-related hazardous material releases at industrial facilities.

The study results indicate that hazardous material releases are a real threat to life and property inside industrial facilities as well as to nearby residential areas. Some of the more significant examples of hazardous material releases triggered by the earthquake include: the air release of 200 metric tons of hazardous anhydrous ammonia to avoid tank over-pressurization due to loss of refrigeration capabilities; the leakage of 6,500 metric tons of toxic acrylonitrile (ACN) into air, soil and water from ruptured tanks; the spill of 50 metric tons of diesel fuel into Izmit Bay from a broken fuel loading arm; the release of 1,200 metric tons of cryogenic liquid oxygen caused by structural failure of concrete support columns in two oxygen storage tanks; and the enormous fires, liquid petroleum gas leakages, and oil spills at the TUPRAS oil refinery.

Several strategies have been identified to make highly populated, industrialized cities safer and more resilient to earthquake threats. These include enforcement of regulations pertaining to seismic-resistant construction codes and other environmental and public safety laws; risk management practices and mitigation measures in industry which account for the possibility of seismic hazards; emergency management programmes in industry and government that take into account the simultaneous effects of the earthquake and possible hazardous materials releases; land use planning as a mitigation strategy to reduce the impact of joint earthquake and hazardous materials releases on urban communities; and the appropriate government structure, organization, and political context in which to effectively manage joint natural and technological emergencies [20].

Hurricane Katrina, USA

Hurricane Katrina (August 2005) was the largest natural catastrophe USA has ever experienced. Although Hurricane Andrew in 1992 made land-fall with far stronger winds, the Galveston Hurricane in 1907 took more lives, and the great Mississippi River flood of 1927 inundated more territory, Hurricane Katrina's strong storm surge along the Gulf Coast of Mississippi and eastern Louisiana, and the failure of New Orleans' levies, combined to devastate a populated and developed area the size of Great Britain. Hurricane Katrina also struck at a time when the nation had far greater expectations of government – and particularly of the federal government – to prepare for and then, after the storm, to assist governments and residents and businesses devastated by the storm. In addition, it struck after four years of investment in preparedness by the government. These investments included a new federal department (the Department of Homeland Security), a new Pentagon command focused on the homeland (U.S. Northern Command, Northcom), a new National Response Plan and National Incident Management System, and billions of dollars of appropriations to help federal, state, and local governments prepare for catastrophic events – albeit primarily catastrophic events caused by terrorism. *Disaster: Hurricane Katrina and the Failure of Homeland Security* vividly documents the failures of this expensive investment in preparedness [21].

Government's recognition and response to Katrina was confused, chaotic, and much too slow. President Bush admits mistakes in handling Hurricane Katrina, saying the storm exposed serious problems in the government's response capability.

Flood Problems in Vietnam

Flooding is the main form of natural disaster in Vietnam; every year floods cause enormous damage to human lives and the economy. The annual rainfall in Vietnam ranges between 1,800 and 2,500 mm and 70-80 per cent falls between July and October. Although beneficial to agriculture, rain creates havoc through flooding [22]. The monsoon winds and typhoons are the main causes of heavy rains. The floods associated with monsoon rains usually occur between July and November. Heavy rains cause flash floods in hilly areas and later inundate the deltas. Typhoons bring heavy rain to the coastal belt, sometimes as much as 800mm per day and the strong winds that accompany them create tidal waves [23]. Loss of human lives and livestock, and damage to economic activities, infrastructure and buildings, are inevitable in the aftermath of flooding.

Floods affect the economic behaviour of the people as well. In flood-prone areas, farmers often grow only one crop a year compared with two or three in flood protected areas of the deltas. In areas where the salinity of the soil has been increased by sea water that comes in with tidal waves, farmers prefer to grow saline-resistant traditional rice varieties instead of high-yielding ones. Despite the scale of the economic losses, no thorough study has yet been conducted to assess the overall impact of floods on the economy of the country [22].

For accurate flood forecasting, good data collection and communication systems are essential. The communication facilities in Vietnam are very poor and worsen during floods, thus necessitating substantial improvements. Vietnam will therefore need external assistance to acquire new technology and experience to reduce the damage caused by floods [22].

5. Results and Conclusions

As can be concluded from many unfortunate events, the losses from miscommunication are much bigger than gains for the companies or the government. Therefore, in order to overcome the problems of miscommunication, preparedness, honesty, coherency, transparency, consistency and being timely are major factors to be considered. Owing to the lessons learned from bad communication examples, many countries have already improved their rules and regulations on risk communication.

The harmful effects can be minimized by informing the population about the hazards and required behaviour under emergency conditions in a vivid and open manner. The communication about risks is a sensitive subject and needs accurate planning. In addition to the well-balanced information contents, the analysis of the social conditions of the population in the neighbourhood has decisive importance for successful communication.

Although the risk communication may create anxiety for the public by engendering some fears, still it is a very important in preventing the negative effects of an emergency. Good communication as a part of risk management is a must in democracy.

References

[1] European Environment Agency (EEA), Multilingual Environment Glossary. http://glossary.eea.europa.eu/EEAGlossary., 2006.

[2] http://www.untj.org/files/minutes/DPM/Annexes/DPM281103(12).pdf

[3] http://www.unisdr.org/unisdr/WSSDdocrevisedsept02.htm

[4] United Nations Development Programme, Bureau for Crisis Prevention and Recovery (UNDP/BCPR), *Reducing Disaster Risk. A Challenge for Development.* A global report, New York, 2004.

[5] United Nations Inter-Agency Secretariat of the International Strategy for Disaster Reduction (UN/ISDR) *Living with Risk: A global review of disaster reduction initiatives.* United Nations, Geneva, 2004.

[6] http://www.unescap.org/enrd/water_mineral/disaster/watdis4.htm

[7] http://www.dphhs.mt.gov/PHSD/risk-communication/risk-comm-index.shtml

[8] http://darwin.nap.edu/books/0309039436/html/

[9] Küçükgül, E., Türkman, A., Uysal, A., Communication Problems During Emergency. NOSHCON 2006, International Risk Management, 45th Conference and Exhibition in Occupational Risk Management, The Lost City Convention Centre, Sun City, South Africa, 144-152, 2006.

[10] http://archive.greenpeace.org/pressreleases/toxics/1999sep15.html

[11] Bogard, William P., The Bhopal Tragedy: Language, Logic, and Politics in the Production of a Hazard. Boulder, Col.: Westview Press, 1989.

[12] http://www.unu.edu/unupress/unupbooks

[13] Prasad, R., and Pandey, R.K., Methyl isocyanate (MIC) hazard to the vegetation in Bhopal, *Journal of Tropical Forestry* **1** (1985), 40-50.

[14] Shrivastava, P. , Bhopal: Anatomy of a Crisis. 2nd edn. London: Paul Chapman, 1992.

[15] Diamond, S. , The Bhopal Disaster: How it happened. New York Times, 28 January., 1985.

[16] Morehouse, W., and Subramaniam, A., The Bhopal Tragedy. New York: Council on International and Public Affairs, 1988.

[17] http://www.tropmed.org/rreh/vol1_10.htm

[18] http://www.world-nuclear.org/info/chernobyl/chernounscear.htm

[19] http://www10.antenna.nl/wise/index.html?http://www10.antenna.nl/wise/385/3760.html

[20] Steinberg, L, Cruz, A., Vardar-Sukan, F, Ersoz, Y. Assessment of Risk Management Practices at Industrial Facilities during the Turkey Earthquake of August 17, 1999, First Annual IIASA-DPRI Meeting “Integrated Disaster Risk Management: Reducing Socio-Economic Vulnerability”, IIASA, Laxemburg, Austria, 1-4 August, 2000.

[21] Abbott, E. B., Review of Disaster: Hurricane Katrina and the Failure of Homeland Security, *Journal of Homeland Security and Emergency Management* **4** (2007), 1-4.

[22] Wickramanayake, E., Flood Mitigation Problems in Vietnam, *Disasters* **18** (1994), 81-86.

[23] UNDP, *Report on the Economy* of *Vietnam.* United Nations Development Programme, Ha Noi., 1990.

I. Apostol et al. (Eds.)
IOS Press, 2009.
doi:10.3233/978-1-58603-948-6-287

Public Participation and Information through the Licensing Phase of Industrial Facilities to Optimize Disaster Forecasting and Prevention Measures

Juliane KNAUL
Legal Department, State Office for Mining, Geology and Minerals, Brandenburg, Germany

Abstract. This paper outlines the key international, EC and German laws that dictate the procedures for informing the public who is adversely affected by industrial development. The paper gives an up-date on the requirements of the relevant regulations for participation by the public and providing information to the public within the license procedure of industrial and mining projects likely to have significant adverse effects on the environment.

Keywords. public participation; information; mining-related industries; legislation

Introduction

The public has an interest in the licensing phase cycle of industrial facilities such as mining-related industries. This interest is based on several accidents which have happened such as some tailing dam bursts in Baia Mare, Romania or Aznalcóllar, Spain. These accidents reflect the general environmental and safety hazards of mining activities which have increased the public awareness. In consideration of this fact, public participation and information to the public through the licensing phase of industrial facilities regulated by international, EC and German national law have an essential role in optimizing disaster forecasting and prevention measures.

1. International Law

UNECE Espoo-Convention on Transboundary Environmental Impact Assessment of 25 February 1991.

The Espoo-Convention is the first general contract of nations which requires an assessment of the trans-boundary environmental impact of certain activities at an early stage of planning. It also lays down the general obligation of States to notify and consult each other on all major projects under consideration that are likely to have a significant adverse environmental impact across boundaries.

Mining activities are also listed in this Convention. The need to give explicit consideration to environmental factors at an early stage in the decision-making process by applying environmental impact assessment requires public participation at all appropriate administrative levels, as a necessary tool to improve the quality of information presented to decision-makers so that environmentally sound decisions can be made, paying careful attention to minimizing significant adverse impact, particularly in a trans-boundary context.

UNECE Århus-Convention on Access to Information, Public, Participation in Decision-Making and Access to Justice in Environmental Matters of June 1998.

Contributing to the protection of the right of every person from present and future generations to live in an environment adequate to his or her health and well-being, the Århus-Convention provides for rights of access to information, public participation in decision-making and lays down the establishment of relevant international minimum standards in these regards.. Aiming thereby to further accountability and transparency in decision-making and to strengthen public support for decisions on the environment, the Convention sets significant standards for protection, prevention and improvement of the state of the environment and for ensuring sustainable and environmentally sound development. This Convention promotes environmental education to further the understanding of the environment and sustainable development and encourages widespread public awareness of, and participation in, decisions affecting the environment and sustainable development. In this context, there is a need for making use of the media and of electronic or other forms of communication.

In summary, the Convention serves an essential role in furthering human wellbeing and the use and enjoyment of basic human rights, including the right to life itself.

2. European Community Legislation

Council Directive 85/337/EEC of 27 June 1985: assessment of the effects of certain public and private projects on the environment.

Already before commencement of the above-named general contracts of nations, the Directive on the assessment of the effects of certain public and private projects on the environment required the involvement and participation of the public in the context of the environmental impact assessment of a large number of economic activities, including mining activities, where such activities are likely to have a significant impact on the environment by virtue of their nature, size or locationThis Directive requires an assessment of the likely environmental effects of certain activities before authorization is given. Such assessment must be reflected in an environmental report that must be taken into account by the competent authority granting authorization. Planned mitigation measures form a particular part of such assessment.

An important factor in the impact assessment procedure is the involvement and participation of the public. Within the environmental impact assessment process, which is extensively required for mining activities, the competent authority is required to ensure that information is made available in an appropriate manner to the public. The resulting comments of the public are to be carefully considered by the competent authority. Such a participatory approach ensures transparency and early involvement of the public and helps to identify and mitigate risks for the environment and optimize disaster forecasting. The participation of the public ensures that all direct and indirect effects of a project are determined, reflected and assessed relating to such factors as humans, fauna and flora, soil, water, air, climate and the landscape, the inter-action between these factors, material assets and the cultural heritage.

The Directive also implements the UNECE Espoo-Convention on trans-boundary impact assessment. In the case of a likely significant trans-boundary environmental impact from a planned industrial or mining project, the affected parties have to be notified and all relevant information on the project, including the environmental report, has to be submitted so that members of the public likely to be affected get the opportunity to comment. The results of such trans-boundary consultation have to be taken into account by the competent authority of the party that is responsible for granting authorization to the project.

Council Directive 1996/61/EC of 24 September 1996; integrated pollution prevention and control.

The purpose of this Directive is to achieve integrated prevention and control of pollution arising from industrial facilities with highly significant negative environmental effects that require the authorization of the facilities. Also, it lays down measures designed to prevent or, where that is not practicable, to reduce emissions to air, water and land from such industrial activities whose production capacities or outputs exceed the threshold and limit values set out in the Directive.

For example, the Directive covers several energy industries such as mineral oil and gas refineries,as well as activities in waste management, like landfills that receive more than 10 tonnes per day or with a total capacity exceeding 25,000 tonnes, excluding landfills for inert waste.

Thereafter, the Directive lays down requirements concerning the permitting of industrial activities. The competent authorities determine the conditions of the permit in order to achieve a high level of protection of the environment taken as a whole, without prejudice to relevant Community provisions.

For the public to be aware of the operation of installations and their potential effect on the environment, and in order to ensure the transparency of the licensing process throughout the Community, the public must have access, before any decision is taken, to information relating to applications for permits for new installations or for substantial changes, and to information on the permits themselves, their updating and the relevant monitoring data. For installations with potential for pollution, and therefore trans-frontier pollution, the applications relating to such proposals, or for substantial changes, must be available to the public of the Member State likely to be affected. During the permitting procedure, the public must be able to comment on the applications for permits before the competent authority reaches its decision and the resulting comments have to be carefully considered by the competent authority.

Directive 2006/21/EC of the European Parliament and Council of 15 March 2006: management of waste from extractive industries.

For waste facilities which hold extractive waste with a substantial risk to the environment and human health such as is found in the case of tailing dams, the Directive on the management of waste from extractive industries, (which also must be consistent with the Århus-Convention) provides a separate and stringent regulating system. In accordance with the objectives of Community policy on the environment, the Directive lays down minimum requirements in order to prevent, or reduce as far as possible, any adverse effects on the environment and human health which are brought about as a result of the management of waste from the extractive industries.

No waste facility must be allowed to operate without a permit granted by the competent authority. Early in the procedure for granting a permit or, at the latest, as soon as the information can reasonably be provided, the public must be informed of the application for a waste management permit. Furthermore, public participation means that the affected members of the public must be consulted prior to the granting of a waste management. The resulting public comments have to be carefully considered by the competent authority. For such waste facilities with potential for pollution, and therefore trans-frontier pollution, the applications must be available to the public of the Member State likely to be affected.

Directive 2001/42/EC of the European Parliament and of the Council of 27 June 2001: assessment of the effects of certain plans and programmes on the environment.

The objective of the Directive on the assessment of the effects of certain industrial plans and programmes on the environment, which also implements the obligations arising under the Århus Convention, is to provide for a high level of protection for the environment. Also, it is to contribute to the integration of environmental considerations into the preparation and adoption of such plans and programmes with a view to promoting sustainable development, by ensuring that an environmental assessment is carried out regarding certain industrial and mining projects which are likely to have significant effects on the environment. For plans and programmes for which the environmental assessment obligation arises simultaneously from this Directive and from other Community legislation, such as the Directive on the assessment of the effects of certain public and private projects on the environment, procedures should be coordinated to avoid duplication of assessment.

An environmental assessment must be carried out for plans and programmes which are likely to have significant environmental effects. Where an environmental assessment is required by this Directive, an environmental report should be prepared in which the likely significant environmental effects of implementing the project, and reasonable alternatives taking into account the objectives and the geographical scope of the plan or programme, are identified, described and evaluated. All plans and programmes including industrial and mining projects, should harmonize with other plans, particularly in their land use, to minimize environmental effects as far as possible.

Where the implementation of a plan or programme is likely to have significant effects on the environment, the Direction lays down the requirements of public participation within the implementation of the plan or programme which are subject to preparation and/or adoption by an authority at national, regional or local level. Where the implementation of a plan or programme is likely to have significant effects on the environment in another Member State, there is also a need for trans-boundary consultations, including public participations and information. The Directive contributes to more transparent decision making to ensure a high level of environmental protection.

In particular, the environmental report must be made available to the public. The environmental report and the opinions expressed by the public, as well as the results of any trans-boundary consultation, must be taken into account during the preparation of the plan or programme and before its adoption or submission to the legislative procedure.

Directive 2003/35/EC of the European Parliament and of the Council of 26 May 2003: providing for public participation in respect of the drawing-up of certain plans and programmes relating to the environment.

The objective of this Directive providing for public participation in the drawing-up of certain plans and programmes relating to the environment is to contribute to the implementation of the obligations arising under the Århus Convention, in particular by:

- providing for public participation in the drawing-up of certain industrial and mining plans and programmes relating to the environment if the public participation is not already required by the Directive on the assessment of the effects of certain plans and programmes on the environment; and
- improving the public participation within the permit procedures of industrial facilities regulated by the Directive concerning integrated pollution prevention and control and by the Directive on the assessment of the effects of certain public and private projects on the environment.

In summary, this Directive sets significant standards for protection, prevention and improvement of the state of the environment and optimization of disaster forecasting as it promotes public participation within the permit procedures of industrial and mining facilities.

Directive 2003/4/EC of the European Parliament and of the Council of 28 January 2003: public access to environmental information.

This Directive increases public access to environmental information and must be consistent with the Århus-Convention. The dissemination of such information contributes to a greater awareness of environmental matters, a free exchange of views, more effective participation by the public in environmental decision-making and, eventually, to a better environment.

It should be noted that this Directive does not address public participation within the authorization procedure of industrial facilities. But the Directive does provide, *inter alia*, the statutory framework for the external emergency plan and obligates the Member States to take necessary measures to ensure that, in the event of an imminent threat to human health or the environment, all information held by or for public authorities (which could enable the public likely to be affected to take measures to prevent or mitigate harm arising from the threat) is disseminated immediately and without delay. Of course, this way of public information is a kind of optimization of disaster forecasting which helps each concerned person to take his individual safety measures.

3. German Law

The detailed arrangements for public participation under the above-named Directives, particularly the Directive on the assessment of the effects of certain industrial projects on the environment, must be determined by every Member State so as to enable the public concerned to prepare and participate effectively.

In German law, the competent authority is required to carry out an environmental impact assessment, which is integrated into the existing licenses procedure of industrial and mining facilities. The environmental impact assessment, including the public participation, must be implemented for projects likely to have significant effects on the environment by virtue, inter alia, of their nature, size or power rating and which are likely to exceed the threshold and limit values set out in the German regulation.

Before beginning the license procedure, the competent authority should prepare to identify the scope of all significant effects on the environment which the project is likely to have on humans, fauna and flora, soil, water, air, climate, the landscape and material assets.

With the subsequent application for licensing, the operator must supply in an appropriate form the information he needs for carrying out the special project. This includes all planning documentation necessary, particularly an environmental impact study, which describes the likely significant effects of the proposed project, and supply details of the measures envisaged to prevent, reduce, and where possible, offset any significant adverse effects on the environment.

Before a license is given, the planning documentation, including the environmental impact study, must be made available to the affected public. This requires that the local public is informed about the following layout of the documentation. Based on this, the public is able to check the documents and establish if they have any concerns about or interests in the planning of the industrial project. Furthermore, the concerned public must get the opportunity to express their opinions and objections to the project at a period of time specified by the authority which will grant the project. The trans-boundary participation is required in case of the environmental effects of the project across boundaries.

Finally, the licensing authority must discuss all opinions and objections particularly those made by the public likely to be concerned by the construction and operation of the industrial and mining facility. The affected public which has expressed objections must also be informed separately about this opportunity. Within this procedure, all opinions, objections and adverse environmental effects of the project should to be substantiated. The discussion procedure serves an essential role for the final result made by the competent authority aiming thereby to further the transparency in decision-making, to strengthen public awareness for industrial and mining projects likely to have adverse effects on the environment, and to optimize disaster forecasting. Only after prior assessment of the likely significant environmental effects and of the measures necessary to prevent, reduce and, where possible, to offset these adverse effects, the licensing authority is able to grant the planned project.

I. Apostol et al. (Eds.)
IOS Press, 2009.
doi:10.3233/978-1-58603-948-6-294

Authorities and Organizations with Security Tasks in the Federal Republic of Germany and Their Legal Basis

Peter PASCALY
Am Blütenhain 5a, D-48163 Münster, Germany

Abstract. The Federal Republic of Germany is built of 16 federal states with police authorities of their own in addition to the Federal Police. Besides that the federation has established law providing guidelines to enable the states to enact regulations to install similar auxiliary attachments such as fire departments. As well as these government organizations there are also various and numerous NGOs. The fire brigades are mostly manned with voluntary personnel. Only major cities and big plants have professional fire-fighters. They work together with NGOs in case of a catastrophe under the control of the police authority. Through Fire Protection Demand Plans the needs of the fire brigades of all regions are focused and perform the catastrophe planning and the civil defence.

Keywords. Police authorities, fire departments, catastrophe, civil defence, GOs, NGOs

Introduction

The Federal Republic of Germany is built as a federal state consisting of 16 states of very different sizes. There are, on the one hand, large states such as Bavaria, Lower Saxony or Mecklenburg-Western Pomerania, and, on the other hand, smaller states such as Saarland or the city states such as Berlin, Hamburg or Bremen, which also differ in size.

This division of singular states was established at the foundation of the Federal Republic in 1949 in the Basic Constitutional Law of Germany (Constitution) but the roots of this structure can be traced back to the Middle Ages. The Basic Constitutional Law also established the separation of powers in Germany as well as the division of the responsibilities in the state structure (i.e. federalism). The Federation is, for example, responsible for foreign affairs and for defence. The states, however, are in charge of the police and cultural affairs, for example. An independent federal police force observes the federal borders (such as at airports) as well as federal-owned equipment and property (such as Deutsche Bahn AG).

In order to manage and coordinate the different interests of the members of the federation, the legislation transferred a framework competence to the Federation. Therefore, the association remits laws for the jurisprudence in all areas. Implementation, as well as supervision, of the laws is the responsibility of the states. Similar regulations apply to authorities with special tasks.

1. The Federation

In principle, the regulation exists that the police authorities take over the control of measures in a catastrophe situation and provide the organization and the implementation of these measures. This control is organized in a decentralized manner but all measures for catastrophe protection have as their goal the defence against danger.

The legal basis for this regulation is the federal civil-defence law which is supplemented through the different fire-protection support-accomplishment laws that each state has implemented slightly differently. These laws allow the use of steering state authorities and the activity of support-organizations such as the Red Cross, for instance.

In principle, the Federation becomes active in the area of danger-defence only in the state of war. Exceptions are catastrophic events which are identified according to certain criteria. Examples for these are the cyclone Kyrill in the winter of 2007 or the Elbe flood in 2002. Catastrophe situations of such dimension can only be managed through the use of all forces available. In such a case, an inter-ministerial coordination group meets with the federal office of defence and catastrophe support under the management of the federal Department of the Interior in the common situation centre. This group can raise the alarm with the federal institution technical relief organization (THW) for federally controlled deployments within the whole country and abroad.

In the THW there are more than 669 local organizations in all federal states, and 80,000 members are organized on a voluntary basis. The THW is equipped with modern appliances and is able to react in all possible damage cases, such as to water, construction/buildings, electrical systems, gas pipes, and to contain oil spills for example. Furthermore, fresh water can be supplied, emergency power and illumination installed, as well as any kind of rescue missions performed.

The Federal government is able to send rapid rescue deployments for catastrophes in foreign countries with the help of the THW (for example, earthquakes in Mexico 1985 or in Iran in December 2003) or for rapid support purposes abroad in the case of disastrous interruption of the water supply, as might be found in the case of earthquakes. The situation centre is also entitled to request the armed forces to provide help in the home country in the case of a catastrophe if all other forces do not have full control of the emergency situation, especially, if the armed forces can make special appliances and teams available.

2. The State of North Rhine-Westphalia

All 16 federal states have self-made regional laws regarding catastrophes and fire-protection as well as a rescue service. In North Rhine-Westphalia, for instance, it is called the law of fire protection and aid (FSHG). Besides that the state also adopted a rescue law (namely, RettG NRW). The Department of the Interior of the state and the Department of Health and Social Affairs are responsible for the implementation of these laws. Normally, the activity of the latter department is limited to control and supervision functions. The ministries in most state authorities delegate their tasks to their branches in the states, the district governments or police headquarters. These institutions, without executive power of their own, only possess coordination responsibilities in a catastrophe, and rely on the help from the government district.

The actual support and rescue activities, and their control, are carried out by members of the local fire brigades of the counties and county-free cities. This authority of the counties and the county-free cities is also incumbent on the planning of fire and catastrophe protection, as well as the protection planning for particularly endangered features. It is this task which becomes more and more important considering the permanent threat of terrorist attacks to exposed and important targets. Such attacks are always spectacular in their effects and can only be repelled through special protection precautions. These precautions must have begun already at the planning phase of the facility, for instance, buildings. Experts are, from a very early stage, involved in fire-protection and other risks.

The fire and catastrophe protection within the Federal Republic is organized by the communities which own the local fire brigade. Usually, honorary firemen serve the local fire brigades; only cities with over 100,000 inhabitants have professional fire-fighters. This cadre of professional fire-fighters in the cities is necessary due to the fact that a large amount of emergency missions are carried out by specialists from the fire brigade. Furthermore, it is not practical for honorary personnel to be alerted and called from their jobs for every small contingency event (for instance a traffic-endangering oil-slick on the road). The fact that catastrophe protection is organized on an honorary basis has the great advantage that a large number of fire-fighters is highly engaged without being employed by the state and for low costs, and it is also attractive to the participants. They can be trained in numerous and frequently-repeated training sessions that disseminate state-of-the-art knowledge. This training takes place in central state facilities.

The professional and voluntary fire brigades – e.g. the public fire brigades – are supplemented at particularly endangered facilities, for instance chemical plants or airports, by so-called plant fire-brigades which are government controlled and recognized, and are formed from full-time and part time forces. Works' fire brigades must fulfil both the protection requirements of their particular businesses and those of the public fire brigade in regard to education and equipment. Works' fire brigades form the fire brigade of the community in combination with the public fire brigades (professional and voluntary fire brigades). The status of education and the cooperation between works' fire brigades and public fire brigades is monitored by the district government's experts regularly.

The communities support the counties with the planning of catastrophe-defence tasks and with catastrophe protection itself. This includes taking part in the defence-planning committees with their regional knowledge and know-how, as well as participating in common catastrophe-protection practices. These hands-on exercises with the special equipment are supplemented by the common training of the staff assuming a catastrophe situation, e.g. crash of an airplane in the outskirts of the community. On such occasions, neighbouring fire brigades exercise a common-danger defence under instruction from experienced experts.

Furthermore, the extensive honorary basis of the fire-fighters' organization has the advantage that many interested participants are ready to support the fire brigade work by their involvement without the fear of professional disadvantages through the cooperation with the fire brigade. This allows the Federal Republic to have 1.3 million honorary fire-fighters available in the event of a catastrophe. Since these personnel come from the whole population, they are very quickly available for an emergency, as follows:

- within 15 minutes: 300,000 people;
- within 30 minutes: 600,000 people;
- within 60 minutes: 900,000 people.

In this way the Federal Republic, with a total population of approximately 82 million, has more than 1.3 million persons in the active fire brigade service. Among these are 95.6% volunteers, and the proportion of women within the fire brigade is 9.8%. In addition, there are 255,000 forces in the youth fire brigade. The fire brigades have 25,213 fire stations which mainly comprise appliance storage buildings. The operational demands on the fire brigades are very high. Annually, they have 3.5 million deployments, 62.8% of which are by the professional fire brigade and 32.7% by the volunteers. Only 4.5% of all emergency deployments are performed by works' fire brigades.

In order to have state-of-the-art fire-fighting skills, the planning of the fire protection demand is the duty of the community in the so-called Fire Protection Demand Plan since it knows best which equipment is necessary for the required purposes. On this basis, the county/county-free cities focus the needs of the communities in their area and perform the catastrophe planning.

The demand registrations of these subordinate catastrophe protectors are implemented in the demand planning by the district governments. The plans are forwarded to the responsible Department of the Interior that, again, assesses the collected demand plans and finally forms a demand plan for the entire country. This planning then forms the basis for the planning of the finance demand for the procurement of tools and the budgeting of the necessary means in the next federal budget. The normal fire cases are usually under the control of the fire brigade from the responsible community. This is also applicable to the rescue tasks with traffic accidents within the community area, unless the accident rescue is delegated to support organizations (NGOs), such as the German Red Cross or Johanniter-Unfallhilfe.

In the case of a catastrophe, the county or the county-free city, in whose area the catastrophe occurred, takes over the control of the operation and of the support forces that are provided by the public fire brigades. For this, the county/county-free city has, in addition to the control centre within its service building, also got a mobile command centre available. Both are inter-connected via radio with the support forces. However, the mobile command centre can act closer to accident events. Beside this operational control, there is an emergency task force that coordinates the administration and reconciles the cooperation with the operation control (Fig. 1).

In the case of a catastrophe the whole emergency task force meets, headed by a headquarter's official. This is the mayor when the catastrophe event occurs in a city; it is the District Chief Executive if the event occurs in a county-affiliated community. The personnel of the task force vary depending on the severity of the catastrophe event. Besides administrative personnel from the corresponding county, necessary experts of various disciplines are also consulted, as are specialists of the companies involved in the event.

An example for the common action in the catastrophe case is the snow catastrophe in November 2005. In that instance, large quantities of wet snow fell in North Rhine-Westphalia in the counties of Borken, Coesfeld and Steinfurt, and caused such a severe increase in weight of the overhead high voltage cables that they broke and caused pylons to bend. As a result, large areas of these counties in North Rhine-Westphalia's most north-western area were without power and with a consequent lack of heat in low

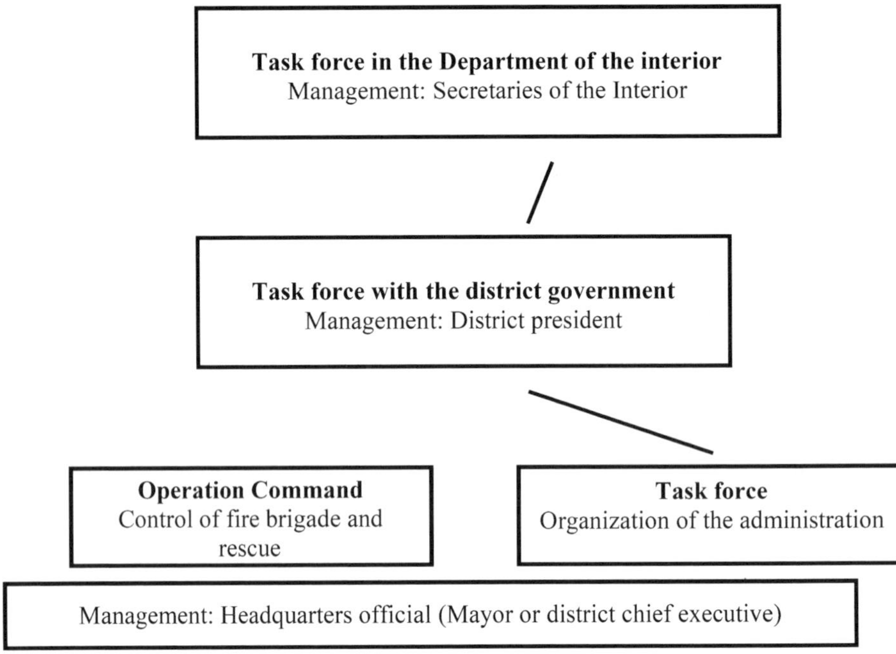

Figure 1. Operation control by emergency task forces.

external wintry temperatures. Thus, farmers could not milk their cows and no warm food could be prepared.

The task forces of the three counties met and, since the snow catastrophe had occurred across county lines, this meeting also included the responsible district government's task force in Münster, which is the district capital.

The local task forces of the involved counties organized the support and rescue forces from within the county area for and at the damage event, as well as providing support goods and appliances from the respective county area. Also, traffic control in the catastrophe area was managed.

The task force of the district president of Münster had central tasks. He coordinated the work of the three county task forces with the three district governments so that they supplemented each other and did not overlap. In addition, he could arrange for help through non-governmental organizations (NGOs), such as the German Red Cross, the Worker-Samaritan-Association, Johanniter-Unfallhilfe and many others, so that helpless people could be brought to a warm shelter, that hospitals could be run, and that field kitchens were managed, for example. In addition, support goods, such as generators for emergency electric power, for the business of electric milking machines, illumination, heaters etc. needed to be moved from other parts of the government district in the concerned county areas. Furthermore, it had the obligation to defend against further dangers and also to organize repair measures of the involved big power-supply companies, as well as local power and utility suppliers.

Even after the Cold War has ended, the extension of the civil defence is continued, but with a different standard. Mobile army surgical hospitals with decontamination facilities are intended for each county in the state of North Rhine-Westphalia for 50 injured as well as contaminated people. These military hospitals possess all necessary facilities for the situation, which normal hospitals are not able to provide due to the large numbers of injured people. The district government's task force can bring further support facilities from the neighbouring counties to the place of the catastrophe at any time.

In areas with water bodies (for example, the Rhine) water-rescue facilities are obtained centrally and are held ready in case of emergencies. The catastrophe protection authorities of the state of North-Rhine-Westfalia hold facilities for water treatment and supply available.

3. Conclusion

The German authority for catastrophe defence is able to manage all incoming events under the aspects of the legal powers they have. The fire brigades are proud of the voluntary structure of their organisation; they are quick and can react effectively to demands in conjunction with the technical and health organisation. Precautions are been taken to keep the existing standard at this high level.

References

Gesetz über den Feuerschutz und die Hilfeleistungen (FSHG) für das Land Nordrhein-Westfalen (Law About Fire Services in North Rhine-Westfalia) vom 10. Februar 1998. – GV. NW. 1998 Seite 122.

Author Index